华侨大学教材建设资助项目

工程流体力学

英汉双语版

主　编　刘晓梅
副主编　李洪友　林添良

机械工业出版社

本书作为工科学生流体力学基础课程的双语教材，内容包括流体力学的基本原理和方程。该书各章节都配有丰富的案例详解、习题及相关知识在工程中的应用案例分析和研究，尤其增加了关于流体传动装置、液压系统及元件的实例分析，可使读者认识并理解如何将流体力学知识应用于工程实践。本书共六章，包括流体的基本性质，流体静力学，流体运动基本内容，流体动力学基本方程及其应用，量纲分析和相似原理，黏性不可压缩流体在孔口、狭缝中的内部流动（泵 - 涡轮系统的能量方程分析及应用，流体传动系统功率和效率的计算等）。

本书可作为普通高等学校工科各专业，特别是机械类相关专业流体力学课程双语教学教材，也可供相关专业的工程技术人员参考。

This book is a bilingual textbook for engineering students on the fundamentals of fluid mechanics, covering fundamental principles and equations of fluid mechanics. Readers will understand how to apply the knowledge of fluid mechanics to the engineering practices through abundant examples and application case analysis and studies, especially on analysis of fluid transmission devices, hydraulic systems and elements. The book consists of six chapters, including fluid properties, fluid statics, basics of fluid kinematics, fundamental equations of fluid dynamics and their application, dimensional analysis and similarity, internal flow of viscous incompressible fluid in orifices and narrow gaps, energy equation and its application to systems involving pump and turbine, calculations of the power and efficiency of fluid transmission systems. The analysis of some problems and derivation of formulas are appropriately simplified in this book, and there are abundant exercises in each chapter.

This book is written as a text book for the course in fluid mechanics for engineering students of colleges and universities, especially for students of mechanical engineering. The book can also be used as reference for engineers and technicians.

图书在版编目（CIP）数据

工程流体力学：英汉双语版 / 刘晓梅主编 . —北京：机械工业出版社，2022.11

ISBN 978-7-111-72563-3

Ⅰ . ①工… Ⅱ . ①刘… Ⅲ . ①工程力学 – 流体力学 – 高等学校 – 教材 – 英、汉 Ⅳ . ① TB126

中国国家版本馆 CIP 数据核字（2023）第 010660 号

机械工业出版社（北京市百万庄大街 22 号 邮政编码 100037）
策划编辑：张秀恩　　　　　　　　　　　责任编辑：王春雨
责任校对：刘本明　陈 越 李 杉　　　　封面设计：马精明
责任印制：邓　博
天津翔远印刷有限公司印刷
2023 年 5 月第 1 版第 1 次印刷
169mm × 239mm ·23.75 印张·476 千字
标准书号：ISBN 978-7-111-72563-3
定价：79.00 元

电话服务　　　　　　　　网络服务
客服电话：010-88361066　机 工 官 网：www.cmpbook.com
　　　　　010-88379833　机 工 官 博：weibo.com/cmp1952
　　　　　010-68326294　金 书 网：www.golden-book.com
封底无防伪标均为盗版　机工教育服务网：www.cmpedu.com

前　　言

"工程流体力学"是工科相关专业的重要基础课程。本书为适用于单学期 30 ～ 40 课时双语教学及境内、境外本科生共同学习使用的简明教材。

本书内容共六章，着重讲述流体力学的基础原理和基础假设，以及怎样应用这些基本原理来解决实际的工程问题。第 1 章介绍流体力学的基本研究内容、压强的表示方法及流体的基本性质，包括流体的密度、黏度、表面张力等；第 2 章为流体静力学相关内容，包括流体上的力，流体静压强分布，测压管的原理与应用，平面壁和曲面壁上的流体静压力以及流体的相对静止等；第 3 章介绍描述流体运动的方法和相关概念，如流线、系统和控制体，以及如何应用雷诺输运定理将物理关系在系统和控制体间转换；第 4 章介绍理想流体动力学的基础方程，包括连续性方程、伯努利方程及其应用和动量方程；第 5 章介绍量纲分析和相似原理的基本思想及其在设计和简化实验系统、分析实验数据方面的优势；第 6 章为黏性不可压缩流体的内部流动，主要介绍实际流体的能量损失和层流及紊流状态，其中还简单分析了黏性流体在圆管、孔口和缝隙中的流动特点。

本书对一些问题的分析和公式推导做了适当简化，增加了很多关于流体传动、液压泵、阀及液压系统的案例分析和例题。特别在第 6 章关于实际流体的能量方程部分，重点增加了包括泵和涡轮系统的能量方程及系统功率、效率的计算，通过这些内容向读者展示作为机械工程师可能处理的与流体相关的问题和应用。通过本课程的学习，对简单的工程问题，读者应该能够独立对待解决的问题进行分析，做出适当的假设或近似，然后运用相关的定理建立控制方程并求解，最后代入数据进行计算。

本书的编者感谢华侨大学教材建设项目的支持和教学团队各位教师的通力协作。本书由刘晓梅任主编，李洪友、林添良任副主编。编写团队其他成员包括闫洁、刘晶峰、顾永华、段闯闯。

本书可作为高等院校相关专业流体力学的教材，特别适用于机械工程类专业，也可供工程技术人员参考。

由于作者水平有限，书中难免有不妥之处，恳请读者批评指正。

Preface

Engineering fluid mechanics is an important basic course for related majors. This is a brief bilingual course matching with one-semester 30~40 class hours for both domestic and overseas under-graduates.

In this book of six chapters, we stress the governing principles, the assumptions and the method to apply the principles to the solution of practical engineering problems. In chapter 1, some fundamental aspects of fluid mechanics, pressure scales and fluid properties are introduced, including density, viscosity and surface tension. Chapter 2 introduces fluid statics, including forces on fluid, distribution of pressure in fluid at absolute and relative rest, theories and applications of manometer, hydrostatic force on a plane surface and a curved surface etc. Chapter 3 explains methods and concepts for the description of fluid motion, including streamline, system and control volume, and Reynolds transport theorem for the transformation from system to control volume. Emphasis is placed on the fundamental equations of ideal fluid dynamics in chapter 4, including the equation of continuity, the Bernoulli equation and the equation of momentum. The advantages of using dimensional analysis and similarity for planning and simplifying experiments and for organizing test data are featured in Chapter 5. Chapter 6 expands on the internal flow of viscous incompressible fluid, includes energy losses of viscous fluid flow, the flow regimes of laminar and turbulent, practical concerns such as pipe flow, orifice and nozzle flow, and narrow gap flow are also discussed in this chapter.

The analysis of some problems and derivation of formulas are appropriately simplified in this book, and there are abundant examples and exercises of fluid transmission, hydraulic pump and valves. Special attention are given to the energy equation of real fluid, calculation of power and efficiency for the systems involving pump and turbine are presented so that the readers can understand and analyze many of the practical problems encountered by the engineer. Through this course, readers should be able to analyze problems independently, make assumptions, establish equilibrium equations, substitute parameters and solve simple engineering problems by using relevant principles.

The authors are very much indebted to the textbook funding project of Huaqiao University. This book is edited by Liu Xiaomei, Li Hongyou and Lin Tianliang. Thanks to Yan Jie, Liu Jingfeng, Gu Yonghua and Duan Chuangchuang for their cooperation in the editing.

The book is written as a bilingual textbook of fluid mechanics for related major students of colleges and universities, especially for students of mechanical engineering. This book can also be used as reference for engineers and technicians.

Some biased points are unavoidable in the book due to the author's limited knowledge and ability level, please feel free to point out the mistakes.

CONTENTS/ 目录

Nomenclature   常用符号表

a	Acceleration	加速度
A	Area	面积
α	Kinetic energy correction factor	动能修正系数
β	Momentum correction factor	动量修正系数
γ	Specific weight	重度
CS	Control surface	控制面
CV	Control volume	控制体
d、D	Diameter	直径
ε	Roughness average height	粗糙度平均高度
g	Gravity	重力加速度，自由落体加速度
h_C	Submergence depth of centroid	形心的淹深
h_D	Submergence depth of center of pressure force	压力中心的淹深
h_f	Frictional head loss	沿程损失
h_j	Minor head loss	局部损失
h_L	Loss head	损失水头
h_P	Head of pump	泵的水头
h_T	Head of turbo	涡轮的水头
λ	Friction loss factor	沿程损失系数
μ	Viscosity	动力黏度
v	Kinematic viscosity	运动黏度
ω	Angular velocity	角速度
Ω	Volume of system	系统的体积
p	Pressure	压强
π	Dimensionless parameter	无量纲量
q	Flow rate　Volume flow rate	体积流量（q 或 q_V）
q_m	mass flow rate	质量流量
r	Radius	半径
ρ	Density	密度
S_g	Specific gravity	比重，相对密度
sys	System	系统
ζ	Minor loss factor	局部损失系数
σ	Surface tension	表面张力
t	Time	时间
τ	Shearing stress	剪切应力
u	Velocity	速度
u,v,w	velocity components in x-, y-, z-direction	x-, y-, z- 方向的速度分量
V	Average velocity	平均流速（用 V 代替 $\bar{V}$）
~~V~~	Volume of control volume	控制体的体积

CHAPTER 1
BASIC CONSIDERATIONS

Mechanics is one of the oldest physical science, the subcategory fluid mechanics is defined as the science that deals with the behavior of fluids both at rest and in motion. A proper understanding of fluid mechanics is extremely important in engineering applications. You can master the fundamentals of fluid mechanics and see how those fundamentals can be applied to practical engineering problems by understanding this book and working many problems.

1.1   Introduction

Fluid mechanics is a discipline which studies the equilibrium and macro mechanical motions of fluids, it is one of the most important branches of mechanics.

The engineering fluid mechanics puts emphasis on solving problems occurring in engineering practice, basic takes of engineering fluid mechanics lie in establishing fundamental equations describing the movement of a fluid, determining velocity and pressure distribution rules for flows in different pipes and immersed bodies, probing calculation methods for energy and various energy losses, and solving interactive problems between fluids and solid bodies.

Fluid mechanics can be divided into three branches: *fluid statics* the study of the mechanics of fluids at rest; *kinematics* deals with velocities and streamlines without considering forces or energy; and *fluid dynamics* is concerned with the relations between velocities and accelerations and the forces exerted by or upon fluids in motion and the interaction of fluids with solids or other fluids at the boundaries.

In this book we present the general equations result from the conservation of mass principle, the conservation of energy theorem, and Newton's second law. From these a number of particular situations will be considered that are of special interest. After studying this book, the readers should be able to apply the basic principles of the mechanics of fluids to new and different situations.

Fluid mechanics is widely used both in everyday activities and in the design of modem engineering systems. Fluid mechanics plays very important role in the design and analysis of civil engineering systems such as the transportation of water and air, sewage treatment, irrigation channels and dams. It is also considered in the design of ocean engineering and chemical-processing equipment. In biomechanics the flow of blood and cerebral fluid are of particular interest. The knowledge of mechanics of fluids are required in the design of an automobile, internal combustion engines, wind turbines, boats, submarines, jet engines, aircraft and rockets.

Mechanical engineers design the automatic transmission system, the pumps and turbines, the hydraulic brakes and power steering, the lubrication system, the air compressors, air-conditioning equipment, pollution-control equipment and the cooling system using a proper understanding of fluid mechanics.

For the long time before seventeenth century, the application of fluid machinery expanded slowly but steadily. People observed fluid phenomena and built masterpieces in abundance. Ancient civilizations had knowledge about fluids, particularly in the areas of irrigation channels and sailing ships. The earliest hydraulic engineering projects such as

第1章
基础知识

力学是物理学最古老的部分之一，流体力学作为其分支研究流体静止和运动状态下的行为特征。正确理解流体力学对其工程应用至关重要。通过本课程的学习和书中大量习题的训练，读者能够掌握流体力学的基础知识，知晓如何应用这些知识去解决工程实践问题。

1.1 引言

流体力学是力学的一个重要分支，主要研究流体的宏观运动与平衡。

工程流体力学则注重解决工程实践中的问题，主要内容包括建立方程描述流体的运动，分析各种管道或淹没在流体中物体内部的速度及压强分布规律，探求能量及能量损失的计算方法，求解流体与固体间的相互作用等。

流体力学可分为三部分内容：流体静力学研究静止流体的力学规律；流体运动学排除力和能量的影响讨论流体的运动速度和趋势；流体动力学分析流体运动速度、加速度与流体受力或施力的关系以及流体与固体界面或流体间的相互作用。

本书从质量守恒定律、能量守恒定律和牛顿第二定律导出基本方程，由此出发分析各种流动情况。通过本书学习，读者将能够运用流体力学基本原理分析和解决各种简单的工程问题。

流体力学广泛应用于日常生活和现代工程设施，并发挥重要作用，如输水供气的市政系统、废水处理系统、灌渠和堤坝等。海洋工程和化工设备的设计需要考虑流体的力学特性；生物力学领域更关注血液和体液的流动；汽车、内燃机、风力发电机、舰船、潜艇、喷气机、飞船和火箭等的设计也需要运用流体力学相关知识。

机械工程师在设计自动运输系统、泵和涡轮、液压制动器和动力转向机构、润滑系统和空气压缩机、空调装置、污染防治装备和冷却装置等时，都需要合理考虑流体力学的相关因素。

十七世纪前很长时间内，人类对流体知识的应用虽然进展缓慢但却持续而坚

Dujiangyan and Lingqu canal were built in the second century BC in China. The Greek mathematician Archimedes (285—212 BC) formulated and applied the buoyancy principle to determine the gold content of the crown of King Hiero I. However the names of almost all those early creators are lost to history, clay pipes for supply of water and sewage disposal even fire prevention, watermills for irrigation and grain grinding, hand bellows for forge metal, clepsydra for time measuring, windmills and sails. The thought of scientific was perfect and the methods of scientific were adopted, but the studies did not yet constitute the fluid mechanics system since fluid properties and parameters were poorly quantified.

From eighteenth century to the beginning of twentieth century was an important stage for establishing the science of fluid mechanics. Newton's law of viscosity, Bernoulli equation for steady incompressible fluid, Euler's describing method of fluid motion and kinetic differential equation for inviscid fluid, Reynolds experiment and Reynolds number, the Navier-Stokes differential equation of viscous fluid motion, are all achievements of this period. The primitive invention of Wright brothers (Wilbur, 1867—1912; Orville, 1871—1948) contained all the major aspects of modem aircraft. Qianxuesen and other scientists ushered China into the aerospace age. These works also explored the links between fluid mechanics, thermodynamics, and heat transfer. New advance supported a huge expansion of the industrial, chemical, aeronautical, and hydraulics; each of which pushed fluid mechanics in new directions.

The development of the digital computer in late twentieth century gave us the ability to solve large complex problems, entirely new field of computational fluid dynamics has developed. The experimental data can be used to verify theory or to provide information supplementary to mathematical analysis. The fluid mechanics has begun to research such as the global climate modeling, ocean, petroleum, chemical engineer, energy sources, environmental protection and construction fields etc. Various numerical methods such as finite elements, boundary elements, and analytic elements are now used to solve advanced problems in fluid mechanics.

There are three approaches for the research of engineering fluid mechanics, the theoretical approach, the experimental approach and the calculating approach.

The theoretical approach analyzes the primary and lesser the factors of question, abstracts theoretical model. Mathematical tools are used to find the general answer about the fluid movement.

The experimental approach summarizes the fluid question by experiment on a similar model, observes the phenomena, determines the data and speculates on the theories from the experiment results according to a certain method.

The calculating approach draws out the experiment scheme according to the theory analysis and experiment observation; simulates the fluid flow under various conditions; calculates the numerical value solution with computer.

All these approaches must combine each other can they advance the development of fluid mechanics.

1.2 Hypothesis of continuum model

A fluid is composed of molecules which may be widely spaced apart, especially in the gas phase. Yet it is convenient to disregard the atomic nature of the fluid and view it as a continuous, homogeneous matter with no holes，that is, a continuum. The continuum idealization allows us to treat properties as point functions and to assume that the properties vary continually in space with no jump discontinuities.

定。人们根据对流体现象的观察创造出了丰富的成果。古人对流体的知识主要用于灌溉和航行，最早的水利工程杰作如都江堰和灵渠，建造于公元前二世纪的中国。希腊数学家阿基米德（公元前 285—212）提出浮力原理和公式，成功验证了希洛一世王冠的含金量。但更多早期发明家的名字已被历史遗忘，用于给排水甚至消防的陶管、灌溉和碾米用的水车、冶炼和锻造时用的皮老虎、计时的滴漏，风车和风帆，等等。虽然由于对流体性质的认识还没有被量化，尚未形成系统的流体力学学科，但科学的思想已经孕育，科学的方法已广为接受。

十八世纪到二十世纪初是流体力学学科建立的重要阶段。牛顿内摩擦定律、不可压缩流体稳定流动的伯努利方程、欧拉法描述无黏流体运动和运动方程、雷诺实验和雷诺数、描述黏性流体运动的纳维 - 斯托克斯方程等，都是这一时期的研究成果。莱特兄弟（威尔伯·莱特，1867—1912；奥维尔·莱特，1871—1948）的创造发明包含了现代飞行器的全部主要原理。钱学森和其他科学家引领中国迈入航天时代。这些成就同时也将流体力学与热力学和传热学等联系起来，新的进展大大促进了工业、化工、航空航天以及水利工程等领域的发展，每个领域都展现出流体力学的新方向。

二十世纪后期，计算机的发展赋予了流体力学强大的计算能力去解决更为复杂和庞大的问题，促生了计算流体力学这一新领域。我们能够运用实验数据来验证理论或为数学分析提供信息和依据。流体力学的研究活跃于全球气候建模、海洋和石化、化工、能源、环保和建筑等领域。多种数字分析方法如有限元、边界元、解析元等方法正被运用于流体力学前沿问题的研究。

流体力学有三种主要研究方法，分别是理论法、实验法和计算法。理论法运用数学工具分析并简化问题，提出精简的模型，发现流体运动的普遍规律。实验法选择合理的方法在模型上进行实验并观察实验现象，根据实验数据推测理论公式。计算法根据理论和实验分析设计实验流程，模拟流体在不同条件下的流动，在计算机上得出数值解。这三种方法联合运用必将促进流体力学的快速发展。

1.2　连续介质假设

流体是由可彼此分离的分子构成的，尤其是气体分子间距更大。但是如果忽略流体的原子属性，把它视作均质且中间无孔隙的物质，即连续物质，会更方便。基于这种理想的连续性，我们假设流体的性质随空间位置连续变化，且是空间位置的函数。

1.2.1　Conception of fluid particle

Density of fluid can also defined by

$$\rho = \lim_{\Delta\Omega \to 0} \frac{\Delta m}{\Delta\Omega} \qquad (1.2.1)$$

where Δm is the incremental mass contained in the incremental volume $\Delta\Omega$.

Physically, we cannot let $\Delta\Omega \to 0$ since, as $\Delta\Omega$ gets extremely small, e.g. smaller than the mean free path λ, that is the average distance a molecule travels before it collides with another molecule, the mass contained in $\Delta\Omega$ would vary discontinuously depending on the number of molecules in $\Delta\Omega$, as shown in Fig.1.2.1a.

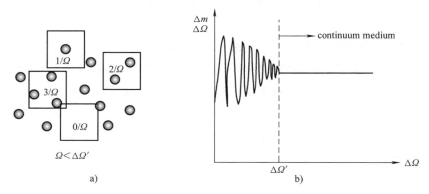

Fig.1.2.1　Density at a point
a) Density at difference points　b) Density at a point in a continuum

Seeing from microstructure, there are big gaps between fluid molecules and which are not continuous, while to fluid mechanics which researching macroscopical rules it is to model the physical bodies and is not necessary to discuss molecular microstructure. Therefore, the zero in the definition of density in Eq.1.2.1 should be replaced by a minimum volume $\Delta\Omega'$, below which $\Delta m/\Delta\Omega$ is going to vary dramatically, a definite ρ is not available and continuum assumption fails. This minimum volume is small enough in macroscopic to be treated mathematically as a point, while large enough in microscopic to no longer consider the individual molecules of the fluid, , as shown in Fig.1.2.1b.

A *fluid particle* is assumed to be the least unit in fluid whose linear scale can be ignored, and it is defined as a physical body in the fluid whose macroscopical scale is very small but microcosmic scale is big enough and has its mass.

For most engineering applications, the minimum volume $\Delta\Omega'$ shown in Fig. 1.2.1 is extremely small. For example, there are 3×10^7 molecules contained in a cubic micrometer of air and 3.3×10^{10} of water at 1 atm pressure and 20℃ . Hence, the scale of a fluid particle is even smaller than a micrometer.

1.2.2　Hypothesis of continuum model

A fluid is composed of fluid particles, which occupy flowing space continuously with inter molecular distance and molecular motion ignored. This is the *hypothesis of continuum model* that Euler put forward in 1753.

1.2.1　流体质点的概念

流体密度的定义为

$$\rho = \lim_{\Delta\Omega\to0} \frac{\Delta m}{\Delta\Omega} \tag{1.2.1}$$

式中，$\Delta\Omega$ 是微小体积增量；Δm 是 $\Delta\Omega$ 内流体的质量增量。

但是从物理的角度，不能使 $\Delta\Omega\to0$。因为当 $\Delta\Omega$ 变得非常小，小于分子平均自由程 λ，即分子与其他分子发生碰撞前可自由移动的平均距离时，那么体积 $\Delta\Omega$ 内的流体质量将非连续变化而是取决于体积 $\Delta\Omega$ 内分子的数量，如图 1.2.1 所示。

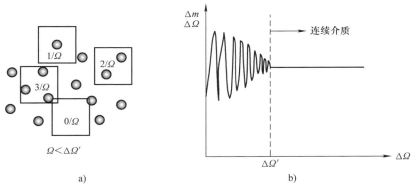

图 1.2.1　空间一点的流体密度

a）空间各点流体密度　b）连续介质的密度

从微观角度看流体是不连续的，分子间存在很大的间隙，但是流体力学研究的是流体的宏观运动规律，可以不必讨论其微观结构。因此可以用最小体积 $\Delta\Omega'$ 代替式 1.2.1 中的体积极限零，小于这一体积时，$\Delta m/\Delta\Omega$ 的值将剧烈变化而无法得到确定的密度 ρ 值，即质量不连续。这个最小体积在宏观上足够小，数学上可以被当做一个点；而微观上又足够大，可以不再考虑单个流体分子的影响，如图 1.2.1b 所示。

假设这种宏观尺度非常小而微观尺度足够大，并具有质量的**流体质点**，是构成流体的最小单元。

多数情况下，图 1.2.1 中的极限体积 $\Delta\Omega'$ 非常小。例如，1 标准大气压 20℃时，一立方微米的体积内可包含 3×10^7 个空气分子 3.3×10^{10} 个水分子。而一个流体质点的尺度甚至小于 $1\mu m$。

1.2.2　连续介质模型假设

忽略分子间距和分子运动，流体由流体质点构成，流体质点连续无间隙地充满流动空间。这就是欧拉于 1753 年提出的**连续介质模型假设**。

With the hypothesis of continuum model, fluid properties can be assumed to apply uniformly at all points in a region at any particular instant in time. For example, the density ρ can be defined at all points in the fluid; it may vary from point to point and from instant to instant; that is, in Cartesian coordinates ρ is a continuous function of x, y, z, and t, written as $\rho(x, y, z, t)$.

All the physical variables not just density, but speed, pressure, shear stress and temperature which represent fluid properties are single values and continuous differentiable functions when fluid flows continuously. So the scalar fields and vector fields of all kinds of physical variables come into being. In this way, we can use continuous function and field theory to study the problems about fluid movement and balance successfully.

1.3 Pressure scales

Pressure is defined as a normal force exerted by a fluid per unit area. We speak of pressure only when we deal with a gas or a liquid. The counterpart of pressure in solids is normal stress. Since pressure is defined as force per unit area, it has the unit of N/m^2, which is called a Pascal (Pa). In engineering practice, kPa ($1kPa = 10^3Pa$) and MPa ($1MPa = 10^6Pa$) are commonly used.

The actual pressure that is measured relative to absolute vacuum (i.e., absolute zero pressure), is called *absolute pressure*. Practically all pressure gages register zero when open to the atmosphere, and so they measure the difference between the pressures of the fluid they are connected to and that of the surrounding air, therefore, a *gage pressure* is defined by measuring pressures relative to the local atmospheric pressure. If a pressure is below that of the atmosphere, it is called a *vacuum pressure*, and its gage value is the amount by which it belows that of the atmosphere. What we call a "high vacuum" is really a low absolute pressure. Absolute, gage, and vacuum pressures are related to each other by

$$p_{\text{gage}} = p_{\text{abs}} - p_{\text{atm}} \tag{1.3.1}$$

$$p_{\text{vac}} = p_{\text{atm}} - p_{\text{abs}} \tag{1.3.2}$$

This is illustrated in Fig.1.3.1.

Here p_{gage} is the gage pressure; p_{abs} is the absolute pressure and p_{atm} represents the local atmospheric pressure; p_{vac} denotes the vacucm pressure. The units of which are all Pa.

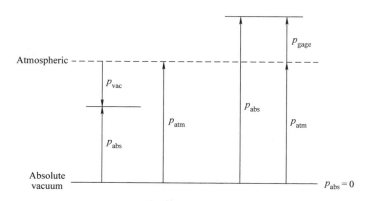

Fig.1.3.1   Absolute, gage, and vacuum pressures

有了连续介质模型，就可以在空间内任意点和任意时刻一致地应用流体的各项性质。例如，可以确定各点的流体密度 ρ，密度随位置和时间的变化，即，在直角坐标系下密度是 x，y，z 和 t 的函数，记作 $\rho(x, y, z, t)$。

对于连续流动的流体，密度以及其他各种描述流体性质的物理量如速度、压强、剪切应力以及温度等也都是有确定值的连续可微函数，这些变量构成了标量场和矢量场。我们可以应用连续函数和场论对流体的运动及受力平衡问题进行有效的研究。

1.3　压强的表示方法

压强是流体单位面积所受的法向力，主要用于液体和气体，固体则称正应力。因为压强是单位面积上的力，所以其单位为 N/m²，也称作帕斯卡 [Pa]。工程中常用 kPa（$1\text{kPa} = 10^3\text{Pa}$）和 MPa（$1\text{MPa} = 10^6\text{Pa}$）。

实际压强以真空（即绝对零压强）为起点，称为绝对压强。而各种压强计在连通大气时指示数为零，所以测量的是与其相连的流体与周围大气间的压强差，因此定义**表压强（相对压强）**为相对于当地大气压的压强。如果某处的绝对压强低于当地大气压，则称其低于当地大气压的数值为**真空度**。所以，我们说某处"真空度高"实际是指这里的绝对压强很低。绝对压强、表压强和真空度之间的关系为

$$p_{\text{gage}} = p_{\text{abs}} - p_{\text{atm}} \qquad (1.3.1)$$

$$p_{\text{vac}} = p_{\text{atm}} - p_{\text{abs}} \qquad (1.3.2)$$

其间关系如图 1.3.1 所示。

以上两式中，p_{gage} 是表压强；p_{abs} 是绝对压强；p_{atm} 是当地大气压；p_{vac} 是真空度，单位均为 Pa。

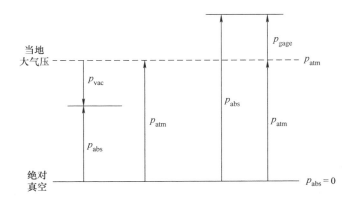

图 1.3.1　绝对压强、表压强和真空度关系

Pressure does not usually much affect the properties of liquids, so we commonly use gage pressures in problems dealing with liquids. Also, we usually find that the atmospheric pressure appears on both sides of an equation, and hence cancels. Thus the value of atmospheric pressure is usually of no significance when dealing with liquids, and, for this reason as well, we almost universally use gage pressures with liquids. Throughout this book, the pressure p will denote gage pressure unless specified otherwise.

Note that the atmospheric pressure is the local atmospheric pressure, which may change with time. However, if the local atmospheric pressure is not given, we use the value of 101325Pa or 101.3kPa, commonly, using standard atmospheric pressure of 100kPa is acceptable, which is within acceptable engineering accuracy.

Example 1.3.1   A vacuum gage connected to a chamber reads 45kPa at a location where the atmospheric pressure is 100kPa. Determine the absolute pressure in the chamber.

Solution   From Eq.1.3.2, $p_{vac} = p_{atm} - p_{abs}$=100kPa$ - p_{abs} = 45$kPa
The result is

$$p_{abs} = 55\text{kPa}$$

Example 1.3.2   A gage is connected to a tank in which the pressure of the fluid is 80kPa above atmospheric (Fig. X1.3.2a), what is the absolute pressure within the tank? If the absolute pressure of the fluid remains unchanged but the gage is in a chamber where the air pressure is reduced to a vacuum of 60kPa (Fig. X1.3.2b), what reading in kPa will then be observed?

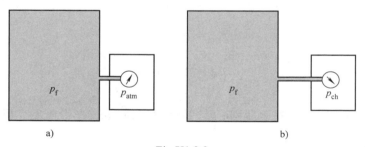

Fig.X1.3.2

Solution a   From Eq.1.3.1, $p_{abs} = p_{gage} + p_{atm} = 80\text{kPa} + 100\text{kPa} = 180\text{kPa}=p_f$

Solution b   In Fig.X1.3.2b, the gage measures the difference between the pressures of the fluid and that of the air in the chamber, which means,

$$p_{abs\ ch} = 100\text{kPa} - 60\text{kPa} = 40\text{kPa}$$

From Eq.1.3.1, gage pressure is calculated from

$$p_{gage} = p_{abs} - p_{atmch} = 180\text{kPa} - 40\text{kPa} = 140\text{kPa}$$

1.4   Fluid Properties

In this section we present several of the more common fluid properties. Some familiar properties are density, specific weight, some are less familiar ones such as viscosity, modulus of elasticity, vapor pressure.

体积膨胀系数 β（1/K）是指在压强不变的条件下，流体密度随温度变化的情况，定义式为

$$\beta = -\frac{\Delta\rho/\rho}{\Delta T}\bigg|_p = -\frac{1}{\rho}\left(\frac{\partial\rho}{\partial T}\right)\bigg|_p \tag{1.4.7}$$

因为密度是随温度增加而减小的，所以式 1.4.7 中用负号使 β 保持正值，其单位是 K^{-1}。

密度随压强和温度的变化率见式 1.4.8。

$$\frac{\mathrm{d}\rho}{\rho} = \frac{1}{\rho}\frac{\partial\rho}{\partial p}\mathrm{d}p + \frac{1}{\rho}\frac{\partial\rho}{\partial T}\mathrm{d}T = \kappa\mathrm{d}p - \beta\mathrm{d}T \tag{1.4.8}$$

式中，$\mathrm{d}\rho$ 是密度随压力和温度的变化量；$\mathrm{d}\rho/\rho$ 是密度的变化率。

大多数气体的密度与压强成正比而与温度成反比，而液体密度受压强的影响通常可以忽略。所以一般认为液体为不可压缩流体，气体的密度变化率小于 3% 时也可以被视为不可压缩流体。

1.4.3　黏性

两个接触的固体相对运动时，其接触面上会产生与运动方向相反的摩擦力。两部分流体做相对运动或流体与固体相对运动时也存在同样的情况。我们用黏性来描述流体运动时的这种内部摩擦阻力。**黏性**可以理解为流体的内部阻力，是流体的一个主要性质，对研究流体运动十分重要。

取两块平行平板中间存在 h 宽的间隙（见图 1.4.1），假设间隙内充满流体且沿 x 方向以不同速度流动，即速度 u 随 y 坐标值变化。保持下平板固定不动，对上平板施加恒定的外力 F，拖动上板以速度 U 平行于下板运动。

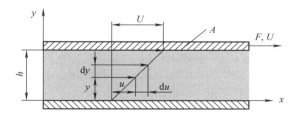

图 1.4.1　两平行平板间流体的运动

首先我们观察到，上平板处的流体会随上平板一起以速度 U 运动，而下平板处的流体则与下板一致保持速度为零。这种现象是由于黏性使得流体附着于固体平面上，称为**壁面不滑移**效应。

当间隙 h 和速度 U 的值都比较小时，平板间的流体做稳定层流流动，流速在 0

If the separation distance h and the velocity U is not too great, a steady laminar flow takes the fluid velocity varies linearly between 0 and U. The behavior of the fluid particles is much as if it consisted of a series of thin layers, each of which slips a little relative to the next.

Experiments have shown that the relationship between the applied force and the change rate of velocity is

$$F \propto A \frac{\mathrm{d}u}{\mathrm{d}y} \tag{1.4.9}$$

here quantity $\mathrm{d}u/\mathrm{d}y$ is a velocity gradient that equals U/h under the linear profile condition, and can be interpreted as a strain rate, describes the rate that a fluid deforms. If a constant of proportionality μ is introduced, we can express the shearing stress τ between any two thin layers of fluid by

$$\tau = \frac{F}{A} = \mu \frac{U}{h} = \mu \frac{\mathrm{d}u}{\mathrm{d}y} \tag{1.4.10}$$

Units of shearing stres are respresented by N/m^2 or Pa.

Eq.1.4.10 is known as *Newton's equation of viscosity* since Sir Isaac Newton (1642—1727) who expressed it first in 1687. In transposed form, Eq.1.4.11 defines the proportionality constant as viscosity μ (Pa · s)

$$\mu = \frac{\tau}{\mathrm{d}u/\mathrm{d}y} \tag{1.4.11}$$

here μ is known as the *coefficient of viscosity* or the *dynamic viscosity*, or simply the *viscosity* of the fluid.

The rate of deformation of a fluid is directly linked to the viscosity of the fluid. For a given stress, a highly viscous fluid deforms at a slower rate than a fluid with a low viscosity.

In many problems involving viscosity the dynamic viscosity is often divided by density, it has become useful to define the ratio as *kinematic viscosity* ν

$$\nu = \frac{\mu}{\rho} \tag{1.4.12}$$

so called the only dimensions being length and time, as in m^2/s.

We usually measure the kinematic viscosity in m^2/s in the SI, and another common unite is stoke (St, $1St = 1cm^2/s = 10^{-4}m^2/s$) or the centi-stoke (cSt, $1cSt = 10^{-6}m^2/s$), a more convenient unit to work with.

The dynamic viscosity μ, of most liquids is practically independent of pressure, and any small variation with pressure is usually disregarded, except at extremely high pressures. The kinematic viscosity ν of gases, however, varies strongly with pressure since density of gas is pressure sensitive.

Kinematic viscosity is useful for comparing the viscosity of two fluids with very different densities. For example, water has a dynamic viscosity of 1.01×10^{-3} Pa·s at 20°C and atmospheric condition, which is 55.8 times of air for 1.81×10^{-5} Pa · s. By kinematic viscosity comparison, that of water is $1.01 \times 10^{-6}m^2/s$ and $1.51 \times 10^{-5}m^2/s$ for air, the air is 15 times "sticky" than water.

至 U 之间线性变化。运动的流体好像被分成许多薄层一样，每一层相对邻层做一点点滑移。

上述实验同样揭示了拖动平板的外力与流体速度变化率之间的关系

$$F \propto A \frac{\mathrm{d}u}{\mathrm{d}y} \qquad (1.4.9)$$

式中，变量 $\mathrm{d}u/\mathrm{d}y$ 称为速度梯度，当流速为线性变化时速度梯度值等于 U/h，也可以将这一变量理解为剪切速率，表示流体的变形率。如果引入比例系数 μ，那么两层流体之间的剪切应力就可以表示为

$$\tau = \frac{F}{A} = \mu \frac{U}{h} = \mu \frac{\mathrm{d}u}{\mathrm{d}y} \qquad (1.4.10)$$

剪切应力的单位是 N/m^2 或 Pa。

式 1.4.10 就是著名的牛顿内摩擦定律，由牛顿（1642—1727）于 1687 年提出。将式 1.4.10 改写为式 1.4.11 的形式来表达比例系数 μ，即为流体的**黏度**

$$\mu = \frac{\tau}{\mathrm{d}u/\mathrm{d}y} \qquad (1.4.11)$$

式中，μ 是流体的黏性系数、动力黏度，或简单地称为黏度。

流体的变形率与其黏度直接相关。一定的剪切力作用下，高黏度流体变形较低黏度流体更慢。

分析与黏度相关问题时也常使用**运动黏度** ν，是将黏度除以流体密度得出的

$$\nu = \frac{\mu}{\rho} \qquad (1.4.12)$$

式中，ν 是运动黏度，其量纲仅由长度和时间构成。

运动黏度的国际单位为 m^2/s，工程中常用斯托克（缩写为 St，$1St = 1cm^2/s = 10^{-4}m^2/s$）或厘斯（cSt，$1cSt = 10^{-6}m^2/s$）更加方便。

压力不是非常大的情况下，绝大多数液体的黏度 μ 几乎不随压强变化，或黏度随压强的微小改变往往可以忽略。而气体的运动黏度 ν 则明显受压强影响，这是因为气体的密度对压力非常敏感。

在比较两种密度明显不同的流体时常用运动黏度。例如大气压下 20℃空气的动力黏度是 $1.81 \times 10^{-5}Pa \cdot s$，而水的黏度是 $1.01 \times 10^{-3}Pa \cdot s$，为空气黏度的 55.8 倍。但如果采用运动黏度进行比较，则水的运动黏度为 $1.01 \times 10^{-6}m^2/s$，空气运动黏度是 $1.51 \times 10^{-5}m^2/s$，比水"粘稠"15 倍。

Viscosity as an internal resistance is caused by the cohesive forces between the molecules and the molecular collisions, and it varies greatly with temperature. The viscosity of liquids decreases with temperature, whereas the viscosity of gases increases with temperature (Fig.1.4.2). This is because cohesive forces play a dominant role in a liquid, the molecules possess more energy at higher temperatures, and they can oppose the large cohesive intermolecular forces more strongly. As a result, the energized liquid molecules can move more freely. In a gas, on the other hand, the intermolecular forces are negligible for lower density and larger molecular distance, and the gas molecules at high temperatures move more actively at higher velocities. This results in more molecular collisions per unit volume per unit time and therefore in greater resistance to flow.

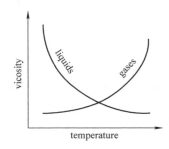

Fig.1.4.2　Trends in viscosity variation with temperature

The viscosity is to be measured by a viscometer shown in Fig.1.4.3. The concept of viscosity and velocity gradients can also be illustrated by this two concentric cylinders with gap filled with viscous liquid.

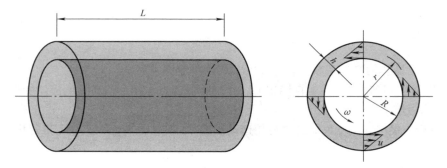

Fig.1.4.3　Rotating concentric cylinders with gap completely filled with liquid

In Fig.1.4.3 two concentric cylinders with gap completely filled with liquid, the inner cylinder rotates at constant rotational speed ω while the outer cylinder remains stationary. When the gap h is very small relative to radius of cylinder R ($h<<R$), flow between the cylinders can be modeled as two parallel plates, the velocity gradient can be approximated as being linear and depends only on r. Therefore

$$\frac{\mathrm{d}u}{\mathrm{d}r}=\frac{\omega R}{h} \qquad (1.4.13)$$

The torque necessary to rotate the inner cylinder at constant rotational speed ω is equal to the resistance torque, and the resistance is due to viscosity of liquid in the gap. The only stress that exists to resist the applied torque is a shear stress, which is observed to depend directly on the velocity gradient

黏性是由分子间内聚力和分子动量交换引起的内部阻力，很容易受温度影响。
液体黏度随温度的升高而减小，而气体黏度随温度升高而增大（见图 1.4.2）。这是
因为液体中分子内聚力影响显著，温度升高时分子内能增加更容易克服内聚力，所以高能量的液体分子运动更加自由。气体由于密度低、分子间距大，所以其分子内聚力可以忽略，温度升高时气体分子更加活跃、运动速度更快，所以单位体积内的气体分子在单位时间内发生碰撞的概率大大提高，动量交换的次数也随之增加，从而形成更大的流动阻力。

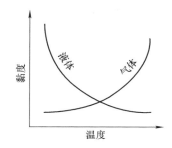

图 1.4.2 黏度随温度的变化趋势

图 1.4.3 所示为测量黏度的旋转黏度计，由两个同心圆筒组成，圆筒间的间隙内充满黏性液体。该图同样可以解释黏度和速度梯度的意义。

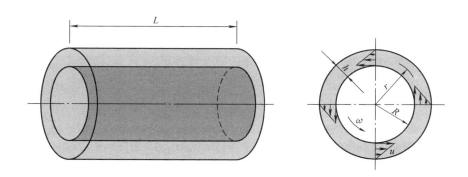

图 1.4.3 间隙内充满黏性液体的旋转同心圆筒

图 1.4.3 中两个同心圆筒间隙内充满黏性液体，外筒保持固定，内筒以恒定角速度 ω 旋转。当间隙 h 远远小于圆筒半径 R 时（$h \ll R$），间隙内的液体流动可以简化为两平行平板间的流动，速度可以视为沿半径 r 线性变化，即

$$\frac{\mathrm{d}u}{\mathrm{d}r} = \frac{\omega R}{h} \tag{1.4.13}$$

保证内筒以恒定角速度 ω 旋转所需的力矩与间隙内液体黏性阻力产生的力矩相等。阻力的作用形式是剪切力，其大小与速度梯度成正比

$$\tau = \mu \frac{du}{dr} = \mu \frac{U}{h} = \mu \frac{\omega R}{h} \qquad (1.4.14)$$

here the tangential velocity is $U = \omega R$ (angular velocity times the radius).

Noting that torque is $T = FR$ (force times the moment arm, which is the radius R of the inner cylinder in this case), taking the wetted surface area of the inner cylinder to be $A = 2\pi RL$ by disregarding the shear stress acting on the two ends of the inner cylinder, torque need to overcome the viscous force can be expressed as

$$T = \mu \frac{\omega R}{h} \cdot 2\pi RL \cdot R = \frac{2\pi R^3 L \omega \mu}{h} \qquad (1.4.15)$$

where L represent the length of rotating cylinders.

Note that in Eq.1.4.15 the torque depends directly on the viscosity, thus the cylinders could be used as a viscometer, a device that measures the viscosity of a liquid. When the torque is measured and the n for rpm (revolutions per minute) of a cylinder viscometer is given, the viscosity of the fluid is determined to be

$$\mu = \frac{60Th}{4\pi^2 R^3 Ln} \qquad (1.4.16)$$

Viscosity and kinematic viscosity of water at different temperature and of other common liquids at 20 ℃ are listed in Tab.1.4.1.

Tab.1.4.1 Viscosity of water at different temperature and of other common liquids at 20 ℃

liquid	Temperature/℃	Density/(kg/m³)	Viscosity/($\times 10^{-3}$Pa·s)	Kinematic viscosity / ($\times 10^{-6}$m²/s)
water	0	999.8	1.781	1.785
	5	1000.0	1.518	1.519
	10	999.7	1.307	1.306
	15	999.1	1.139	1.139
	20	998.2	1.002	1.003
	25	997.0	0.890	0.893
	30	995.7	0.798	0.800
	40	992.2	0.653	0.658
	50	988.0	0.547	0.553
	60	983.2	0.466	0.474
	70	977.8	0.404	0.413
	80	971.8	0.354	0.364
	90	965.3	0.315	0.326
	100	958.4	0.282	0.294
Sea water	20	1023	1.07	1.046
CTC	20	1588	0.97	0.611
Gasoline	20	680	0.29	0.426
Glycerin	20	1258	1494	1188
Kerosene	20	808	1.92	2.376
SAE l0 oil	20	918	82	89.325
SAE 30 oil	20	918	440	479.3

$$\tau = \mu \frac{du}{dr} = \mu \frac{U}{h} = \mu \frac{\omega R}{h} \quad (1.4.14)$$

式中，切向线速度等于角速度与内筒半径的乘积 $U = \omega R$。

力矩等于力与力臂的乘积，即，$T = FR$，这里力臂为内筒的半径，受力面积为内筒的侧面积。忽略作用在内筒两端面上的剪切力，转动内筒所需的力矩为

$$T = \mu \frac{\omega R}{h} \cdot 2\pi RL \cdot R = \frac{2\pi R^3 L \omega \mu}{h} \quad (1.4.15)$$

式中，L 是内筒的长度。

从式 1.4.15 可以看出，液体黏度决定转动力矩，所以该装置可用于测量液体黏度，称为旋转黏度计。如果力矩可以测量得出，内筒转速 n（每分钟转数）已知，则液体黏度可由下式计算

$$\mu = \frac{60Th}{4\pi^2 R^3 Ln} \quad (1.4.16)$$

表 1.4.1 列出了不同温度下水和 20℃其他几种常见液体在 20℃的黏度和运动黏度。

表 1.4.1 不同温度下水和 20℃其他几种常见液体的黏度和运动黏度

液体	温度 /℃	密度 /（kg/m³）	黏度 /（×10⁻³Pa·s）	运动黏度 /（×10⁻⁶m²/s）
水	0	999.8	1.781	1.785
	5	1000.0	1.518	1.519
	10	999.7	1.307	1.306
	15	999.1	1.139	1.139
	20	998.2	1.002	1.003
	25	997.0	0.890	0.893
	30	995.7	0.798	0.800
	40	992.2	0.653	0.658
	50	988.0	0.547	0.553
	60	983.2	0.466	0.474
	70	977.8	0.404	0.413
	80	971.8	0.354	0.364
	90	965.3	0.315	0.326
	100	958.4	0.282	0.294
海水	20	1023	1.07	1.046
四氯化碳	20	1588	0.97	0.611
汽油	20	680	0.29	0.426
甘油	20	1258	1494	1188
煤油	20	808	1.92	2.376
SAE 10 机油	20	918	82	89.325
SAE 30 机油	20	918	440	479.3

Fluids for which the rate of deformation (or strain rate, represented by velocity gradient) is linearly proportional to the shear stress are called *Newtonian fluids*, most common fluids such as water, air, gasoline, and oils are Newtonian fluids. *Non-Newtonian fluids*, the relationship between shear stress and rate of deformation is not linear, shown in Fig.1.4.4, often have a complex molecular composition, blood and liquid plastics are examples of non-Newtonian fluids. We will restrict the remainder of this book to the common fluids that under normal conditions obey Newton's equation of viscosity.

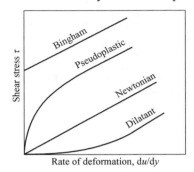

Fig.1.4.4　Variation for Newtonian and non-Newtonian fluids

Example 1.4.1　A 5mm wide space between two horizontal plane surfaces is filled with lubricating oil which viscosity is 40cSt and density is 900kg/m^3. What force is required to drag a very thin plate of 1m^2 area through the oil at a velocity of 1m/s if the plate is 2mm from upper surface?

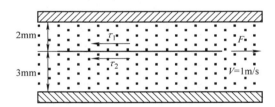

Fig.X.1.4.1　Move a plate that immerged by viscous fluid

Solution　The viscosity of oil is, $\mu = \nu\rho = 40 \times 10^{-6} \times 900 \text{Pa} \cdot \text{s} = 0.036 \text{Pa} \cdot \text{s}$.
The shearing stress on both sides of the thin plate

$$\tau_1 = \mu \frac{V}{h_1} = 0.036 \times 500 \text{N/m}^2 = 18 \text{N/m}^2$$

$$\tau_2 = \mu \frac{V}{h_2} = 0.036 \times \frac{1}{0.003} \text{N/m}^2 = 12 \text{N/m}^2$$

The force required to drag the thin plate is

$$F = F_1 + F_2 = \tau_1 A + \tau_2 A = 30 \text{N}$$

变形率（或应变率，用速度梯度表示）与剪切应力成线性关系的流体称为**牛顿流体**，常见流体中水、空气、汽油和食用油等都属于牛顿流体。**非牛顿流体**的应力 - 应变关系为非线性，如图 1.4.4 所示，通常具有复杂结构，血液和液体塑料等都是典型的非牛顿流体。本书中分析的流体均为牛顿流体。

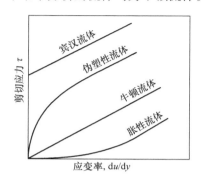

图 1.4.4　牛顿流体与非牛顿流体的区别

例 1.4.1　两平行平板间的间隙为 5mm，间隙内充满黏度为 40cSt、密度为 900kg/m³ 的润滑油。间隙内距上平板 2mm 处平行放置一面积为 1m² 的薄板。计算以 1m/s 的速度拖动薄板匀速移动所需力的大小。

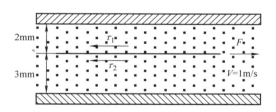

图 X.1.4.1　移动淹没在黏性液体中的平板

解　油液黏度 $\mu = \nu\rho = 40 \times 10^{-6} \times 900 \mathrm{Pa \cdot s} = 0.036 \mathrm{Pa \cdot s}$。
薄板所受剪切应力

$$\tau_1 = \mu \frac{V}{h_1} = 0.036 \times 500 \mathrm{N/m^2} = 18 \mathrm{N/m^2}$$

$$\tau_2 = \mu \frac{V}{h_2} = 0.036 \times \frac{1}{0.003} \mathrm{N/m^2} = 12 \mathrm{N/m^2}$$

拖动薄板所需的力为

$$F = F_1 + F_2 = \tau_1 A + \tau_2 A = 30 \mathrm{N}$$

Example 1.4.2 A viscometer is constructed with two 300-mm-long concentric cylinders, one 200mm in diameter and the other 202mm in diameter. A torque of 0.18N·m is required to rotate the inner cylinder at 300rpm, calculate the viscosity of fluid.

Solution The radius is $R = d/2 = 100$mm, gap $h = (D - d)/2 = 1$mm, Eq.1.4.16 provides

$$
\begin{aligned}
\mu &= \frac{60Th}{4\pi^2 R^3 Ln} \\
&= \frac{0.18 \times 0.001 \times 60}{4 \times 3.14^2 \times 0.1^3 \times 0.3 \times 300} \text{Pa} \cdot \text{s} \\
&= 0.003043 \text{Pa} \cdot \text{s}
\end{aligned}
$$

Note: All lengths are in meters so that the desired units on μ, are obtained.

Example 1.4.3 The clutch system shown in Fig.X1.4.3 is used to transmit torque through a 1.2-mm-thick oil film with $\mu = 0.38$Pa·s between two identical 300-mm-diameter disks. When the driving shaft rotates at a speed of 1450rpm, the driven shaft is observed to rotate at 1390rpm. Assuming a linear velocity profile for the oil film, determine the transmitted torque.

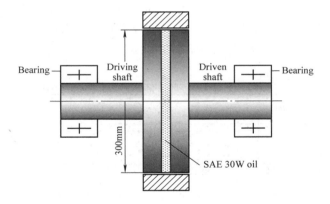

Fig.X1.4.3 An oil transmitted clutch

Solution The relative angular velocity between the driving disk and driven disk is

$$\Delta\omega = 2\pi(1450 - 1390)\text{r/min} = 2\pi 60\text{r/min} = 2\pi \text{rad/s}$$

Linear velocity distribution in direction of radius is, $u(r) = \Delta\omega r = 2\pi r$.
Strain rate of oil inside the gap is

$$\frac{\mathrm{d}u}{\mathrm{d}y} = \frac{u(r)}{h} = \frac{\omega r}{h} = \frac{2\pi r}{1.2 \times 10^{-3}}$$

Shear stress inside the oil film is

例 1.4.2 旋转黏度计筒长 300mm，内筒直径 200mm，外筒直径 202mm。当内筒以 300r/min 转动时所需转矩为 0.18N·m，确定被测液体的黏度。

解 内筒半径 $R = d/2 = 100$mm，间隙宽 $h = (D-d)/2 = 1$mm，根据式 1.4.16 可得

$$\mu = \frac{60Th}{4\pi^2 R^3 Ln}$$
$$= \frac{0.18 \times 0.001 \times 60}{4 \times 3.14^2 \times 0.1^3 \times 0.3 \times 300} \text{Pa} \cdot \text{s}$$
$$= 0.003043 \text{Pa} \cdot \text{s}$$

注意：各长度尺寸单位均为米（m）。

例 1.4.3 如图 X1.4.3 所示，黏性液体离合器采用厚度为 1.2mm 的油膜传递转矩，油液黏度 $\mu = 0.38$Pa·s，两转盘直径 300mm。主动盘以转速 1450r/min 转动时，测得从动盘转速为 1390r/min。假设油膜内速度线性分布，计算该离合器可传递的转矩。

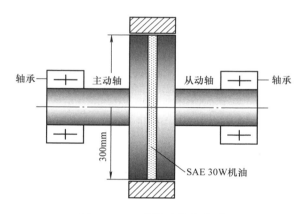

图 X1.4.3 黏性液体离合器

解 主动盘和从动盘间角速度差为

$$\Delta\omega = 2\pi(1450-1390)\text{r/min} = 2\pi 60\text{r/min} = 2\pi\text{rad/s}$$

沿半径方向线速度分布规律为 $u(r) = \Delta\omega r = 2\pi r$
间隙内油液的剪切速率为

$$\frac{\mathrm{d}u}{\mathrm{d}y} = \frac{u(r)}{h} = \frac{\omega r}{h} = \frac{2\pi r}{1.2 \times 10^{-3}}$$

油膜内剪切应力

$$\tau = \mu \frac{\mathrm{d}u}{\mathrm{d}y} = \mu \frac{2\pi r}{h}$$

An element of shearing force is

$$\mathrm{d}F_\tau = \tau \mathrm{d}A = \tau 2\pi r \mathrm{d}r = \mu \frac{4\pi^2}{h} r^2 \mathrm{d}r$$

Transmitted torque can be determined by

$$T = \int_0^R r\mathrm{d}F_\tau = \int_0^R \frac{4\mu\pi^2}{h} r^3 \mathrm{d}r = \int_0^R \frac{4\mu\pi^2}{1.2\times10^{-3}} r^3 \mathrm{d}r = 1.58\mathrm{N}\cdot\mathrm{m}$$

1.4.4　Engineering applications

The viscosity of some fluids changes when a strong electric field is applied on them. This phenomenon is known as the electrorheological (ER) effect, and fluids that exhibit such behavior are known as ER fluids. The Bingham plastic model for shear stress, which is expressed as $\tau = \tau_y + \mu(\mathrm{d}u/\mathrm{d}y)$ is widely used to describe ER fluid behavior because of its simplicity(Fig.X1.4.4a). Where τ_y is defined as yield stress, which is adjustable by changing the applied electric field. The fluid does not deform, i.e. $\mathrm{d}u/\mathrm{d}y$ has been to zero when yield stress is not reached, the ER fluid flows and exhibits the behavior similar to Newtonian fluids until the shear stress exceeds the yield stress, i.e. $\tau > \tau_y$.

One of the most promising applications of ER fluids is the ER clutch. A typical multi-disk ER clutch consists of several equally spaced steel disks of inner radius R_1 and outer radius R_2, N of them attached to the input shaft. The gap h between the parallel disks is filled with ER fluid, shown in Fig.X1.4.4b. The relationship for the torque generated by the clutch through the output shaft is adjustable by the applied electric field.

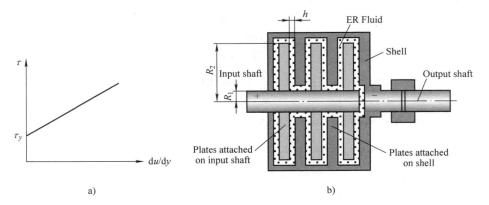

Fig.A1.4.4　Schematic diagram of multi-plate-cluth

a) Bingham plastic model of ER fluid　b) a multi-plate-clutch applied the ER fluid

A strain rate in the ER fluid filled in the gap due to the rotative speed difference $\Delta\omega$ between the input and output shaft is essential to generate a transmitting torque.

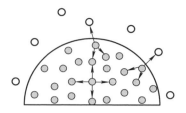

图 1.4.5 半球形液滴的表面张力

我们来分析一下水滴内的压强。如图 1.4.6a 所示的水滴，内部压强为 $p_1\pi R^2$，外部压力为 $p_2\pi R^2$，内外压力与表面张力平衡，所以 $(p_1 - p_2)\pi R^2 = 2\pi R\sigma$，化简为

$$p_1 - p_2 = \frac{2\sigma}{R} \tag{1.4.17}$$

液体薄膜泡泡同样是内外压力与表面层两侧张力平衡的结果，假设气泡膜的厚度极小，取膜的内、外表层半径均为 R，如图 1.4.6.b 所示。

$$\pi R^2 (p_1 - p_2) = 2 \times 2\pi R\sigma$$

$$p_1 - p_2 = \frac{4\sigma}{R} \tag{1.4.18}$$

我们看到泡泡内部的压强是同样大小的液滴内部压强的二倍。

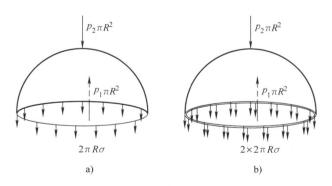

图 1.4.6 水滴和气泡的表面张力

a）水滴 b）泡泡内的压强

不同物质的表面张力差别很明显。表 1.4.2 列出了几种常见液体在 1 标准大气压下的表面张力值。总的来说，液体的表面张力随温度的升高而减小，并且在临界温度（物质以液态存在的最高温度）时下降至零。压力对表面张力的影响不明显，可以忽略不计。

Tab.1.4.2　Surface tension of some fluids in air at 1atm

Fluid	temprature/℃	Surface tension σ /(N/m)
Water	0	0.076
	20	0.073
	100	0.059
	300	0.014
Mercury	20	0.440
SAE 30 oil	20	0.035
Gasoline	20	0.022
Kerosene	20	0.028
Soap solution	20	0.025
Ammonia	20	0.021
Glycerin	20	0.063
Ethyl alcohol	20	0.023
Blood	37	0.058

Another interesting consequence of surface tension is the *capillary effect* or *capillarity*, which is the rise or fall of a liquid in a small-diameter tube inserted into the liquid, such narrow tubes are called *capillary tubes*. It is commonly observed that capillarity makes water rise in a glass tube, while mercury depresses below the true level. This effect is usually expressed by saying that water wets the glass while mercury does not, and can be considered due to both cohesion and adhesion. When the cohesion between water molecules for example, is of less effect than the adhesion between water and glass tube, the water will wet a glass surface it touches and rise at the point of contact (capillary rise); when cohesion within mercury predominates, the liquid surface will depress at the point of contact (capillary drop).

Figure 1.4.7a shows the rise of a liquid in a clean glass capillary tube due to surface tension. If h is the capillary rise, d the diameter，ρ the density, and σ the surface tension, h can be determined from equating the vertical component of the surface tension force to the weight of the liquid column.

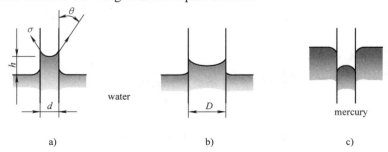

Fig.1.4.7　Capillary rise and capillary drop
a)、b) Capillary rise of water in tubes of diameter d and D ($d < D$)　c) Capillary drop of mercury

表 1.4.2　几种液体在 1 标准大气压时的表面张力值

液体	温度 /℃	表面张力 σ /（N/m）
水	0	0.076
	20	0.073
	100	0.059
	300	0.014
水银	20	0.440
SAE 30 机油	20	0.035
汽油	20	0.022
煤油	20	0.028
皂液	20	0.025
氨水	20	0.021
甘油	20	0.063
无水乙醇	20	0.023
血液	37	0.058

　　表面张力的另一个有趣表现是**毛细现象**，是指插入液体内部非常细的小管内液面高度的上升或下降，这样的小管称为**毛细管**。最常见的毛细现象是玻璃管内水柱高于真实液面而水银柱低于真实液面，我们通常把这两种情况分别称为水对玻璃浸润而水银对玻璃不浸润。以上两种情况可以通过内聚力和附着力解释，例如，当水分子间的内聚力小于水对玻璃管壁的附着力时，水就会浸润玻璃而上升一定高度；而水银分子内聚力占优势时，水银液柱则下降（不浸润）。

　　图 1.4.7a 所示液体在干净玻璃管内由于表面张力而浸润的情况，平衡状态下表面张力的竖直分量与液柱重量相等，用 d 为管径，ρ 为液体密度，σ 为表面张力，可计算液柱浸润高度 h。

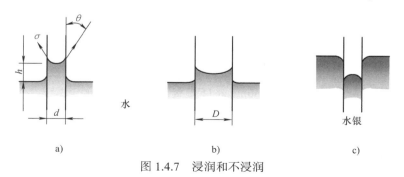

a)　　　　　　　　　　b)　　　　　　　　　c)

图 1.4.7　浸润和不浸润

　　a）水在毛细管内浸润　b）水在较大毛细管内浸润　c）水银在毛细管内不浸润

$$\sigma \pi d \cos \theta = \rho g \frac{\pi d^2}{4} h \tag{1.4.19}$$

therefore

$$h = \frac{4\sigma \cos \theta}{\rho g d} \tag{1.4.20}$$

here, θ is a contact angle the liquid makes with the glass tube and it will be very small if the tube is clean，$\theta = 0°$ for water and about 140° for mercury. When the contact angle is bigger than 90°, then $\cos\theta < 0$, it makes $h < 0$, which is shown as Fig1.4.7c, called capillary drop, or we can say that mercury is not wet the glass.

Also note that the capillary rise is inversely proportional to the diameter of the tube. Therefore, the thinner the tube is, the greater the rise (or fall) of the liquid in the tube. When we use small tubes to measure fluid properties, such as pressures, we must take the readings while aware of the surface tension effects. When pressure measurements are made using manometers, it is important to use sufficiently large tubes to minimize the capillary effect (e.g. Fig.1.4.7b). In practice, the capillary effect for water is usually negligible in tubes whose diameter is greater than 10mm.

Example 1.4.5 Water at 5℃ stands in a clean glass tube of 2-mm diameter at a height of 40 mm. What is the true static height? (Surface tension of water at 5℃ is 0.0749N/m.)

Solution For clean glass tube $\theta = 0°$, Surface tension at 5℃ is 0.0749N/m.

Eq.1.5.20 $h = \dfrac{4\sigma \cos \theta}{\rho g d} = \dfrac{4 \times 0.0749}{9800 \times 0.002} = 0.0153\text{mm} = 15.3\text{mm}$

True static height = 40mm − 15.3mm = 24.7mm

1.4.6 Vapor Pressure and Cavitation

The *vapor pressure* of a pure substance is the pressure exerted by its vapor molecules when the system is in phase equilibrium with its liquid molecules at a given temperature. The vapor pressure is the pressure resulting from liquid vaporize. In other words, the vapor pressure, at a given temperature, is the phase-change pressure between the liquid and vapor phases.

A transition from the liquid state to the gaseous state occurs if the local absolute pressure is less than the vapor pressure of the liquid. The vapor pressure of water at 15℃ is 1.71kPa absolute, at this temperature the water is boiling, that is, the liquid state of the water can no longer be sustained because the attractive forces are not sufficient to contain the molecules in a liquid phase. Vaporization will terminate when equilibrium is reached between the liquid and gaseous states of water.

The vapor pressure is a property of pure substance, p_v is different from one liquid to another, the vapor pressure of ammonia at 15℃ is 33.8kPa absolute.

The vapor pressure is highly dependent on temperature, it increases significantly when the temperature increases. For example, the vapor pressure of water at 20℃ is 2.34kPa absolute. Vapor pressure increases with temperature, thus, a substance at higher pressure boils at higher temperature. For example, water boils at 134℃ in a pressure cooker operating at 3atm absolute pressure. At high elevations where the atmospheric pressure is relatively low, boiling occurs at temperatures less than 100℃ . At an elevation of 2000m, boiling would occur at approximately 93℃ where the atmospheric pressure is 0.8atm, and at 90℃ of a 3000m elevation. An abridged table for water

$$\sigma \pi d \cos\theta = \rho g \frac{\pi d^2}{4} h \qquad (1.4.19)$$

因此

$$h = \frac{4\sigma \cos\theta}{\rho g d} \qquad (1.4.20)$$

式中，θ 是液体与玻璃管壁的接触角，水与干净玻璃管的接触角很小，可假设 $\theta = 0°$，水银与玻璃的接触角约 140°。当接触角大于 90° 时，$\cos\theta < 0$，所以对于水银 $h<0$，如图 1.4.7c 所示，为不浸润。

由式 1.4.20 可知，浸润高度与毛细管直径成反比，管径越小浸润（或不浸润）高度越大。当使用玻璃管测量流体属性，如测量压强时，需考虑毛细现象的影响。测压计中使用的玻璃管直径应取得适当大一些以减小毛细现象的影响（见图 1.4.7.b）。工程上使用的玻璃管直径大于 10mm 时，毛细现象的影响可以忽略。

例 1.4.4　直径 2mm 的玻璃管内 5℃ 水柱高 40mm，计算水柱的真实高度。（5℃ 水的表面张力为 0.0749N/m）

解　干净玻璃管与水的接触角 $\theta = 0°$，5℃ 时水的表面张力为 0.0749N/m。

由式 1.4.20 得，$h = \dfrac{4\sigma \cos\theta}{\rho g d} = \dfrac{4 \times 0.0749}{9800 \times 0.002}$mm=0.0153mm=15.3mm

水柱真实高度 = 40mm − 15.3mm = 24.7mm

1.4.6　饱和蒸汽压和气穴现象

纯净物质的**饱和蒸汽压**是指在一定温度下，与液体处于相平衡的蒸汽的压强。饱和蒸汽压是液体气化时的压强，换句话说，是在一定温度下发生气 - 液相变的压强。

当地绝对压强低于饱和蒸汽压时，物质就会由液相转化为气相。15℃ 水的饱和蒸汽压为 1.71kPa 绝对压强。在此压强和温度下，水分子间的引力无法维持其液态，即水发生沸腾而气化，当气 - 液达到平衡时气化过程停止。

饱和蒸汽压是纯净物质的属性，用 p_v 表示，不同液体具有不同的饱和蒸汽压，5℃ 氨的饱和蒸汽压为 33.8kPa 绝对压强。

饱和蒸汽压对温度极其敏感，温度升高时饱和蒸汽压也明显升高。如 20℃ 水的饱和蒸汽压可达 2.34kPa 绝对压强。因为饱和蒸汽压随温度升高，反过来高压下物质的沸点也会升高。如工作压力为 3 个大气压的高压锅内水的沸点为 134℃。大气压较低的高海拔地区水的沸点会低于 100℃。海拔 2000m 大气压为 0.8atm 的位置，水的沸点约为 93℃，海拔高度升至 3000m 时，沸点降至 90℃。表 1.4.3 节选了一些不同温度下水的饱和蒸汽压，也可参照图 1.4.8 找出更具体温度下的蒸汽压数值。表 1.4.4 选取了一些常用液体 16 ~ 21℃ 间的饱和蒸汽压。

is given in Table 1.4.3 and figure 1.4.8 for easy reference. Vapor pressure of some common liquids at approximately 16 to 21℃ are listed in Table.1.4.4.

Table 1.4.3　Vapor pressure of water at various temperatures

Temperature $T/℃$	−10	−5	0	5	10	15	20
Vapor pressure p_v/kPa	0.26	0.403	0.611	0.872	1.23	1.71	2.34
Temperature $T/℃$	25	30	40	50	60	70	80
Vapor pressure p_v/kPa	3.17	4.25	7.38	12.35	19.9	31.2	47.3
Temperature $T/℃$	90	100	150	200	250	300	
Vapor pressure p_v/kPa	70.1	101.3	475.8	1554	3973	8581	

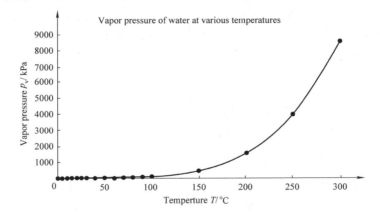

Fig.1.4.8　Vapor pressure of water at various temperatures

Table 1.4.4　Vapor pressure of common liquids at approximately 16 to 21℃

liquid	Vapor pressure p_v/kPa
Benzene	10.3
Carbon tetrachloride	86.2
Glycerin	1.4×10^{-5}
Mercury	1.59×10^{-4}
Turpentine	5.31×10^{-2}
Gasoline	37.1
Kerosene	46.5

In liquid flows, conditions can be created that lead to a pressure below the vapor pres-

表 1.4.3　不同温度下水的饱和蒸汽压

温度 $T/℃$	−10	−5	0	5	10	15	20
饱和蒸汽压 p_v/kPa	0.26	0.403	0.611	0.872	1.23	1.71	2.34
温度 $T/℃$	25	30	40	50	60	70	80
饱和蒸汽压 p_v/kPa	3.17	4.25	7.38	12.35	19.9	31.2	47.3
温度 $T/℃$	90	100	150	200	250	300	
饱和蒸汽压 p_v/kPa	70.1	101.3	475.8	1554	3973	8581	

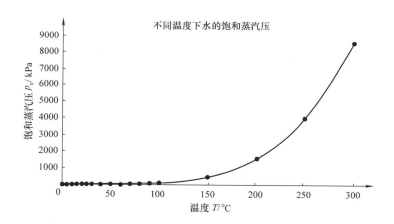

图 1.4.8　不同温度下水的饱和蒸汽压

表 1.4.4　常用液体 16 ~ 21℃的饱和蒸汽压

液体	饱和蒸汽压 p_v/kPa
苯	10.3
四氯化碳	86.2
甘油	1.4×10^{-5}
水银	1.59×10^{-4}
松节油	5.31×10^{-2}
汽油	37.1
煤油	46.5

　　液体在某些流动条件下，可能发生压强低于饱和蒸汽压的情况。我们之所以要

sure of the liquid. The reason for our interest in vapor pressure is the possibility of the liquid pressure in liquid-flow systems dropping below the vapor pressure at some locations, and the resulting unplanned vaporization. When this happens, bubbles are formed locally. This phenomenon, called *cavitation*, can be very damaging when these bubbles are transported by the flow to higher-pressure regions. What happens is that the bubbles collapse upon entering the higher-pressure region, and this collapse produces local pressure spikes which have the potential of damaging a pipe wall or other working elements.

For example, water at 10℃ may vaporize and form bubbles at locations (such as suction sides of pumps or the tip regions of impellers) where the pressure drops below 1.23kPa. The vapor bubbles collapse as they are swept away from the low-pressure regions, generating highly destructive, extremely high-pressure waves. This phenomenon, which is a common cause for drop in performance and even the erosion of hydraulic components, and it is an important consideration in the design of hydraulic pumps and turbines.

Cavitation must be avoided, or at least minimized, in most flow systems since it reduces performance, generates annoying vibrations and noise, and causes damage to equipment. The presence of cavitation in a flow system can be sensed by its characteristic tumbling sound. Additional discussion on cavitation is included in Section 4.4.

Example 1.4.5 Estimating the boiling temperature of water in a pressure cooker operating at 1.8 atm absolute pressure. Using Fig.1.4.8.

Solution 1.8 atm absolute pressure = 182.385kPa

Water boils at 116.93℃ at 1.8 atm absolute pressure.

Example 1.4.6 A pump is used to drain water from lake, the temperature of water is observed to be as high as 25℃. Determine the minimum pressure allowed in the pumping system to avoid cavitation.

Solution The vapor pressure of water at 25 ℃ is 3.17kPa. To avoid cavitation, the pressure anywhere in the system should not be allowed to drop below the vapor pressure. Therefore, the pressure should be maintained above 3.17kPa everywhere in the flow.

Example 1.4.7 Calculate the vacuum necessary to cause cavitation in a water flow at a temperature of 80℃ somewhere in Kunlun Mountain where the elevation is 2500m, the atmospheric pressure is about 75kPa.

Solution The vapor pressure of water at 80℃ is given in Table 1.4.1. It is 47.3kPa absolute. The atmospheric pressure is 75kPa.

The required pressure is then

$$p = 47.3\text{kPa} - 75\text{kPa} = -27.7\text{kPa} \quad \text{or} \quad p_{\text{vac}} = 27.7\text{kPa}$$

1.5 Forces on a Fluid

Each fluid particle is acted by some kinds of forces no matter it keeps moving or balance. According to the behaviors of the forces there are two types of forces: mass force and surface force.

Mass force is a force exerting on all particles in a fluid by certain force field, the magnitude of *mass force* is in direct proportion to mass of the fluid. Gravity, inertia force are known as mass force or field force, body force.

The mass force acting on unit mass fluid is called unit mass force, usually denoted by

对这一现象进行分析，就是为了防止由于液体流动系统内局部压力骤降至低于饱和蒸汽压时可能突发的气化问题。气化发生的局部会产生气泡，气泡被流动液体运输到高压区可能造成严重破坏，这种现象称为**气穴现象**。进入高压区的气泡可能发生破裂，导致相应位置压强陡升甚至可能损坏管壁和其他元件。

例如，10℃的水在压强低于 1.23kPa 的区域（例如泵的吸入口或叶轮尖部等）就会气化并产生气泡。这些气泡离开低压区后就可能发生破裂从而形成破坏力极大的高压冲击波，这是导致液压元件工作性能降低甚至发生锈蚀的原因之一，也是设计液压泵和涡轮机时必须要考虑的因素。

多数流动系统内的气穴现象因其发生可导致系统性能下降、振动和噪声甚至损坏内部元件，所以必须被消除或抑制。液体流动时的不正常的翻滚声会提示可能的气穴现象，4.4 节还会继续讨论气穴现象。

例 1.4.5　估算工作压力为 1.8atm 绝对压强的高压锅内水的沸点。可根据图 1.4.8 查找相关数据。

解　1.8atm 绝对压强 = 182.385kPa

在此压强下水的气化温度为 116.93℃。

例 1.4.6　用一台水泵从湖中抽水，水温约 25℃。为避免发生气穴现象，水泵系统的最低压力为多少？

解　25℃水的气化压力为 3.17kPa。为避免气穴现象，水泵系统内各处的压强都应不低于此气化压力值。所以流动系统内各处压力均应高于 3.17kPa。

例 1.4.7　昆仑山脉某处海拔高度 2500m，大气压约 75kPa，欲将 80℃的水气化，计算所需的压力。

解　由表 1.4.1 查得，80℃水的饱和蒸气压为 47.3kPa 绝对压强。当地大气压为 75kPa，

则所需压强为

$$p = 47.3\text{kPa} - 75\text{kPa} = -27.7\text{kPa} \qquad 或 \qquad p_{\text{vac}} = 27.7\text{kPa}$$

1.5　流体受的力

任何流体质点，无论是运动还是静止，都受到力的作用。这些力根据其特点可以分为两类：质量力和表面力。

流场内作用于流体质点上的**质量力**，其大小与流体的质量成正比。重力、惯性力都是质量力或称为场力。

单位质量流体所受的质量力称为单位质量力，通常表示为

$$\boldsymbol{f} = f_x \boldsymbol{i} + f_y \boldsymbol{j} + f_z \boldsymbol{k} \tag{1.5.1}$$

where f_x, f_y and f_z are components in x, y, and z direction.

In field of gravity, the unit mass force is gravitational acceleration g, can be expressed as

$$\boldsymbol{f} = -g\boldsymbol{k} \tag{1.5.2}$$

here the negative sign is due to the direction of gravity is vertical downward, opposites to the direction of z axis.

A force that is directly proportional to surface area of a fluid and distributes on the fluid surface is called *surface force*. The press force and the friction or shearing force are all surface forces.

Classified according to the direction of acting, the surface force can be resolved into the normal force and the tangential force (Fig.1.5.1). Force in outward normal direction is the tensile force, and is positive; force in the opposite direction of outward normal vector, is the press force, and is negative.

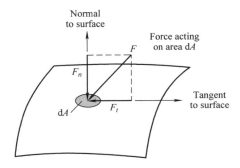

Fig.1.5.1   The normal stress and shear stress at the surface of a fluid element

Force per unit area is defined as *stress* and is determined by dividing the force by the area upon which it acts. The normal component of a force acting on a surface per unit area is called the normal stress, i.e., $\sigma = F_n / \mathrm{d}A$, the one is called pressure if it points to inward normal direction. The tangential component of a force acting on a surface per unit area is called shear stress, i.e., $\tau = F_t / \mathrm{d}A$.

A fluid at rest is at a state of zero shear stress, whereas a fluid deforms continuously under the influence of a shear stress, no matter how small, and the stress is proportional to strain rate. When a constant shear stress is applied, a fluid never stops deforming and approaches a constant rate of strain.

Exercises 1

1.1   What is specific gravity? How is it related to specific weight?

1.2   What does the coefficient of volumetric compressibility of a fluid represent? What does the bulk modulus of elasticity of a fluid represent?

1.3   What does the coefficient of volume expansion of a fluid represent?

1.4   What is the no-slip condition? What causes it?

$$f = f_x i + f_y j + f_z k \tag{1.5.1}$$

式中，f_x，f_y 和 f_z 是质量力在 x，y 和 z 方向的分力。

重力场中，单位质量力是重力加速度 g，表示为

$$f = -gk \tag{1.5.2}$$

式中负号表示重力的方向竖直向下，与 z 轴正方向相反。

大小与流体受力面的面积成正比的力称为**表面力**。压强、摩擦力或剪切力都是表面力。

力根据其作用方向可分为法向力和切向力（见图 1.5.1）。沿外法向的力为拉力，为正；与外法线方向相反的力为压力，为负。

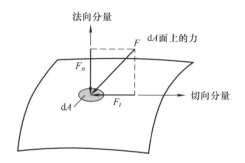

图 1.5.1　流体微元表面的法向力和切向力

作用于单位面积上的力为应力，其大小为力与受力面积之比。单位面积上的法向力称为正应力（或法向应力），即，$\sigma = F_n/\mathrm{d}A$，内法向的正应力称为压强。单位面积上的切向力称为剪切应力，即，$\tau = F_t/\mathrm{d}A$。静止流体不能承受剪切力，受到剪切力作用的流体，无论该剪切力多么小，都会产生变形，且变形率与剪切力的大小成正比。当一恒定的剪切力作用于流体时，流体也会以一定的速率持续变形。

习题一

1.1　什么是比重？比重与重度的关系如何？

1.2　如何表示流体的体积压缩系数？如何表示流体的弹性模量？

1.3　如何表示流体的体积膨胀系数？

1.4　什么是壁面不滑移？引起壁面不滑移的原因是什么？

1.5   How does the viscosity of liquids and gases vary with temperature?

1.6   What is a Newtonian fluid? Is water a Newtonian fluid? Is ER fluid a Newtonian fluid?

1.7   What is viscosity? What is the cause of it in liquids and in gases? Are liquids stickier than gases?

1.8   Express the dynamic viscosity with the units of kg, m and s.

1.9   What is surface tension? What is it caused by?

1.10   Is the pressure inside a soap bubble higher or lower than the pressure outside?

1.11   What is the capillary effect? What is it caused by?

1.12   How is the capillary rise or capillary down affected by the contact angle?

1.13   What is vapor pressure? Does water boil at higher temperatures at higher pressures?

1.14   What is cavitation? What causes it?

1.15   Define stress, shear stress, and pressure.

1.16   Define mass force and surface force.

1.17   The pressure at Typhoon center 97500Pa can be written as
A. 97.5hPa   B. 97.5kPa   C. 97.5MPa   D. 97.5bar

1.18   A 5000N force acts on a $250cm^2$ area at an angle of 30° to the normal. The shear stress acting on the area is
A. 10kPa   B. 100kPa   C. 1MPa   D. 10MPa

1.19   The mass of 5 liters of certain kind of liquid is 4.3kg. The density of this liquid is:
A. $0.86kg/m^3$ B. $86kg/m^3$ C. $860kg/m^3$ D. $8600kg/m^3$

1.20   The viscosity of the liquid in exercise 1.19 is measured of $2\times10^{-3}Pa\cdot s$ at 30°C, the kinematic viscosity of this liquid is nearest:
A. $1.72mm^2/s$ B. $17.2mm^2/s$ C. $2.33mm^2/s$ D. $23.3mm^2/s$

1.21   The specific weight of a liquid which 48kg occupies 30 Liters is nearest:
A. $1.6N/m^3$ B. $1600N/m^3$ C. $15.969N/m^3$ D. $15969N/m^3$

1.22   The specific gravity of mercury can be usually taken as
A. $13.6kg/m^3$ B. $13600kg/m^3$ C. 13.6   D. 13600

1.23   A liquid with a specific gravity of 1.2 fills a volume. If the mass in the volume is 144 kg, the magnitude of the volume is
A. $0.12m^3$ B. $1.2m^3$ C. $12m^3$ D. $120m^3$

1.24   The specific weight of an unknown liquid is $9310N/m^3$. The mass of the liquid contained in a volume of $1500cm^3$ is
A. 13.965kg   B. 1396.5kg   C. 1.425kg   D. 1425kg

1.25   The measured viscosity of a liquid with a density of $900kg/m^3$ is $45\times10^{-3}Pa\cdot s$, the kinematic viscosity of this fluid is:
A. 40.5cSt   B. 4050cSt   C. 50cSt   D. 0.05cSt

1.26   The velocity distribution in a 50-mm-diameter pipe transporting 20°C water is given by $u(r)=2.5m/s-4000r^2(m/s)$. The shearing stress at the wall is nearest:
A. 0.01Pa   B. 0.1Pa   C. 1Pa   D. 10Pa

1.27   The distance 5°C water would climb in a long 3-mm-diameter, clean glass tube is nearest:

1.5　液体和气体黏度如何随温度变化？

1.6　什么是牛顿流体？水是否为牛顿流体？电流变液是否为牛顿流体？

1.7　什么是黏度？导致液体和气体黏性的主要原因是什么？液体一定比气体更黏稠吗？

1.8　请用物理量单位千克、米和秒表示动力黏度。

1.9　什么是表面张力？引起表面张力的原因是什么？

1.10　肥皂泡内部的压强较外界压强高还是低？

1.11　什么是毛细现象？引起毛细现象的原因是什么？

1.12　接触角如何影响液体对某物体浸润或不浸润？

1.13　什么是饱和蒸气压？高压下水的沸点也会相应提高吗？

1.14　什么是气穴现象？引起气穴现象的原因是什么？

1.15　说出应力、剪切应力和压强的定义。

1.16　说出质量力和表面力的定义。

1.17　台风中心气压为 97500Pa，也可以表示为

A. 97.5hPa　　　　B. 97.5kPa　　　　C. 97.5MPa　　　　D. 97.5bar

1.18　5000N 力作用于 $250cm^2$ 的平面上且与平面法向夹角 30°，则平面上的剪切应力为

A. 10kPa　　　　B. 100kPa　　　　C. 1MPa　　　　D. 10MPa

1.19　某种液体 5L 的质量为 4.3kg，则该液体的密度为

A. $0.86kg/m^3$　　B. $86kg/m^3$　　C. $860kg/m^3$　　D. 8600kg/m

1.20　19 题中的液体 30℃时黏度为 $2\times10^{-3}Pa\cdot s$，则其运动黏度约为

A. $1.72mm^2/s$　　B. $17.2mm^2/s$　　C. $2.33mm^2/s$　　D. $23.3mm^2/s$

1.21　质量为 48kg、体积为 30L 的液体，其重度约为

A. $1.6N/m^3$　　B. $1600N/m^3$　　C. $15.969N/m^3$　　D. $15969N/m^3$

1.22　水银的比重通常取值为

A. $13.6kg/m^3$　　B. $13600kg/m^3$　　C. 13.6　　　　D. 13600

1.23　某比重为 1.2 的液体质量为 144kg，则其体积为

A. $0.12m^3$　　B. $1.2m^3$　　C. $12m^3$　　D. $120m^3$

1.24　某种未知液体重度为 $9310N/m^3$，则体积 $1500cm^3$ 的该液体质量为

A. 13.965kg　　B. 1396.5kg　　C. 1.425kg　　D. 1425kg

1.25　密度为 $900kg/m^3$ 的液体黏度为 $45\times10^{-3}Pa\cdot s$，则其运动黏度为

A. 40.5cSt　　　B. 4050cSt　　　C. 50cSt　　　　D. 0.05cSt

1.26　20℃的水在直径 50mm 的管道内流动，流速分布规律为 $u(r)=2.5m/s-4000r^2$（m/s）。则管壁处的剪切应力约为

A. 0.01Pa　　　B. 0.1Pa　　　C. 1Pa　　　　D. 10Pa

A. 0.1mm   B. 1mm   C. 10mm   D. 100mm

1.28   A pressure gage attached to a rigid tank measures a vacuum of 30kPa inside the tank which is situated at a site in Tibet Plateau where the atmospheric pressure is 78.9kPa. Determine the absolute pressure inside the tank.

1.29   A fluid that occupies a volume of 32L weighs 280N. Determine the mass of this fluid and its density.

1.30   A gage pressure of 51.3kPa is read on a gage, find the absolute pressure.

1.31   A vacuum of 28kPa is measured in an airflow, find the absolute pressure.

1.32   For a constant-temperature atmosphere, the pressure as a function of elevation is given by $p(z) = p_0 e^{-gz/RT}$, where g is gravity, $R = 287J/(kg \cdot K)$, and T (K) is the absolute temperature. Use this equation and estimate the pressure at 5827m the height peak of Chi-lien Mountains , assuming that $p_0 = 101kPa$ and $T = -35℃$.

1.33   An applied force of 200kN is distributed uniformly over a 150-cm^2 area; however, it acts at an angle of 60° with respect to a normal vector (see Fig.E1.33). Calculate the resulting pressure and tangential stress.

1.34   The force on an area of 60cm^2 is due to a pressure of 120kPa and a shear stress of 900Pa, as shown in Fig.E1.34. Calculate the magnitude of the force acting on the area and the angle of the force with respect to a normal coordinate.

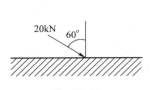

Fig.E1.33

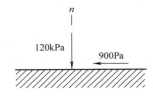

Fig.E1.34

1.35   In a hydraulic cylinder a piston without rings is designed to slide freely inside the vertical cylinder. Hydraulic oil between the piston and cylinder is maintained by a thin oil film. Determine the constant velocity with which the 120-mm-diameter piston will fall inside the 120.2-mm-diameter cylinder. The 350-g piston is 100mm long. The hydraulic oil is HM46 ($S_g = 0.85$, v = 46cSt at 40℃).

1.36   For flow over a plate, the variation of velocity with vertical distance y from the plate is given as $u(y) = ay - by^2$ where a and b are constants. Obtain a relation for the wall shear stress in terms of a, b, and μ.

1.37   A velocity distribution in a 50-mm-diameter pipe is measured to be $u(r) = 5(1 - r^2/r_0^2)$m/s, where r_0 is the radius of the pipe. Calculate the shear stress at the centerline, at $r = 20mm$, and at the wall if water at 20℃ is flowing.

1.38   A rotating viscometer consists of two 0.2-m-long rotating concentric cylinders, the velocity distribution in the gap is given by $u(r) = (0.484/r - 1000r)$ m/s. If the diameters of the cylinders are 40mm and 44mm, respectively, calculate the fluid viscosity if the torque on the inner cylinder is measured to be 0.00444N·m.

1.27 5℃水在直径 3mm 洁净的长玻璃管内可能上升的高度约为

A. 0.1mm   B. 1mm   C. 10mm   D. 100mm

1.28 青藏高原某处大气压为 78.9kPa，测得该位置一密封容器内真空压强为 30kPa，试确定容器内的绝对压强。

1.29 某流体的体积为 32L，重量为 280N，试确定该流体的质量和密度。

1.30 表压强为 51.3kPa 的位置绝对压强是多少？

1.31 气流内真空度为 28kPa，则绝对压强是多少？

1.32 恒温大气层内压强与高度的函数关系为 $p(z) = p_0 e^{-gz/RT}$，式中 g 是重力加速度，$R = 287J/(kg \cdot K)$，T 为绝对温度（K）。应用该公式估算祁连山脉登峰海拔 5827m 处的大气压。假设温度为 $T = -35℃$，标准大气压 $p_0 = 101kPa$。

1.33 200kN 力均布于 150cm² 面积上，且与平面的法向夹角 60°（见图 E1.33）。计算该平面所受压强和切应力。

1.34 图 E1.34 所示面积为 60cm² 的平面，受到 120kPa 压应力和 900Pa 剪切应力的作用，计算平面所受合力的大小及合力与平面法向的夹角。

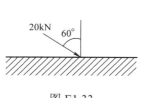

图 E1.33

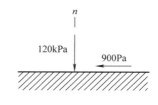

图 E1.34

1.35 某液压缸体中无活塞环的活塞可在竖直放置的缸筒内自由滑动。缸筒和活塞间被油膜分隔。活塞直径为 120mm，缸筒内径为 120.2mm。当质量为 350g 长度为 100mm 的活塞在缸筒内能够匀速滑落时，计算活塞运动速度。液压油牌号为 HM46（40℃时 $S_g = 0.85$，$v = 46cSt$）。

1.36 液体在一平板表面上流过，流速随流体质点距平板表面的竖直距离 y 的分布规律为：$u(y) = ay - by^2$，式中 a 和 b 均为常数。推导平面表面剪切应力与 a、b 和 μ 的关系式。

1.37 直径为 50mm 管道内水的流速为 $u(r) = 5(1 - r^2/r_0^2)$ m/s，式中 r_0 是管道半径。计算 20℃水分别在管道中心、$r = 20mm$ 处和管道内壁处的剪切应力。

1.38 旋转黏度计两同心筒长为 0.2m，已知间隙内液体速度分布 $u(r) = (0.484/r - 1000r)$ m/s。若内、外筒直径分别为 40mm 和 44mm，旋转内筒所需转矩

1.39   A 0.4-m-long, 25.4-mm-diameter shaft rotates inside an equally long concentric cylinder that is 26mm in diameter. Calculate the torque required to rotate the inner shaft at 1500rpm if an oil of 50cSt kinematic viscosity and 860kg/m^3 density fills the gap.

1.40   A 0.64m^2 horizontal plate moves paralled to solil surface at a distance of 2mm above it. Water at 20℃ fills the gap. Estimate the force required to move the plate at 2m/s velocity. Neglect the mass of the plate.

1.41   Consider a fluid flow between two parallel fixed plates 5mm apart, as shown in Fig.E1.41. The velocity distribution for the flow is given by $u(y) = 80(0.05y^{-10} - y^2)$ m/s where y is in meters. The fluid is water at 30℃ (the density is 995.7kg/m^3 and the kinematic viscosity is 0.804×10^{-6}m^2/s). Calculate the magnitude of the shear stress acting on each of the plates.

1.42   Calculate the torque needed to rotate the cone shown in Fig.E1.42 at 2000rpm if SAE-30 oil at 40℃ (about 11×10^{-3}Pa·s viscosity) fills the gap. Assume a linear velocity profile between the cone and the fixed wall.

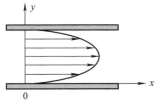

Fig.E1.41

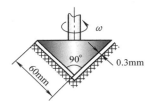

Fig.E1.42

1.43   A plate of 1m^2 area and 2mm distance to a parallel fixed plate, moves at parallel to the fixed plate 1m/s with 2N force. Determine the viscosity of the liquid fills the gap. Neglect the mass of the plate.

1.44   A 300×200×50 (mm) block weighing 230N is to be moved at a constant velocity of 1m/s on an inclined surface with a friction coefficient of 0.25, shown in Fig.E1.44. (a) Determine the force F that needs to be applied in the horizontal direction. (b) If a 0.2-mm-thick oil film with a viscosity of 0.012Pa·s is applied between the block and inclined surface, determine the percent reduction in the required force.

1.45   Consider the flow of a fluid with viscosity μ, through a circular pipe, shown in Fig.E1.45. The velocity profile in the pipe is given as $u(r) = u_{max}(1 - r^2/R^2)$, where u_{max} is the maximum flow velocity, which occurs at the centerline; r is the radial distance from the centerline; and $u(r)$ is the flow velocity at any position r. Develop a relation for the drag force exerted on the pipe wall by the fluid in the flow direction per unit length of the pipe. Obtain the value of the drag force for water flow at 20℃ with R=50mm, L=10m, $u_{max} = 9$m/s, and $\mu = 0.001$Pa·s.

为 0.00444N·m，计算被测液体的黏度。

1.39 直径 25.4mm 长度 0.4m 的轴在相同长度直径 26mm 的同心筒内转动，转速为 1500r/min。若轴和筒间隙内充满运动黏度 50cSt（50×10^{-2}cm^2/s）、密度 860kg/m^3 的油液，计算转动轴所需的转矩。

1.40 面积 0.64m^2 的水平平板在固体平面上方 2mm 处平行移动，间隙内充满 20℃水。若平板移动速度为 2m/s，忽略平板自身重量，估算维持运动所需的力。

1.41 30℃水（密度 995.7kg/m^3，运动黏度 0.804×10^{-6}m^2/s）在间隙 5mm 的平行平板间流动，如图 E1.41 所示，流速分布 $u(y) = 80(0.05y^{-10} - y^2)$。计算每个平板表面剪切应力的大小。

1.42 如 E1.42 图所示，锥体与固体壁面间隙内充满 40℃的 SAE-30 机油（黏度约 11×10^{-3}Pa·s），假设间隙内机油流速线性分布，计算锥体以转速 2000r/min 转动时所需的转矩。

图 E1.41　　　　　　　图 E1.42

1.43 动板与固定平板间有 2mm 缝隙，缝隙内充满液体。动板面积为 1m^2，以 2N 力拖动其以 1m/s 速度平行于固定平板移动。计算缝隙内液体的黏度。忽略动板质量。

1.44 如图 E1.44 所示，尺寸 $300 \times 200 \times 50$（mm×mm×mm）的物块重 230N，在斜面上以 1m/s 匀速运动，物块与斜面的摩擦因数为 0.25。（1）计算移动物块所需的力水平 F；（2）若物块和斜面间存在 0.2mm 厚油膜，油液黏度 0.012Pa·s，计算移动所需力减小的百分比。

1.45 如图 E1.45 所示，圆管内黏度为 μ 的流体流速分布 $u(r) = u_{max}(1 - r^2/R^2)$，式中 u_{max} 是管道中心处最大流速；r 为任意点至中心的半径，$u(r)$ 为半径 r 处的流速。推导单位长度管壁所受流体的拖动力表达式。当流体为 20℃水，$\mu = 0.001$Pa·s，管道半径和长度分别为 $R = 50$mm、$L = 10$m，且 $u_{max} = 9$m/s 时，计算该拖动力的大小。

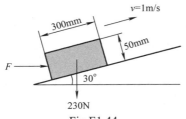

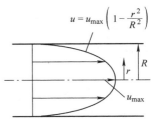

Fig.E1.44　　　　　　　　　　　Fig.E1.45

1.46　A thin 200mm × 200mm plate moves between two parallel, horizontal, stationary flat surfaces at a constant velocity of 5m/s (Fig.E1.46). The two stationary surfaces are spaced 32mm apart, and the medium between them is filled with oil whose viscosity is 0.6Pa · s. The plate is 20mm from the bottom surface (h_2) and 12mm from the top surface (h_1). Determine the force required to maintain this motion.

1.47　Reconsider Exercise 1.46. If the viscosity of the oil above the moving plate is 3 times that of the oil below the plate, determine the distance of the plate from the bottom surface (h_2) that will minimize the force needed to pull the plate between the two oils at constant velocity.

1.48　A thin 200mm×200mm flat plate is pulled at 2m/s horizontally through a 3.2-mm-thick oil layer sandwiched between two plates, one stationary and the other moving at a constant velocity of 0.5m/s, as shown in Fig.E1.48. The viscosity of oil is 0.032Pa · s. Assuming the velocity in each oil layer to vary linearly, determine the force that needs to be applied on the plate to maintain this motion.

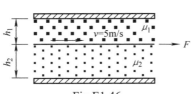

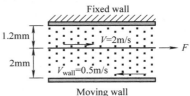

Fig.E1.46　　　　　　　　　　　Fig.E1.48

1.49　In some damping systems, a circular disk immersed in oil is used as a damper, as shown in Fig.E1.49. Derive the relation of the damping torque in terms of angular velocity ω, gap spaces h_1 and h_2, the viscosity of oil μ and the radius of the disk R.

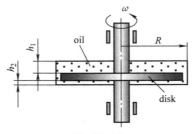

Fig.E1.49

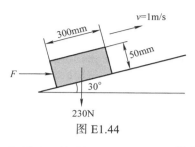

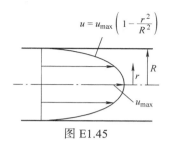

图 E1.44

图 E1.45

1.46　一块 200mm×200mm 的薄板在两块平行的固定平板间以 5m/s 匀速移动，如图 E1.46 所示，两固定平板间距离为 32mm，中间充满黏度为 0.6Pa·s 的油液。动薄板距下平板（h_2）20mm，距上平板（h_1）12mm。计算拖动平板运动所需的力。

1.47　习题 1.46 中的情况，现动板上方油液黏度是其下方油液黏度的 3 倍。若仍需维持薄板匀速运动，确定薄板与下板的合适距离 h_2，可以使以维持平板运动的力最小。

1.48　两平板间距 3.2mm 且中间充满黏度 0.032Pa·s 的油液。现上平板固定不动，下平板以 0.5m/s 匀速向左移动，有一 200mm×200mm 的薄板在间隙内以 2m/s 匀速反向水平移动，如图 E1.48 所示。假设油液流速线性变化，计算拖动薄板运动所需的力。

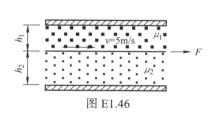

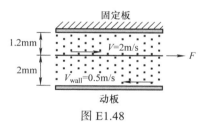

图 E1.46

图 E1.48

1.49　某些阻尼系统使用油液中的转动圆盘作为阻尼器，如图 E1.49 所示。圆盘半径为 R，转动角速度为 ω，与壳体间隙分别为 h_1 和 h_2，油液黏度为 μ，推导阻尼力矩与这些参数间的关系式。

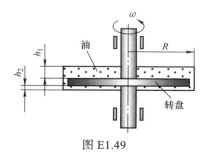

图 E1.49

1.50 Consider a 120mm-long journal bearing that is lubricated with oil whose viscosity is 0.1Pa·s at 20℃ at the beginning of operation and 0.012Pa·s at the anticipated steady operating temperature of 80℃. The diameter of the shaft is 60mm, and the average gap between the shaft and the journal is 0.2mm. Determine the torque needed to overcome the bearing friction initially and during steady operation when the shaft is rotated at 750r/min.

1.51 A viscometer is constructed with two 200mm-long concentric cylinders, one 150mm in diameter and the other 152mm in diameter. A torque of 0.24N·m is required to rotate the inner cylinder at 500r/min. Calculate the viscosity.

1.52 Determine the height that 20℃ water would climb in a vertical 2-mm-diameter tube if it attaches to the wall with an angle of 30° to the vertical.

1.53 Find an expression for the rise of liquid between two parallel plates a distance t apart. Use a contact angle θ and surface tension σ.

1.54 A 3-mm-diameter clean glass tube is inserted in water at 20℃. Determine the height that the water will climb up the tube. The water makes a contact angle of 0° with the clean glass.

1.55 In a piping system，the water temperature remains below 30℃. Determine the minimum pressure allowed in the system to avoid cavitation.

1.56 A 1.2-mm-diameter tube is inserted into an unknown liquid whose density is 910kg/m^3, and it is observed that the liquid rises 5mm in the tube, making a contact angle of 10°. Determine the surface tension of the liquid.

1.57 Water are carried to upper parts of plants by tiny tubes because of the capillary effect. Determine how high the water will rise in a tree in a 2-μm-diameter tube as a result of the capillary effect. Take the water solution with nutrition at 20℃ with a contact angle of 15°.

1.58 A 3-mm-diameter glass tube is inserted into mercury, which makes a contact angle of 140° with glass. Determine the capillary drop of mercury in the tube at 20℃.

1.50　长度 120mm 的轴颈和轴承，轴颈直径 60mm，与轴承间平均间隙 0.2mm，以油液润滑。工作初始温度 20℃，油液黏度 0.1Pa·s，工作稳定后温度升至 80℃，润滑油黏度降至 0.012Pa·s。轴以 750r/min 的转速转动，计算转动初始和稳定后所需的转矩。

1.51　一旋转黏度计，双筒长为 200mm，内筒直径为 150mm，外筒直径为 152mm。内筒以 500r/min 的转速转动时，所需力矩为 0.24N·m。计算被测液体黏度。

1.52　若水与固体管壁竖直方向的接触角为 30°，水温为 20℃，管径为 2mm，计算水在管内上升的高度。

1.53　两平行平板间距为 t，与某液体接触角为 θ，液体表面张力为 σ，导出液体在缝隙内上升高度与这些参数间的关系。

1.54　直径 3mm 的洁净玻璃管插入 20℃水中，水与洁净玻璃间的接触角为 0°，计算水在管中的浸润高度。

1.55　某管网内水温保持在 30℃，计算不发生气穴现象的最小压强。

1.56　直径 1.2mm 的管插入某未知液体中，测得液体密度为 910kg/m³，管内液体上升高度为 5mm，管壁与液体接触角为 10°。计算该未知液体的表面张力。

1.57　植物依靠毛细效应通过细小管道从根部向高处输送水分。如果植物内部管道直径为 2μm，20℃营养水分与管道接触角为 15°，计算水分可送达的高度。

1.58　把直径 3mm 的玻璃管插入水银中，已知水银与玻璃的接触角为 140°，计算 20℃时管内水银下降的高度。

CHAPTER 2
FLUID STATICS

2.1 Introduction

Fluid statics is the study of fluids in which there is no relative motion between fluid particles. If there is no relative motion, no shearing stresses exist, the only surface stress that exists is a normal stress, the pressure. The fluid property responsible for those forces is pressure, so it is the pressure that is of primary interest in fluid statics. Two situations involving fluid statics will be investigated. These include fluids at absolute rest, such as water pushing against a dam, and fluids at relative rest, such as fluids contained in devices that undergo linear acceleration; and fluids contained in rotating cylinders. In each of these two situations the fluid is in static equilibrium with respect to a reference frame attached to the boundary surrounding the fluid. In addition to the examples shown for fluids at rest, we also consider instruments called manometers.

2.2 Pressure at a point

In a fluid at rest, shear stress does not exist, the only forces are pressure forces normal to the surfaces. Therefore the pressure at any point in a fluid is the same in all directions, that is, the pressure has magnitude but not a specific direction. This can be demonstrated even for fluids in motion with no shear, by considering a small wedge-shaped fluid element of unit length ($\Delta z = 1$ into the page) in equilibrium, as shown in Fig.2.2.1.

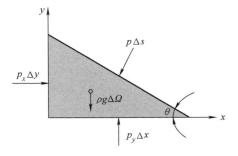

Fig.2.2.1   A fluid element in equilibrium

We define the pressure at a point as being the infinitesimal normal compressive force divided by the infinitesimal area over which it acts:

$$p = \lim_{\Delta A \to 0} \frac{\Delta F_n}{\Delta A} \tag{2.2.1}$$

Assume that a pressure p acts on the hypotenuse and that a different pressure acts on each of the other areas, Newton's second law is applied to the element, for both the x-

第 2 章
流体静力学

2.1 引言

流体静力学的研究对象是质点间没有相对运动处于静止状态的流体。没有相对运动就没有剪切应力，作用于流体的表面力仅为正应力，即压强。流体静力学主要研究静止流体内压强分布及在静压力作用下流体特性的反映。流体的静止可分为两种情况，一种是绝对静止，如堤坝内不流动的水；另一种情况是相对静止，如旋转容器内的流体。这两种情况下的流体相对其所依附的界面建立的坐标系都是处于静止状态。本章介绍流体静力学原理的应用以及测压装置的工作原理。

2.2 静止流体内一点的压强

静止流体内部不存在剪切应力，仅有与受力面垂直的压力。所以流体内部任一点所受的压强在各个方向都是相同的，也就是说，任一点处压强等值地作用于各个方向，即使是处于运动中但无剪切力的流体。我们可以通过对图 2.2.1 中的楔形流体微元受力平衡来分析证明，假设微元的深度边长为单位值（$\Delta z = 1$，垂直页面方向）。

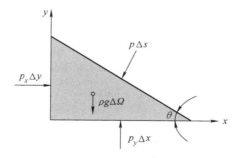

图 2.2.1　流体微元受力平衡

定义压强 p 为法向压力 F_n 与受力面积之比的极限，可表示为

$$p = \lim_{\Delta A \to 0} \frac{\Delta F_n}{\Delta A} \qquad (2.2.1)$$

and y-directions:

$$\Sigma F_x = ma_x: \qquad p_x\Delta y - p\Delta s \sin\theta = \rho\frac{\Delta x\Delta y}{2}a_x$$

$$\Sigma F_y = ma_y: \qquad p_y\Delta x - \rho g\frac{\Delta x\Delta y}{2} - p\Delta s\cos\theta = \rho\frac{\Delta x\Delta y}{2}a_y$$

(2.2.2)

where we have used the volume of fluid element $\Delta\Omega = \Delta x\Delta y/2$ (we could include Δz in each term to account for the depth).

Substituting $\Delta s\sin\theta = \Delta y$, $\Delta s\cos\theta = \Delta x$, that Eq.2.2.2 take the forms:

$$p_x - p = \frac{\rho a_x}{2}\Delta x$$

$$p_y - p = \frac{\rho(a_y + g)}{2}\Delta y$$

(2.2.3)

Note that in the limit as the element shrinks to a point, $\Delta x \rightarrow 0$, and $\Delta y \rightarrow 0$.

Hence the right-hand sides in the equations above go to zero, even for fluids in motion, providing with the result that, at a point $p_x = p_y = p$. We can also prove that by considering a three-dimension case, and we have

$$p_x = p_y = p_z = p$$

(2.2.4)

Since θ is arbitrary, the results are independent of θ, thus we conclude that the pressure in a fluid is constant at a point, it acts equally in all directions at a given point for both a static fluid and a fluid that is in motion in the absence of shear stress.

2.3 Pressure Variation in a Fluid at rest

Consider the infinitesimal element of static fluid shown in Fig.2.3.1. Since the element is very small, we can assume that the density of the fluid within the element is constant. Assume that the pressure at the center of the element is p and that the dimensions of the element are dx, dy and dz. The forces acting on the fluid element in the vertical direction are the mass fore, i.e. the action of gravity on the mass within the element, and the surface forces, transmitted from the surrounding fluid and acting against the top, bottom, and sides of the element. Because the fluid is at rest, the element is in equilibrium and the summation of forces acting on the element in any direction must be in equillibrium.

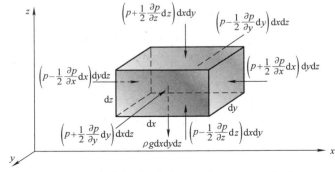

Fig.2.3.1   Forces acting on an infinitesimal element that is at rest in the xyz-reference frame

假设流体微元通过斜边和直角边的面所受压强均不相同，根据牛顿第二定律对 x 和 y 方向建立受力平衡式：

$$\Sigma F_x = ma_x: \qquad p_x \Delta y - p \Delta s \sin\theta = \rho \frac{\Delta x \Delta y}{2} a_x$$

$$\Sigma F_y = ma_y: \qquad p_y \Delta x - \rho g \frac{\Delta x \Delta y}{2} - p \Delta s \cos\theta = \rho \frac{\Delta x \Delta y}{2} a_y \tag{2.2.2}$$

这里用 $\Delta\Omega = \Delta x \Delta y/2$ 表示流体微元的体积（也可以计入表示深度的 Δz）。

代入 $\Delta s \sin\theta = \Delta y$，$\Delta s \cos\theta = \Delta x$，式 2.2.2 可改写为

$$p_x - p = \frac{\rho a_x}{2} \Delta x$$

$$p_y - p = \frac{\rho(a_y + g)}{2} \Delta y \tag{2.2.3}$$

取边长无限小至一点的极限，即，$\Delta x \rightarrow 0$ 和 $\Delta y \rightarrow 0$。

所以上式右边的项为零，证明了静止流体内任一点处 $p_x = p_y = p$。这一结论对运动但无剪切的流体同样成立，也适用于三维流动，即

$$p_x = p_y = p_z = p \tag{2.2.4}$$

由于 θ 角为任意值，所以以上结论与 θ 角无关。由此我们可以确定，静止流体和无剪切力的运动流体内部任一点的压强，在各个方向上的大小均相等。

2.3 静止流体内的压强分布

分析图 2.3.1 所示的静止流体微元。因为微元体积非常小，所以我们可以认为微元内流体密度没有变化为常数。假设微元边长分别为 dx、dy 和 dz 且中心处压强为 p。中心处沿竖直方向向下的质量力为重力，周围流体对微元各表面产生表面力。由于微元处于静止状态，所以各方向的受力均平衡。

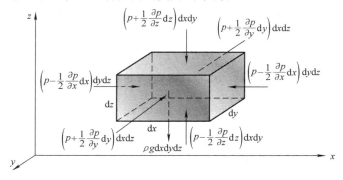

图 2.3.1 直角坐标系内静止流体微元受力

Since that the pressure at the center of the element is p, the pressures at each of the sides can be expressed by using the chain rule from calculus with $p(x, y, z)$:

$$dp = \frac{\partial p}{\partial x}dx + \frac{\partial p}{\partial y}dy + \frac{\partial p}{\partial z}dz \qquad (2.3.1)$$

If we move from the center to the faces a distance $(dx/2)$ away we see that the pressures are

$$p(x - \frac{dx}{2}, y, z) = p - \frac{\partial p}{\partial x}\frac{dx}{2} \qquad (2.3.2)$$

and

$$p(x + \frac{dx}{2}, y, z) = p + \frac{\partial p}{\partial x}\frac{dx}{2} \qquad (2.3.3)$$

The pressures at all faces are expressed in this manner, as shown in Fig. 2.3.1.

Summing forces in the x-direction and setting the sum equal to ma_x by applying Newton's second law and using the mass as $\rho dxdydz$,

$$\sum F_x = \left(p - \frac{\partial p}{\partial x}\frac{dx}{2}\right)dydz - \left(p + \frac{\partial p}{\partial x}\frac{dx}{2}\right)dydz = -\frac{\partial p}{\partial x}dxdydz = \rho dxdydz a_x \qquad (2.3.4)$$

Eq.2.3.4 can be reduced and results in the three component equations, we have

$$\frac{\partial p}{\partial x} = -\rho a_x$$

$$\frac{\partial p}{\partial y} = -\rho a_y \qquad (2.3.5)$$

$$\frac{\partial p}{\partial z} = -\rho(a_z + g)$$

The pressure differential in any direction can now be determined from Eq.2.3.1 as

$$dp = -\rho a_x dx - \rho a_y dy - \rho(a_z + g)dz \qquad (2.3.6)$$

Since a fluid at absolute rest does not undergo any acceleration, therefore set $a_x = 0$, $a_y = 0$ and $a_z = 0$, Eq.2.3.6 reduces to

$$dp = -\rho g dz \qquad (2.3.7)$$

If the density can be assumed constant, Eq. 2.3.7 is integrated to yield

$$p + \gamma z = \text{Constant} \quad \text{or} \quad p + \rho gz = \text{Constant} \quad \text{or} \quad \frac{p}{\rho g} + z = \text{Constant} \qquad (2.3.8)$$

where z denotes the height from the position of a certain fluid particle to the datum plane.

The quantity $(p/\rho g + z)$ is often referred to as static total water head, it has components of pressure head $(p/\rho g)$ and position head z, and remains constant. We may also consider the static total head as the total mechanical energy of a fluid at rest, and it is conserved.

　　流体微元中心处的压强为 p，此压强为坐标位置的函数 $p(x, y, z)$，且沿各方向的变化率可以写为

$$\mathrm{d}p = \frac{\partial p}{\partial x}\mathrm{d}x + \frac{\partial p}{\partial y}\mathrm{d}y + \frac{\partial p}{\partial z}\mathrm{d}z \qquad (2.3.1)$$

则距中心点 $\mathrm{d}x/2$ 的左、右侧面压强可分别表示为

$$p(x - \frac{\mathrm{d}x}{2}, y, z) = p - \frac{\partial p}{\partial x}\frac{\mathrm{d}x}{2} \qquad (2.3.2)$$

和

$$p(x + \frac{\mathrm{d}x}{2}, y, z) = p + \frac{\partial p}{\partial x}\frac{\mathrm{d}x}{2} \qquad (2.3.3)$$

　　其他各表面的压强都可以用这种形式表示，如图 2.3.1 所示。
　　流体微元质量为 $\rho\mathrm{d}x\mathrm{d}y\mathrm{d}z$，根据牛顿第二定律，$x$ 方向的合力 $\sum F_x$ 等于 ma_x

$$\sum F_x = \left(p - \frac{\partial p}{\partial x}\frac{\mathrm{d}x}{2} \right)\mathrm{d}y\mathrm{d}z - \left(p + \frac{\partial p}{\partial x}\frac{\mathrm{d}x}{2} \right)\mathrm{d}y\mathrm{d}z = -\frac{\partial p}{\partial x}\mathrm{d}x\mathrm{d}y\mathrm{d}z = \rho\mathrm{d}x\mathrm{d}y\mathrm{d}za_x \qquad (2.3.4)$$

化简式 2.3.4，并在另外两个方向上建立同样的等式，可以得到

$$\frac{\partial p}{\partial x} = -\rho a_x$$
$$\frac{\partial p}{\partial y} = -\rho a_y \qquad (2.3.5)$$
$$\frac{\partial p}{\partial z} = -\rho(a_z + g)$$

把上式代入式 2.3.1 可得出压强在任意方向的变化率

$$\mathrm{d}p = -\rho a_x\mathrm{d}x - \rho a_y\mathrm{d}y - \rho(a_z + g)\mathrm{d}z \qquad (2.3.6)$$

绝对静止流体各方向的加速度均为零，即令 $a_x = 0$，$a_y = 0$ 和 $a_z = 0$，式 2.3.6 化简为

$$\mathrm{d}p = -\rho g\mathrm{d}z \qquad (2.3.7)$$

假设密度为常数，则对式 2.3.7 积分，可得出

$$p + \gamma z = C \quad \text{或} \quad p + \rho gz = C \quad \text{或} \quad \frac{p}{\rho g} + z = C \qquad (2.3.8)$$

式中，z 表示某流体质点相对于基准面的高度。
　　（$p/\rho g + z$）常称为静压头，由压强水头（$p/\rho g$）和位置水头 z 组成，其值恒定不变。我们可以把静压头理解为静止流体所含的机械能，机械能是守恒不变的。总

The magnitude of total mechanical energy is illustrated by static total head which has a demission of length. While z denotes the height from the position of point that of interested to the datum plane, represents the position potential. Pressure head measures the pressure potential by height of liquid column ($p/\rho g$) which is acted by point pressure p.

Note that z is positive in the upward direction, and $\mathrm{d}p$ is negative when $\mathrm{d}z$ is positive since pressure decreases in an upward direction. If A is a point which vertical coordinate is z_1, has a distance h below the free surface (a surface separating a gas from a fluid), as shown in Fig.2.3.2, the quantity ($p + \rho gz$) at the free surface equals to that of point A, this yields

$$p_0 + \rho gz_0 = p_1 + \rho gz_1 \qquad (2.3.9)$$

we can also write Eq.2.3.9 as

$$p_1 - p_0 = \rho g\left(z_0 - z_1\right) = \rho gh \qquad (2.3.10)$$

Obviously, the left side of Eq.2.3.10 is the gage pressure at point A, a gage pressure of any point bellows the free surface is given by

$$p = \gamma h \quad \text{or} \quad p = \rho gh \qquad (2.3.11)$$

So that pressure increases with depth h, that is defined as depth of immersion. Often we find it more convenient to express pressure in terms of a height of a column of fluid $p/\rho g$ rather than in pressure per unit area.

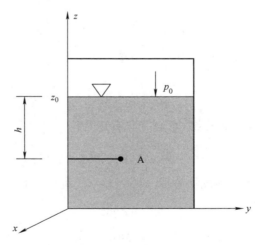

Fig.2.3.2　Pressure below a free surface

Pressure in a fluid at rest is independent of horizontal coordinate or the shape or cross section of the container, it changes with depth, as shown in Fig.2.3.3. A consequence of the pressure in a fluid remaining constant in the horizontal direction is that *the pressure applied to a confined fluid increases the pressure throughout by the same amount*. This is called Pascal's law, after Blaise Pascal (1623—1662).

机械能以具有长度量纲的静压头形式体现。z 表示被分析流体质点距基准面的高度，代表位势能；该点压强 p 的值体现为其产生的液柱高度（$p/\rho g$），代表该处压强势能即压强水头的大小。

　　因为坐标 z 向上方向为正，而压强沿向上方向是递减的，所以当 dz 为正时 dp 为负值。如果某点 A 竖直方向坐标值为 z_1，位于自由液面（液体与气体分隔的界面）下方 h 处，如图 2.3.2 所示，若自由液面上某点的静压头与 A 点处静压头相等，则应该有

$$p_0 + \rho g z_0 = p_1 + \rho g z_1 \qquad （2.3.9）$$

式 2.3.9 也可以写作

$$p_1 - p_0 = \rho g (z_0 - z_1) = \rho g h \qquad （2.3.10）$$

　　显然，式 2.3.10 的左边为点 A 处的表压强，则自由液面下任意一点的表压强都可以表示为

$$p = \gamma h \quad 或 \quad p = \rho g h \qquad （2.3.11）$$

　　上式中压强随深度 h 增大，h 称为淹深。很多时候，用液柱高度 $p/\rho g$ 来表示压强的大小较使用单位面积上的压力更加方便。

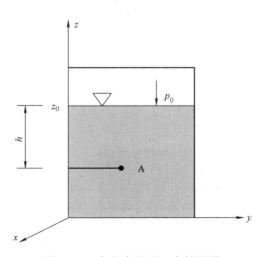

图 2.3.2　自由液面下 A 点的压强

　　静止流体内部压强由深度决定，如图 2.3.3 所示，与水平坐标值和容器截面形状无关。由此，我们可以得出结论：**流体内部任意一点的压强增量，都会等值地传递到流体内部各处**。这就是由帕斯卡（1623—1662）提出的**帕斯卡定律**。

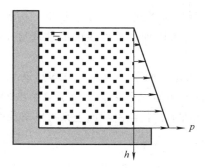

Fig.2.3.3　The pressure of a fluid at rest increases with depth

Example 2.3.1　An open tank (Fig.X2.3.1) contains water 1.5m deep covered by a 2-m-thick layer of oil ($S_g = 0.85$). What is the (a) absolute pressure and (b) gage pressure at the bottom of the tank, (c) gage pressure in terms of a water column?

Solution　(1) At free surface, $p_{free} = p_a = 100\text{kPa}$

Density of oil, $\rho_{oil} = S_g \rho_{water} = 850\text{kg/m}^3$

For interface between oil and water, $p_{oil} = \rho g h_{oil} = 850 \times 9.8 \times 2\text{kPa} = 16.66\text{kPa}$

Pressure as result of weight of water, $p_{water} = \rho_{water} g h_{water} = 1000 \times 9.8 \times 1.5\text{kPa} = 14.7\text{kPa}$

Absolute pressure at the bottom of tank, $p_{abs} = 100\text{kPa} + 16.66\text{kPa} + 14.7\text{kPa} = 131.36\text{kPa}$

(2) Gage pressure at the bottom of tank, $p_{gage} = 16.66\text{kPa} + 14.7\text{kPa} = 31.36\text{kPa}$

(3) Equivalent water head at bottom of tank, $h = \dfrac{p}{\rho_{water} g} = \dfrac{31.36}{9.8}\text{m of water} = 3.2\text{m of water}$

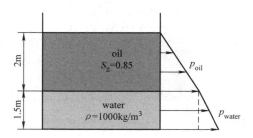

Fig.X2.3.1　Tank contains of oil and water

Example 2.3.2　The atmospheric pressure is given as 608mm Hg at a mountain location. Convert this to kilopascals and meters of water.

Solution　Density of mercury, $\rho_{Hg} = 13.6 \times 10^3 \text{kg/m}^3$.

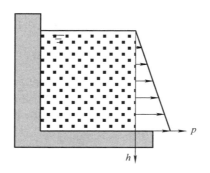

图 2.3.3　静止流体内部压强随深度增加

例 2.3.1　一开放容器（图 X2.3.1）内有 1.5m 深的水，水上方为 2m 深的油液（$S_g = 0.85$）。计算（1）容器底部绝对压强；（2）容器底部表压强；（3）用水柱高表示容器底部表压强。

解　（1）自由液面压强 $p_{\text{free}} = p_a = 100\text{kPa}$

油液密度 $\rho_{\text{oil}} = S_g \rho_{\text{水}} = 850\text{kg/m}^3$

油水分界面处，$p_{\text{油}} = \rho g h_{\text{油}} = 850 \times 9.8 \times 2\text{kPa} = 16.66\text{kPa}$

水重量形成的压强，$p_{\text{水}} = \rho_{\text{水}} g h_{\text{水}} = 1000 \times 9.8 \times 1.5\text{kPa} = 14.7\text{kPa}$

罐底绝对压强，$p_{\text{abs}} = 100\text{kPa} + 16.66\text{kPa} + 14.7\text{kPa} = 131.36\text{kPa}$

（2）罐底表压强，$p_{\text{gage}} = 16.66\text{kPa} + 14.7\text{kPa} = 31.36\text{kPa}$

（3）罐底表压强的等效水头 $h = \dfrac{p}{\rho_{\text{水}} g} = \dfrac{31.36}{9.8}\text{m 水柱} = 3.2\text{m 水柱}$

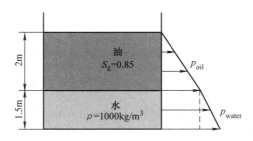

图 X2.3.1　油 - 水容器内压强

例 2.3.2　山区某处的大气压为 608mm 汞柱，用千帕和水柱高度来表示该大气压。

解　水银密度，$\rho_{\text{Hg}} = 13.6 \times 10^3 \text{kg/m}^3$。

Pressure in kilopascals, $p = \rho_{Hg}gh = 13.6\times10^3 \times 9.8 \times 0.608\text{kPa} = 81\text{kPa}$

Pressure in meters of water, $h = \dfrac{p}{\rho_{water}g} = \dfrac{81}{9.8}\text{m} = 8.265\text{m}$

Example 2.3.3 A sealed tank for alcohol ($\rho = 900\text{kg/m}^3$), Fig.X2.3.3a, a gage is connected to the tank in which the air pressure is 35kPa above atmospheric, (a) what is the pressure within the tank? (b) What is the pressure at the bottom of the tank?

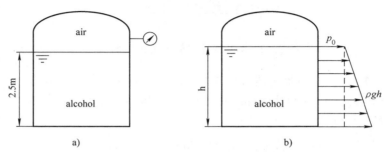

Fig.X2.3.3   A sealed tank for alcohol

Solution   (1) Gage pressure of air at top of the tank is $p_0 = 35\text{kPa}$, and according to the Pascal's law, this part of pressure applied throughout the alcohol by the same amount.

The other part of pressure increases with the depth of alcohol, $p_1 = \rho gh$

Therefore the total pressure within the alcohol is the summation of this two parts, Fig. X2.3.3b, $p = p_0 + \rho gh$

(2) Pressure at the bottom of the tank,

$$p = p_0 + \rho gh = (35 + 0.9\times9.8\times2.5)\text{kPa} = 57.05\text{kPa}$$

2.4   Manometers

We notice from Eq.2.3.8 and Eq.2.3.11 that an elevation change of z or h in a fluid at rest corresponds to $p/\rho g$, which suggests that the height of the liquid column will give the value of the pressure head. Instruments that use columns of liquids to measure pressures are manometers.

Fig.2.4.1a displays a manometer used to measure relatively small pressures. The pressure in the container can be determined by defining a point 1 at the center of the container and a point 2 at the surface of the right column. By using Eq.2.3.11

$$p_1 + \rho gz_1 = p_2 + \rho gz_2 \tag{2.4.1}$$

Since gage pressure is selected, $p_2 = p_{atm} = 0$, where the datum from which z_1 and z_2 are measured is located at any desired position, and $z_2 - z_1 = h$, here is

$$p_1 = p_{1'}$$
$$p_1 = p_A + \rho g \left(h_0 - h_1 \right)$$
$$p_{1'} = p_B + \rho g \left(h_0 - h_2 \right) + \rho' g \left(h_2 - h_1 \right)$$

（2.4.5）

对上式进行整理并用 Δh 代替（$h_2 - h_1$），则

$$p_A - p_B = \left(\rho' - \rho \right) g \Delta h \qquad （2.4.6）$$

式 2.4.6 仅用于 A、B 两容器内流体密度相同的情况。若两容器内流体密度不同，我们可以按照例 2.4.1 的步骤，列出自 A 到 B 各段流体的压强，再得出压差。

测量较大压差时，压差计中应选用密度大的液体，如水银。对于压差不大的情况，应该选用密度较小的液体，如油液甚至是空气，这时压差计可以按图 2.4.2b 所示布置，此时称为 Π 形管。当然，无论如何选择，工作流体都不应与 A 或 B 内流体混合。按照之前的方法，我们可以测得图 2.4.2b 中 A、B 为同种流体时的压差

$$p_1 = p_{1'}$$
$$p_1 = p_A - \rho g h_1$$
$$p_{1'} = p_B - \rho g h_2 - \rho' g \left(h_2 - h_1 \right)$$
$$p_A - p_B = \left(\rho + \rho' \right) g \Delta h$$

（2.4.7）

测量液体间压强差时常常会选用空气或其他气体作为工作流体并按图 2.4.2b 形式布置，然后泵入一定压力的空气直至形成适当的液柱高度差。由于空气密度较液体小很多，其重量可以忽略，所以上式中点 1′ 与点 2 处压强相等，因此

$$p_A - p_B = \rho g \left(h_1 - h_2 \right) = \rho g \Delta h \qquad （2.4.8）$$

被测压差非常小时可以用对压差更加敏感的倾斜微压差计（图 2.4.3a），将竖直高度差 Δh 放大为 $l/\sin\theta$，如式 2.4.9，式中 θ 为斜管与水平方向的夹角，且 θ 角越小，l 长度越大。

$$p_1 - p_2 = \rho' g \left(l \sin\theta + \Delta h \right) \qquad （2.4.9）$$

容器中液面高度变化量 $\Delta h = A_0 l / A$，由于斜管截面积 A_0 与容器截面积 A 相比 $A_0 \ll A$，所以 Δh 可以忽略不计。

因此图 2.4.3a 中压差可由下式得出

$$p_1 - p_2 = \rho' g l \sin\theta \qquad （2.4.10）$$

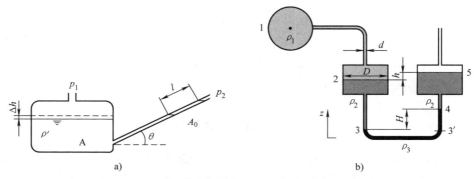

Fig.2.4.3　Micromanometers

a) inclined micromanometer　b) double-cup micromanometer

Figure 2.4.3b also shows a micromanometer that is used to measure very small pressure changes. Introducing the points indicated, requiring that $p_3 = p_{3'}$, we can write

$$\begin{aligned} p_3 &= p_1 + \rho_1 g \left(z_1 - z_2\right) + \rho_2 g \left(z_2 - z_3\right) \\ p_{3'} &= \rho_2 g \left(z_5 - z_4\right) + \rho_3 g \left(z_4 - z_3\right) + p_5 \end{aligned} \tag{2.4.11}$$

observe that $z_2 - z_3 + h = H + z_5 - z_4$ and set $p_5 = 0$, then

$$p_1 = \left(\rho_3 - \rho_2\right) gH + \rho_2 gh - \rho_1 g \left(z_1 - z_2\right) \tag{2.4.12}$$

The micromanometer is capable of measuring small pressure changes because a small pressure change in p_1 results in a relatively large deflection H. The change in H due to a change in p_1 can be determined using Eq. 2.4.12. Suppose that p_1 increases by Δp_1 and, as a result, z_2 decreases by Δz; then h and H also change. Using the fact that a decrease in z_2 is accompanied by an increase in z_5 leads to an increase in h of $2\Delta z$ and, similarly, assuming that the volumes are conserved, it can be shown that H increases by $2\Delta z D^2/d^2$. Hence the reading of H is amplified because that $D^2 >> d^2$.

Many engineering problems and some manometers involve multiple immiscible fluids of different densities stacked on top of each other. Such systems can be analyzed easily by remembering that:

(1) The pressure change across a fluid column of height h is $\Delta p = \rho g h$.

(2) Pressure increases downward in a given fluid and decreases upward.

(3) Two points at the same elevation in a continuous fluid column (same type of fluid) are at the same pressure.

(4) One of the points is selected at the interface of different fluids.

Example2.4.1　The water in a tank is pressurized by air, and the pressure is measured by a multi-fluid manometer as shown in Fig.X2.4.1. The tank is located on a mountain where the atmospheric pressure is 85kPa. Determine the air pressure in the tank if $h_1 = 0.3$m, $h_2 = 0.1$m, and $h_3 = 0.2$m. Take the densities of water, oil, and mercury to be 1000kg/m³, 850kg/m³, and 13600kg/m³, respectively.

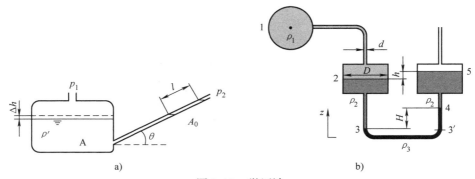

图 2.4.3　微压计

a）倾斜微压计　b）双杯微压计

图 2.4.3b 同样测量微小压强，U 形管中两点 $p_3 = p_{3'}$，所以

$$p_3 = p_1 + \rho_1 g(z_1 - z_2) + \rho_2 g(z_2 - z_3)$$
$$p_{3'} = \rho_2 g(z_5 - z_4) + \rho_3 g(z_4 - z_3) + p_5$$

（2.4.11）

因为 $z_2 - z_3 + h = H + z_5 - z_4$，且取 $p_5 = 0$，则

$$p_1 = (\rho_3 - \rho_2)gH + \rho_2 gh - \rho_1 g(z_1 - z_2)$$

（2.4.12）

双杯微压差计把 p_1 的微小变化转化为较大高度差 H 值，可以测量很小的压强变化量。高度差 H 随压强 p_1 的变化见式 2.4.12。压强 p_1 增加 Δp_1，z_2 相应减小 Δz，h 和 H 随之改变。z_2 减小使 z_5 增大并导致 h 值增加 $2\Delta z$，因为杯内液体的体积不变，所以 H 高度增加 $2\Delta z D^2/d^2$。双杯截面积和管截面积相比 $D^2 >> d^2$，所以 H 被放大。

实际使用的测压管中常常会有几种不同密度、不混溶的流体叠加的情况，这时可按以下提示处理：

（1）每段高度为 h 的液柱产生的压强为 $\Delta p = \rho gh$。

（2）沿液柱向下压强增大、向上则减小。

（3）同一液柱内处于相同高度位置的同种液体压强相等。

（4）至少有一点应该选在不同流体的分界面处。

例 2.4.1　如图 X2.4.1 所示的密封水箱上方有空气加压，用测压管测量压强。已知水箱放置于大气压 85kPa 的山区。若测得 $h_1 = 0.3$m，$h_2 = 0.1$m 和 $h_3 = 0.2$m，计算水箱上方空气压强。水、油和水银的密度分别取 1000kg/m³、850kg/m³ 和 13600kg/m³。

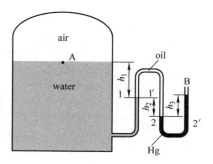

Fig.X2.4.1

Solution Assume that the air pressure in the tank is uniform (i.e., its variation with elevation is negligible due to its low density), and thus we can determine the pressure at the air-water interface. Starting with the pressure at point 1 at the oil- water interface and point 1′ at the same elevation in a continuous fluid, set an equation of $p_1 = p_{1'}$. Moving along the tube to the oil-mercury interface and set the equation of $p_2 = p_{2'}$ by adding or subtracting the ρgh terms, keep moving until we reach point B, and setting the result equal to p_{atm} since the tube is open to the atmosphere gives.

$$p_1 = p_A + \rho_1 gh_1 \,, p_1 = p_{1'} = p_2 - \rho_2 gh_2$$

$$p_2 = p_{2'} = p_{atm} + \rho_3 gh_3$$

ρ_1, ρ_2, ρ_3 and h_1, h_2, h_3 are given, solving for p_A and substituting,

$$
\begin{aligned}
p_A &= p_{atm} + \rho_3 gh_3 - \rho_2 gh_2 - \rho_1 gh_1 \\
&= (85 + 13.6 \times 9.8 \times 0.2 - 0.85 \times 9.8 \times 0.1 - 9.8 \times 0.3)\text{kPa} \\
&= 108\text{kPa}
\end{aligned}
$$

Example 2.4.2 In Fig.X2.4.2, liquid A weights 850kg/m^3, liquid B weights 1240kg/m^3, manometer fluid is mercury, $h_1 = 100\text{mm}$, $h_2 = 500\text{mm}$, $h_3 = 750\text{mm}$, find the pressure difference between A and B.

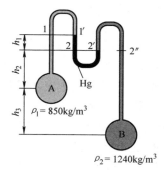

Fig.X2.4.2

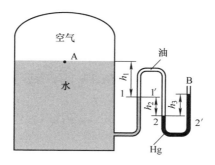

图 X2.4.1

解　假设气压均匀分布（由于空气密度低，所以忽略压强随高度的变化），计算空气与水分界面处 A 点的压强。在液柱上取水 - 油分界面点 1 和与之同高的连通液柱内另一点 1′，则有 $p_1 = p_{1'}$。沿液柱加或减相应的 ρgh，移至油 - 水银分界面，取等压点 $p_2 = p_{2'}$，继续沿液柱移动到开放的 B 点，其压强为大气压。

$$p_1 = p_A + \rho_1 gh_1, \quad p_1 = p_{1'} = p_2 - \rho_2 gh_2$$

$$p_2 = p_{2'} = p_{atm} + \rho_3 gh_3$$

ρ_1、ρ_2、ρ_3 和 h_1、h_2、h_3 均已知，代入可解 p_A 得

$$\begin{aligned}
p_A &= p_{atm} + \rho_3 gh_3 - \rho_2 gh_2 - \rho_1 gh_1 \\
&= (85 + 13.6 \times 9.8 \times 0.2 - 0.85 \times 9.8 \times 0.1 - 9.8 \times 0.3) \text{kPa} \\
&= 108 \text{kPa}
\end{aligned}$$

例 2.4.2　图 X2.4.2 所示容器 A 中液体密度为 850kg/m^3，容器 B 中液体密度为 1240kg/m^3，测压液体为水银。测得 $h_1 = 100 \text{mm}$，$h_2 = 500 \text{mm}$，$h_3 = 750 \text{mm}$。计算容器 A、B 的压强差。

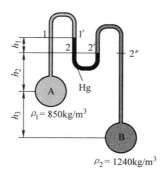

图 X2.4.2

Solution   Starting with the pressure at point $1'$ at the interface and point 1 at the same elevation in a continuous fluid, set an equation of $p_1 = p_{1'}$. Repeat the same steps at points $2' - 2$, and $2' - 2''$.

$$p_1 = p_{1'} = p_A - \rho_1 g\left(h_1 + h_2\right) = p_2 - \rho_3 g h_1$$

$$p_2 = p_{2'} = p_{2''} = p_B - \rho_2 g\left(h_2 + h_3\right)$$

$$p_B - p_A = \rho_3 g h_1 + \rho_2 g\left(h_2 + h_3\right) - \rho_1 g\left(h_1 + h_2\right) = 23.5\text{kPa}$$

Example 2.4.3   Alcohol is contained in A and B, density of the manometer fluid is 1600kg/m^3, as shown in Fig.X2.4.3, what is the pressure difference between A and B?

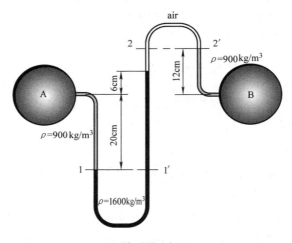

Fig.X2.4.3

Solution   Starting with the pressure p_A at center of container A, moving along the tube by adding or subtracting the $\rho g h$ terms until we reach center of container B, skip at interface to the point at the same elevation in a continuous fluid, and setting the result equal to p_B, gives,

$$p_A + \rho g h_1 - \rho' g\left(h_1 + h_2\right) + \rho g h_3 = p_B$$
$$p_A - p_B = \rho' g\left(h_1 + h_2\right) - \rho g\left(h_1 + h_3\right)$$
$$p_A - p_B = [1.6 \times 9.8 \times \left(0.2 + 0.06\right) - 0.9 \times 9.8 \times \left(0.2 + 0.12\right)]\text{kPa} = 1.25\text{kPa}$$

Example 2.4.4   Measured fluid contained in A and B, density of manometer fluid in the double-U-tube manometer is ρ', shown in FigX2.4.4, calculate the pressure difference between A and B.

解　取分界面处点 $1'$ 及其等压点 1，建立等式 $p_1 = p_{1'}$。对点 $2'{-}2$ 和 $2'{-}2''$ 重复上述步骤。

$$p_1 = p_{1'} = p_A - \rho_1 g\left(h_1 + h_2\right) = p_2 - \rho_3 g h_1$$

$$p_2 = p_{2'} = p_{2''} = p_B - \rho_2 g\left(h_2 + h_3\right)$$

$$p_B - p_A = \rho_3 g h_1 + \rho_2 g\left(h_2 + h_3\right) - \rho_1 g\left(h_1 + h_2\right) = 23.5\,\text{kPa}$$

例 2.4.3　如图 X2.4.3 所示容器 A、B 内液体为酒精，测压管工作液体密度 $1600\,\text{kg/m}^3$，计算容器 A、B 的压强差。

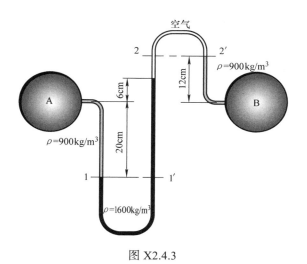

图 X2.4.3

解　从容器 A 中心开始，沿测压管下行时加相应 $\rho g h$ 项，上行时减去相应 $\rho g h$ 项，遇流体分界面则跃至同液柱内等高的等压点，直至容器 B 中心处压强为 p_B，列出等式：

$$p_A + \rho g h_1 - \rho' g\left(h_1 + h_2\right) + \rho g h_3 = p_B$$
$$p_A - p_B = \rho' g\left(h_1 + h_2\right) - \rho g\left(h_1 + h_3\right)$$
$$p_A - p_B = [1.6 \times 9.8 \times \left(0.2 + 0.06\right) - 0.9 \times 9.8 \times \left(0.2 + 0.12\right)]\,\text{kPa} = 1.25\,\text{kPa}$$

例 2.4.4　测量图 X2.4.4 所示容器 A、B 压差，复式测压管内工作液体密度为 ρ'，写出 A、B 压差表达式。

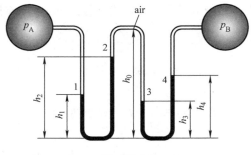

Fig.X2.4.4

Solution　Firstly identify the relevant points as shown in Fig.X2.4.4. Start from point A (A container), moving along the tube by add pressure when the elevation decreases and subtract pressure when the elevation increases, "jump" at interface points from one fluid column to the next point which is in the same continuous fluid, until point B (B container) is reached, note that by neglecting the weight of the air, the pressure at point 2 is equal to the pressure at point 3.

$$p_A + \rho g(h_0 - h_1) - \rho'g(h_2 - h_1) - \rho'g(h_4 - h_3) - \rho g(h_0 - h_4) = p_B$$

$$p_A - p_B = \rho'g\left[(h_2 - h_1) + (h_4 - h_3)\right] - \rho g(h_4 - h_1)$$
$$p_A - p_B = \rho'g(\Delta h_1 + \Delta h_2) - \rho g(h_4 - h_1)$$

We can also start at point B and end at point A, then

$$p_B + \rho g(h_0 - h_4) + \rho'g(h_4 - h_3) + \rho'g(h_2 - h_1) - \rho g(h_0 - h_1) = p_A$$
$$\rho g(h_0 - h_4) + \rho'g(h_4 - h_3) + \rho'g(h_2 - h_1) - \rho g(h_0 - h_1) = p_A - p_B$$
$$p_A - p_B = \rho'g(\Delta h_1 + \Delta h_2) - \rho g(h_4 - h_1)$$

2.5　Forces on Plane Surfaces

A plane surface (such as a gate valve in a dam, the wall of a liquid storage tank, or the hull of a ship at rest) is subjected to fluid pressure distributed over its surface when exposed to a liquid. On a plane surface, the hydrostatic forces form a system of parallel forces that are all normal to the surfaces, and we often need to determine the magnitude of the force and its point of application, which is called the center of pressure. In most cases atmospheric pressure acts on both sides of the plate, yielding a zero resultant, it is convenient to subtract atmospheric pressure and work with the gage pressure only.

2.5.1　Magnitude of pressure on plane surfaces

Consider the top surface of a flat plate of arbitrary shape completely submerged in a liquid, as shown in Fig.2.5.1 together with its normal view. The plane of this surface (normal to the page) intersects the horizontal free surface at angle θ. The absolute pressure above the liquid is p_{atm}, which can be ignored in most force calculations since it acts on both sides of the plate. Then the gage pressure at any point on the plate is

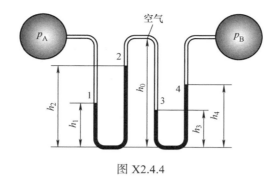

图 X2.4.4

解　首先如图 X2.4.4 所示标记各点。由 A 点（A 容器）起始，液柱下降时加上该段液柱形成的压强；上升时减去此段液柱产生的压强，遇到流体分界面则跳跃至同一液柱高度相等的另一侧，直至终点 B（B 容器）。注意忽略空气重量，即，认为点 2 和点 3 压强相等。

$$p_A + \rho g(h_0 - h_1) - \rho' g(h_2 - h_1) - \rho' g(h_4 - h_3) - \rho g(h_0 - h_4) = p_B$$

$$p_A - p_B = \rho' g\left[(h_2 - h_1) + (h_4 - h_3)\right] - \rho g(h_4 - h_1)$$

$$p_A - p_B = \rho' g(\Delta h_1 + \Delta h_2) - \rho g(h_4 - h_1)$$

也可以反向，从 B 点起始至 A 点终此，列出等式

$$p_B + \rho g(h_0 - h_4) + \rho' g(h_4 - h_3) + \rho' g(h_2 - h_1) - \rho g(h_0 - h_1) = p_A$$

$$\rho g(h_0 - h_4) + \rho' g(h_4 - h_3) + \rho' g(h_2 - h_1) - \rho g(h_0 - h_1) = p_A - p_B$$

$$p_A - p_B = \rho' g(\Delta h_1 + \Delta h_2) - \rho g(h_4 - h_1)$$

2.5　平面壁上的流体静压力

平面壁（如堤坝的闸门、储液罐的壁、静止的船体等）置于液体中时，其被淹面都会受到流体静压力的作用。平面上的流体静压力由一系列垂直于受压面的平行分力构成，需要计算其合力的大小和作用点位置，即静压力中心点的位置。绝大多数情况下作用于物体两侧的大气压可互相抵消，所以通常忽略大气压仅用表压强。

2.5.1　平面壁流体静压力大小

如图 2.5.1 所示一任意形状完全淹没于液体中的平板，取与平板垂直的视图。平板（本视图中垂直于页面）与水平自由液面夹角为 θ。液面上方大气压为 p_{atm}，由于平板两侧大气压平衡故可以忽略。所以平板上各处的表压强为

$$p = \rho g h = \rho h y \sin \theta \qquad (2.5.1)$$

$$p = \rho gh = \rho hy \sin \theta \tag{2.5.1}$$

where h is the vertical distance of the point from the free surface and y is the distance of the point from the x-axis (from point O in Fig. 2.5.1). The resultant hydrostatic force F acting on the surface is determined by integrating the force $p\mathrm{d}A$ acting on a differential area $\mathrm{d}A$ over the entire surface area

$$F = \int_A p\mathrm{d}A = \int_A \rho gy \sin \theta \mathrm{d}A = \rho g \sin \theta \int_A y\mathrm{d}A \tag{2.5.2}$$

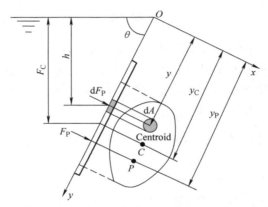

Fig.2.5.1　Hydrostatic force on an inclined plane surface completely submerged in a liquid

The distance from Ox along the plane to the centroid (point C) is defined as

$$y_C = \frac{1}{A} \int_A y\mathrm{d}A \tag{2.5.3}$$

Substituting

$$F = \rho g y_C \sin \theta A = \rho g h_C A \tag{2.5.4}$$

where ρgh_C is the pressure at the centroid of the surface, which is equivalent to the average pressure on the surface, and $h_C = y_C \sin\theta$ is the vertical distance of the centroid from the free surface of the liquid, is called *depth of immersion* of *the centroid*.

Thus we conclude that: the pressure acts normal to the surface, and the magnitude of the resultant force acting on a plane surface equals the pressure p_C at the centroid of the surface multiplied by the surface area A. But the force does not, in general, act at the centroid.

2.5.2　Center of pressure on plane surfaces

The point of action of the resultant force is the center of pressure. We note that the sum of the moments of all the infinitesimal pressure forces acting on the area A must equal the moment of the resultant force. Let the force F act at the point $P(x_P, y_P)$, the center of pressure, the value of y_P can be determined by equating the moment of the resultant force to the moment of the distributed pressure force about the x-axis

式中 h 为面上任意点至自由液面的垂直距离，y 为各点沿平板表面至 x 轴（原点，为图 2.5.1 中 O 点）的距离。对面积微元 dA 上的压力 pdA 在整个平板表面积分可得到静压力的合力 F

$$F = \int_A p\,dA = \int_A \rho g y \sin\theta\,dA = \rho g \sin\theta \int_A y\,dA \qquad (2.5.2)$$

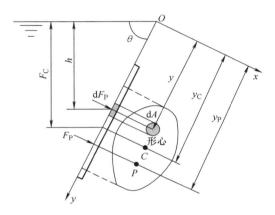

图 2.5.1　完全淹没于液体中的倾斜平板所受流体静压力

定义 Ox 轴至平板形心（C 点）沿平板表面的距离为

$$y_C = \frac{1}{A}\int_A y\,dA \qquad (2.5.3)$$

代入式 2.5.2 得

$$F = \rho g y_C \sin\theta A = \rho g h_C A \qquad (2.5.4)$$

式中，$\rho g h_C$ 是平板形心处的静压强，等于平板上各点压强的平均值；$h_C = y_C \sin\theta$ 为形心至自由液面的垂直距离，称为**形心的淹深**。

可以得出结论：平面壁所受流体静压力垂直于壁面且其大小等于壁面形心处的压强 p_C 乘以平面壁的受压面积 A。但静压力的作用点通常不在形心处。

2.5.2　静压力的中心

合力的作用点即为静压力的中心。各微元面积上的压力 dF 对面积 A 的力矩之和，一定等于合力 F 对该面积的力矩。假设静压力中心点坐标 $P(x_P, y_P)$，由各分力对 x 轴的力矩之和必等于合力对 x 轴的力矩，可以确定坐标值 y_P

$$y_P F = \int_A yp\,\mathrm{d}A = \rho g \sin\theta \int_A y^2 \mathrm{d}A = \rho g \sin\theta I_{Ox} \qquad (2.5.5)$$

where the second moment of the area (also called the area moment of inertia) about the x-axis is

$$I_{Ox} = \int_A y^2 \mathrm{d}A \qquad (2.5.6)$$

The second moments of area about two parallel axes are related to each other by the parallel axis theorem, which in this case is expressed as

$$I_{Ox} = I_{Cx'} + y_C^2 A \qquad (2.5.7)$$

where $I_{Cx'}$ is the second moment of area about the x'-axis passing through the centroid of the area and y_C (the y-coordinate of the centroid C) is the distance between the two parallel axes.

Substitute Eq.2.5.7 into Eq.2.5.5, and obtain

$$y_P = y_C + \frac{I_{Cx'}}{Ay_C} \qquad (2.5.8)$$

From Eq.2.5.8, we can see that the location of the center of pressure y_P is independent of the angle θ, that is, we can rotate the plane area about x-axis without affecting the location of P. Also, we see that the center of pressure is always below the centroid (except on a horizontal area for which the center of pressure and the centroid coincide).

The $I_{Cx'}$ values for some common areas are given in Fig.2.5.2.

Similarly, to locate the x-coordinate x_P of the center of pressure, we write

$$x_P F = \int_A xp\,\mathrm{d}A = \rho g \sin\theta \int_A xy\,\mathrm{d}A \qquad (2.5.9)$$

where the product of inertia of the area A is

$$I_{xy} = \int_A xy\,\mathrm{d}A \qquad (2.5.10)$$

Using the transfer theorem for the product of inertia

$$I_{xy} = I_{Cx'y'} + Ax_C y_C \qquad (2.5.11)$$

Eq.2.5.9 becomes

$$x_P = x_C + \frac{I_{Cx'y'}}{Ay_C} \qquad (2.5.12)$$

For areas that are symmetric about either the y-axis or the x-axis, $I_{Cx'y'} = 0$, it gives $x_P = x_C$, we see that the center of pressure lies directly below the centroid.

For areas that possess symmetry about the y-axis, the center of pressure lies on the y-axis directly below the centroid.

$$y_{\mathrm{P}}F = \int_A y p \mathrm{d}A = \rho g \sin\theta \int_A y^2 \mathrm{d}A = \rho g \sin\theta I_{Ox} \tag{2.5.5}$$

式中微元面积对 x 轴的二阶矩（也称为截面惯性矩）定义为

$$I_{Ox} = \int_A y^2 \mathrm{d}A \tag{2.5.6}$$

根据移轴定理，两平行轴惯性矩之间的关系可以表示为

$$I_{Ox} = I_{Cx'} + y_{\mathrm{C}}{}^2 A \tag{2.5.7}$$

式中，$I_{Cx'}$ 是关于 x' 轴的惯性矩，x' 轴平行于 x 轴且两轴之间的距离为 y_{C}（形心 C 的 y 坐标值）。

将式 2.5.7 代入式 2.5.5，可得

$$y_{\mathrm{P}} = y_{\mathrm{C}} + \frac{I_{Cx'}}{A y_{\mathrm{C}}} \tag{2.5.8}$$

由式 2.5.8 可知，压力中心的坐标 y_{P} 与平板倾斜角度 θ 无关，即平板绕 x 轴任意转动不会影响流体静压力中心点 P 的位置。同样可知，压力中心总是低于形心（除非水平放置的平板，其力心与形心重合）。

图 2.5.2 列出了几种常见几何形状的 $I_{Cx'}$。

同样方法可以得出压力中心的 x 坐标值 x_{P}，

$$x_{\mathrm{P}}F = \int_A x p \mathrm{d}A = \rho g \sin\theta \int_A xy \mathrm{d}A \tag{2.5.9}$$

式中对面积 A 的惯量积为

$$I_{xy} = \int_A xy \mathrm{d}A \tag{2.5.10}$$

同样应用移轴定理，得

$$I_{xy} = I_{Cx'y'} + A x_{\mathrm{C}} y_{\mathrm{C}} \tag{2.5.11}$$

则式 2.5.9 改写为

$$x_{\mathrm{P}} = x_{\mathrm{C}} + \frac{I_{Cx'y'}}{A y_{\mathrm{C}}} \tag{2.5.12}$$

对于任何关于 y 轴或 x 轴对称的平面，都有 $I_{Cx'y'} = 0$，所以 $x_{\mathrm{P}} = x_{\mathrm{C}}$，即力心在形心下方。

关于 y 轴对称的几何形状，其力心一定位于 y 轴上、且在形心下方。

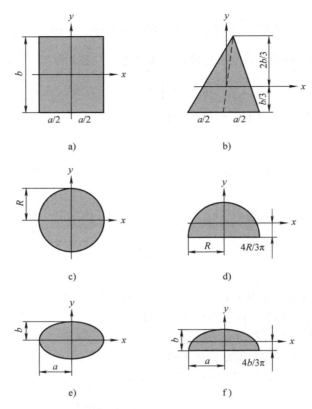

Fig.2.5.2　The centroid and the centroidal moments of inertia for some common geometries

a) Rectangle　$A = ab$　$I_{Cx'} = ab^3/12$　b) Triangle　$A = ab/2$　$I_{Cx'} = ab^3/36$　c) Circle　$A = \pi R^2$　$I_{Cx'} = \pi R^4/4$

d) Semi-circle　$A = \pi R^2/2$　$I_{Cx'} = 0.11R^4$　e) Ellipse　$A = \pi ab$　$I_{Cx'} = \pi ab^3/4$

f) Semi-ellipse　$A = \pi ab/2$　$I_{Cx'} = 0.11ab^3$

Using the expression above, we can show that the force on a rectangular plate with the top edge even with the liquid surface, as shown in Fig.2.5.3, acts two-thirds of the way down through the centroid.

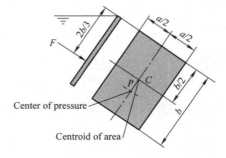

Fig.2.5.3　Force on a rectangle plane area with top edge even with a free surface

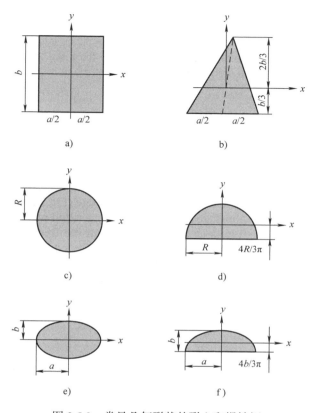

图 2.5.2　常见几何形状的形心和惯性矩

a）矩形　$A = ab$　$I_{Cx'} = ab^3/12$　b）三角形　$A = ab/2$　$I_{Cx'} = ab^3/36$　c）圆　$A = \pi R^2$　$I_{Cx'} = \pi R^4/4$

d）半圆　$A = \pi R^2/2$　$I_{Cx'} = 0.11R^4$　e）椭圆　$A = \pi ab$　$I_{Cx'} = \pi ab^3/4$

f）半椭圆　$A = \pi ab/2$　$I_{Cx'} = 0.11ab^3$

由以上分析可知，对于如图 2.5.3 所示的矩形平面顶边与自由液面平齐时，其压力中心刚好位于顶点沿对称中心线向下 2/3 处。

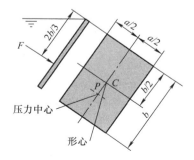

图 2.5.3　顶部与自由液面平齐的矩形平面所受静压力

2.5.3 Submerged rectangular plate

The pressure distribution on a submerged horizontal surface is uniform, and its magnitude is $F=\rho gHA$, where H is the distance of the surface from the free surface. Therefore, the hydrostatic force acting on a horizontal rectangular surface is

$$F = \rho gHab \qquad (2.5.13)$$

and it acts through the centroid of the plate(Fig.2.5.4a).

The force on a vertical rectangular plate, with the top edge even with the liquid surface, as shown in Fig. 2.5.4b, is obviously considered the triangular pressure distribution and acts two-thirds of the way down.

The force on a submerged rectangular plate could be divided into two composite forces: a rectangular distributed force with centroid at its center, and a triangular distributed force with centroid two-third the distance from the top edge (Fig.2.5.4c).

Consider a completely submerged rectangular plate tilted at an angle θ from the horizontal and whose top edge is at a distance H from the free surface, as shown in Fig.2.5.4d. The two composite forces: a rectangular distributed force with centroid at its center, and a triangular distributed force with centroid two-third of the length from the top along the plate, the location of the force is then found by locating the centroid of the composite forces.

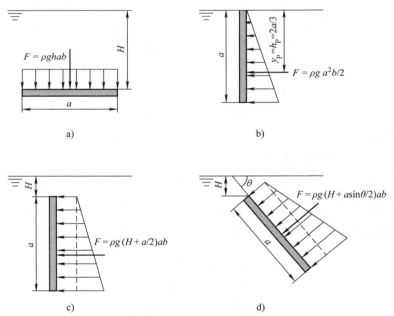

a) b)

c) d)

Fig.2.5.4   Hydrostatic force acting on a rectangular plate for horizontal, vertical, and tilted cases

a) Horizontal plate   b) Vertical plate with top edge even with a free surface

c) Submerged Vertical plate   d) Submerged tilted plate

Example 2.5.1   A rectangular flat plate completely submerged under the water (as shown in Fig.2.5.4d), the plate of length 3m and width (normal to the page) 2m tilted at an angle 30°from the horizontal and the top edge is 2m below the free surface. Deter-

2.5.3　完全浸没的矩形平板

完全浸没的水平平板表面压力均匀分布，其值大小为 $F=\rho gHA$，式中 H 是平板距自由液面的距离。所以完全淹没在自由液面下水平放置的矩形平板所受流体静压力为

$$F = \rho gHab \tag{2.5.13}$$

且静压力通过平板形心（图 2.5.4a）。

顶点与自由液面平齐的竖直平板（见图 2.5.4b）所受静压力为三角形分布且力心位于顶点下方 2/3 高度处。

完全淹没在液面下一定深度的平板所受流体静压力可以看作由两部分组成：力心与形心重合且矩形分布的静压力；力心在顶点下方 2/3 处，且为三角形分布的静压力（见图 2.5.4c）。

现在分析图 2.5.4d 所示完全淹没在液面下且与水平方向夹角 θ 的矩形平板，其顶部至自由液面的距离为 H。静压力由两部分构成：力心在形心处的矩形载荷和力心在平板顶部沿平板向下 2/3 处的三角形载荷，合力的作用点可通过两部分载荷来确定。

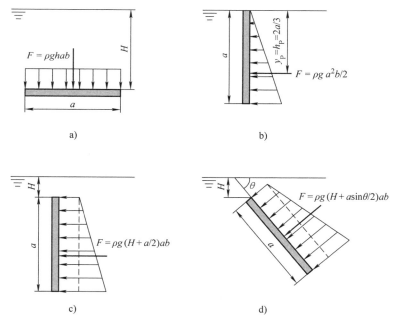

a)　　　　　　　　　　　　　　　　　b)

c)　　　　　　　　　　　　　　　　　d)

图 2.5.4　流体静压力作用于水平、竖直及倾斜的矩形平面壁

a) 水平矩形平面　b) 竖直且顶点与自由液面平齐的矩形平面

c）完全淹没的竖直矩形平面　d）完全淹没的倾斜矩形平面

例 2.5.1　完全淹没的矩形平板（见图 2.5.4d）长 3m 宽 2m，与水平方向成 30° 角倾斜放置于水下。平板顶部淹没深度为 2m。（1）计算流体静压力并确定力的

mine the hydrostatic force on the plate and the location of the pressure center as well as the depth of immergence of centroid of the plate, and compare to the locations when the submerge depth of top edge is 10m.

Solution (1) The rectangular distributed force with centroid at its center is

$$F_1 = \rho g H A = 117.6\text{kN}$$

The triangular distributed force with centroid two-third of the length from the top along the plate is

$$F_2 = \rho g \frac{a}{2} \sin\theta A = 44.1\text{kN}$$

The resultant force is

$$F = F_1 + F_2 = 161.7\text{kN}$$

Let the center of pressure is $(1, y_P)$, equating the moment of the two composite forces and the resultant force about the top edge of the plate.

$$F_1 \frac{a}{2} + F_2 \frac{2}{3} a = F y_P$$
$$y_P = 1.64\text{m}$$

The depth of center of pressure is

$$h_P = H + y_P \sin 30° = 2.82\text{m}$$

The depth of immergence of centroid of plate is $h_C = H + y_C \sin 30° = 2.75\text{m}$.

(2) The rectangular distributed force when the submergence depth of top is 10m is

$$F_1 = \rho g H A = 588\text{kN}$$

Location of center of pressure along the plate is

$$F_1 \frac{a}{2} + F_2 \frac{2}{3} a = F y_P$$
$$y_P = 1.53\text{m}$$

The depth of center of pressure is

$$h_P = H + y_P \sin 30° = 10.7515\text{m}$$

The depth of immergence of centroid of plate is 10.75m.

Note that, as the depth of submersion is increased, the depth of immersion of centroid increases and therefore center of the force approaches the centroid of area.

Example 2.5.2 A plane area of 1m×1m acts as an escape hatch on a submersible in a lake as shown in Fig.X2.5.2. If it is on a 45°angle with the horizontal, what force applied normal to the hatch at the bottom edge is needed to just open the hatch, if it is hinged at the top edge when the top edge is 10m below the surface? The pressure inside the submersible is assumed to be atmospheric.

中心;（2）若顶点淹深为 10m 时，比较形心和力心的位置。

解　（1）中心在形心处的矩形均布载荷大小为

$$F_1 = \rho g H A = 117.6 \text{kN}$$

力心位于顶点下 2/3 处的三角形分布载荷大小为

$$F_2 = \rho g \frac{a}{2} \sin\theta A = 44.1 \text{kN}$$

合力大小等于

$$F = F_1 + F_2 = 161.7 \text{kN}$$

设力心坐标为（1，y_P），两分力至平板顶部力矩之和与合力至顶部的力矩相等。

$$F_1 \frac{a}{2} + F_2 \frac{2}{3} a = F y_P$$
$$y_P = 1.64 \text{m}$$

力心的淹深为

$$h_P = H + y_P \sin 30° = 2.82 \text{m}$$

形心的淹深为 $h_C = H + y_C \sin 30° = 2.75 \text{m}$。

（2）顶点淹深为 10m 时，矩形均布载荷的大小为

$$F_1 = \rho g H A = 588 \text{kN}$$

流体静压力作用中心位置 y_P

$$F_1 \frac{a}{2} + F_2 \frac{2}{3} a = F y_P$$
$$y_P = 1.53 \text{m}$$

压力中心淹深

$$h_P = H + y_P \sin 30° = 10.7515 \text{m}$$

形心的淹深为 10.75m。

可见，固体壁面淹没深度越大，形心和力心的位置越接近。

例 2.5.2　图 X2.5.2 所示淹没于湖水中的潜水舱表面有一个 1m×1m 与水平方向成 45° 的舱门。若舱门上方铰链淹没于水下的深度为 10m，计算开启舱门所需的力。假设潜水舱内部压强为大气压。

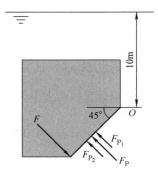

Fig.X2.5.2

Solution   The force on the plane could be considered has two components, the rectangular distributed pressure by the 10m deep water and the pressure changes with the depth of water from the hinge downward along the hatch.

$$F_{p_1} = \rho g H A$$

$$F_{p_2} = \rho g h'_C A, \ h'_C = \frac{l}{2} \sin 45°$$

When the escape hatch is open, the torque by force applied equating with the torque by hydrostatic pressure.

$$Fl = F_{p_1} \frac{l}{2} + F_{p_2} \frac{2}{3} l$$

$$F = 51.3 \text{kN}$$

Example2.5.3   A gate represented in Fig.X2.5.3 of 1m width perpendicular to the page, it is on a 60° angular with the horizontal and pivoted at hinge O. For what value of hinge position x will the gate open automatically when the water depth H and h are 2m and 0.4m relatively.

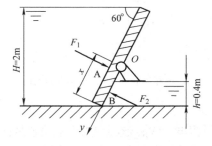

Fig.X2.5.3

Solution   When the gate is open, moments about hinge O by water pressure from left and right are even.

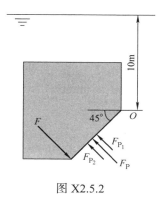

图 X2.5.2

解　舱门表面的流体静压力可以看作由两部分组成，一部分是 10m 深的湖水形成的均布载荷，另一部分是自铰链向下随舱门表面各点深度变化的三角形载荷

$$F_{P_1} = \rho g H A$$

$$F_{P_2} = \rho g h_C' A \qquad h_C' = \frac{l}{2}\sin 45°$$

潜水舱门开启时，开启舱门的力对铰链的力矩等于流体静压力对铰链的力矩和

$$Fl = F_{P_1}\frac{l}{2} + F_{P_2}\frac{2}{3}l$$

$$F = 51.3\text{kN}$$

例 2.5.3　一倾斜 60° 宽 1m 的水闸，如图 X2.5.3 所示，可绕 O 点铰链转动。设计铰链 O 的位置 x，保证当上下游水位 H 和 h 分别达到 2m 和 0.4m 时，闸门可自动开启。

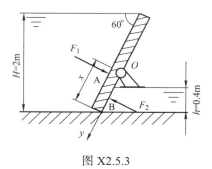

图 X2.5.3

解　闸门开启时，上、下游水静压力对铰链 O 的力矩应该相等。

Left side of the gate:

$$F_1 = \rho g h_{C1} A_1 = \rho g \frac{H}{2} \frac{H}{\sin \theta}$$

The moment arm of F_1 with respect to hinge O is $x - y_{P1}$, and y_{P1} is one-third of the length from the bottom along the gate

$$T_1 = F_1 \left(x - \frac{1}{3} \frac{H}{\sin \theta} \right)$$

Right side of the gate:

$$F_2 = \rho g h_{C2} A_2 = \rho g \frac{h}{2} \frac{h}{\sin \theta}$$

The moment arm of F_2 with respect to hinge O is $x - y_{P2}$, and y_{P2} is one-third of the wet length from the bottom along the gate

$$T_2 = F_2 \left(x - \frac{1}{3} \frac{h}{\sin \theta} \right)$$

When the gate is about to open, moments of these two forces about O are equal.

$$F_1 \left(x - \frac{H}{3 \sin 60°} \right) = F_2 \left(x - \frac{h}{3 \sin 60°} \right)$$

The result is: $x = 0.8$m

Example 2.5.4 Fig.X2.5.4a. represents a gate, 1m wide perpendicular to the sketch. It is pivoted at hinge H. The gate weighs 500kg. Its center of gravity is 0.4m to the right of and 0.3m above H. For what values of water depth x above H will the gate remain closed? Neglect friction at the pivot and neglect the thickness of the gate.

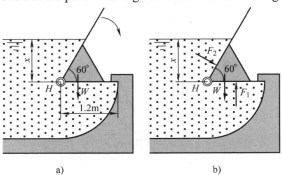

a)   b)

Fig.X2.5.4

a) Structure of the gate b) force diagram of the gate

Solution There are three forces acting on the gate: its weight W, the vertical hydrostatic force F_1 upward on the bottom of the gate, and the slanting hydrostatic force F_2

闸门左侧：

$$F_1 = \rho g h_{C1} A_1 = \rho g \frac{H}{2} \frac{H}{\sin\theta}$$

力 F_1 至铰链 O 的力臂为 $x - y_{P1}$，式中 y_{P1} 等于闸门底部沿闸门左侧受淹长度的 1/3

$$T_1 = F_1 \left(x - \frac{1}{3} \frac{H}{\sin\theta} \right)$$

闸门右侧：

$$F_2 = \rho g h_{C2} A_2 = \rho g \frac{h}{2} \frac{h}{\sin\theta}$$

力 F_2 至铰链 O 的力臂为 $x - y_{P2}$，式中 y_{P2} 等于闸门底部沿闸门右侧受淹长度的 1/3

$$T_2 = F_2 \left(x - \frac{1}{3} \frac{h}{\sin\theta} \right)$$

闸门刚好开启时，两个力对 O 点的力矩相等。

$$F_1 \left(x - \frac{H}{3\sin 60°} \right) = F_2 \left(x - \frac{h}{3\sin 60°} \right)$$

解得 $x = 0.8\text{m}$

例 2.5.4　图 X2.5.4a 所示闸门宽 1m（垂直图面方向），可绕铰链 H 转动。闸门质量 500kg，其重心位于 H 右侧 0.4m、上方 0.3m 处。计算可使闸门保持关闭的水深 x 值。忽略铰链处摩擦和闸门厚度。

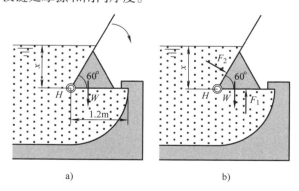

a)　　　　　　　　　　b)

图 X2.5.4

a）闸门结构　b）闸门受力

解　闸门受三部分力作用：自重 W；闸门底部竖直向上的静压力 F_1；与闸门

acting at right angles to the sloping portion of the gate. The vertical hydrostatic force F_1 tends to make the gate keep closed while the weight and slanting force F_2 make attempt to open it. A diagram showing these three forces is as Fig.X2.5.4b, the magnitudes of the latter three forces are:

Weight is given as $W=500g$

$$F_1 = \rho gxA_1 = 1.2\rho gx$$

$$F_2 = \rho g\frac{x}{2}A_2 = \rho g\frac{x}{2}\frac{x}{2\sin 60°}b = \frac{\sqrt{3}}{3}\rho gx^2$$

The moment arms of weight and F_1 respect to H are 0.4m and 0.6m, respectively. The line of action of F_2 for the triangular distributed load is at the one-third of the length along the slopping surface of the gate downward to H, and the location is calculated from

$$y_{P_2} = \frac{1}{3}\frac{x}{\sin 60°} = \frac{2\sqrt{3}}{9}x$$

When the gate is about to open (incipient rotation), the moments of all forces about H are balanced, that is

$$F_2 y_{P_2} + 0.4W = 0.6F_1 \quad \rightarrow \quad \frac{\sqrt{3}}{3}\rho gx^2\frac{2\sqrt{3}}{9}x + 0.4\times 500g - 0.6\times 1.2\rho gx = 0$$

It is reduced to

$$\frac{2}{9}x^3 - 0.72x + 0.2 = 0$$

There are three roots are $x = 0.285, 1.641, -1.926$, the negative one is meaningless. Therefore, the gate will remain closed when 0.285m<x<1.641m.

Disscusion: Example 2.5.5

A heavy car plunges into a lake during an accident and lands at the bottom of the lake on its wheels (Fig. X2.5.5). The door is 1.2m high and 1m wide, and the top edge of the door is 6m below the free surface of the water. The passenger cabin is well-sealed so that no water leaks inside, and the pressure in the passenger cabin remains at atmospheric. Determine the hydrostatic force on the door and the location of the pressure center, and discuss whether the driver can open the door.

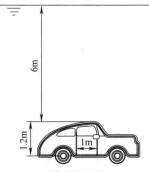

Fig.X2.5.5

斜面垂直的静压力 F_2。压力 F_1 使闸门关闭，自重 W 和压力 F_2 使闸门开启。三部分力对闸门的作用如图 X2.5.4b 所示，其大小为

自重 $W=500g$

$$F_1 = \rho gxA_1 = 1.2\rho gx$$

$$F_2 = \rho g \frac{x}{2} A_2 = \rho g \frac{x}{2}\frac{x}{\sin 60°} b = \frac{\sqrt{3}}{3}\rho gx^2$$

自重 W 和力 F_1 至 H 的力臂分别为 0.4m 和 0.6m。力 F_2 的作用点沿闸门斜面向下至 H 点距离为其长度的 1/3，其值为

$$y_{P_2} = \frac{1}{3}\frac{x}{\sin 60°} = \frac{2\sqrt{3}}{9}x$$

闸门即将开启时，H 点的力矩平衡，即

$$F_2 y_{P_2} + 0.4W = 0.6F_1 \quad \rightarrow \quad \frac{\sqrt{3}}{3}\rho gx^2 \frac{2\sqrt{3}}{9}x + 0.4\times 500g - 0.6\times 1.2\rho gx = 0$$

整理为

$$\frac{2}{9}x^3 - 0.72x + 0.2 = 0$$

解一元三次方程得 $x=0.285$，1.641，-1.926，负值解无意义。

所以，当水深 x 在 $0.285\sim 1.641$m 之间时，闸门可自动处于关闭状态。

讨论：例 2.5.5

一较重的小车由于事故落入水中，如图 X2.5.5 所示。车门高约 1.2m 宽约 1m，门顶部至水面距离约为 6m。车体密封良好未进水，所以车内气压仍为大气压。估算车门所受流体静压力及车门力心位置，讨论车内的人是否有能力开启车门。

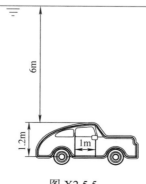

图 X2.5.5

Solution The resultant hydrostatic force on the door is determined to be

$$F_\mathrm{P} = \rho g h_\mathrm{C} A = 9.8 \times 6.6 \times 1.2 \mathrm{kN} = 77.6 \mathrm{kN}$$

A strong man can push 100kg, whose weight is 980N, so the driver can apply the force at a point 1m from the hinges for maximum and generate a moment of 0.98kN·m. The resultant hydrostatic force acts at a distance of 0.5m from the hinges, and this generates a moment of 38.8kN·m, which is about 40 times the moment the driver can possibly generate. The driver should open the window and wait till the car is filled with water since at that point the pressures on both sides of the door are nearly the same.

2.6 Forces on Curved Surfaces

For a submerged curved surface, the determination of the resultant hydrostatic force is more involved since the force on the various elementary areas that make up the curved surface are different in direction and magnitude. The easiest way to determine the resultant hydrostatic force F acting on a two-dimensional curved surface is to determine the horizontal and vertical components F_x and F_z separately.

Consider a free-body diagram of a liquid block enclosed by a two-dimensional curved surface MN and the two plane surfaces (one horizontal and one vertical) passing through the two ends of the curved surface, as shown in Fig2.6.1. A coordinate system is built, in which Oxy plane overlaps with the liquid free surface, and Oz axis downwards. Note that the resultant force F acting on the curved solid surface is equal with and opposite to the counterforce F' acting on the detached fluid block by the curved surface (Newton's third law). So we actually need to calculate are the forces exerted on the fluid block in x- and z-direction.

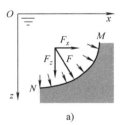 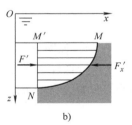 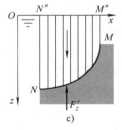

a)   b)   c)

Fig.2.6.1 Hydrostatic forces on curved surfaces

a) Hydrostatic force on a curved surface b) Horizontal force c) Vertical force

2.6.1 Horizontal force on curved surface

The projecting elements, which are all horizontal, enclose a volume whose ends are the vertical plane through N and the irregular area MN, as shown in Fig2.6.1b. This volume of liquid is in static equilibrium. The only horizontal forces on MNM' are the pressure force from surrounding fluid F' and force exerted by the curved surface itself F_x' which is the counterforce of x-component of hydrostatic force F_x, therefore the horizontal force component

$$F_x = F' = -F_x' \qquad (2.6.1)$$

解 车门所受流体静压力大小约为

$$F_P = \rho g h_C A = 9.8 \times 6.6 \times 1.2 \text{kN} = 77.6 \text{kN}$$

一个强壮的成年人可发出约 100kgf 的推力，即大约 980N。驾驶人员在距铰链约 1m 处施加推力可产生的力矩约 0.98kN·m。水的静压力中心距车门铰链的距离为 0.5m，静压力对铰链的力矩约为 38.8kN·m，大约是人力可产生推力力矩的 40 倍。所以驾驶人员暂时无法开启车门，应该开启车窗等待车内几乎充满水时，车门两侧静压力平衡后开启车门。

2.6 曲面壁上的流体静压力

曲面壁各微元面积上的流体静压力方向和大小均不相同，所以淹没在液面下的曲面壁所受流体静压力的计算较平面壁复杂。求解二维曲面上流体静压力的最佳方法就是分别计算水平方向分力 F_x 和竖直方向分力 F_z。

二维曲面 MN 以及通过其两端的水平面和竖直平面封闭成一个"液体块"，其受力情况如图 2.6.1 所示。取自由液面为直角坐标系的 Oxy 平面，z 轴竖直向下。根据牛顿第三定律，曲面壁所受的流体静压力 F 与其反作用力 F' 大小相等方向相反。所以，需要求解的是反作用力在 x 和 z 方向的分量。

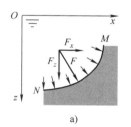 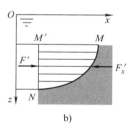 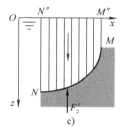

      a)   b)   c)

图 2.6.1 曲面壁上的流体静压力

a）曲面壁上的流体静压力 b）水平方向分力 c）竖直方向分力

2.6.1 曲面壁上水平方向分力

曲面 MN 及其各点沿水平方向的投影线与过 N 的竖直平面围成一个如图 2.6.1b 所示的空间体积，这一体积处于静力平衡状态。体积 MNM' 内的流体在水平方向仅受到来自周围流体的压力 F' 和来自曲面的 x 方向作用力 F_x'。曲面作用于流体的力 F_x' 和曲面在 x 方向所受的流体静压力 F_x 是一对作用力和反作用力，所以流体静压力 x 方向的分量等于

$$F_x = F' = -F_x' \tag{2.6.1}$$

The pressure force F' by surrounding fluid acts on the vertical surface $M'N$ which is the horizontal projection of the curved surface whose area is expressed as A_x can be calculated by

$$F' = F_x = \rho g h_{Cx} A_x \tag{2.6.2}$$

Hence the horizontal force on any curved area is equal to the force on an imaginary horizontal projection of that area onto a vertical plane. The magnitude of the horizontal force is equal to the pressure at center of the projection multiplied by the projection area, as discussed in Section 2.5 and expressed as Eq.2.6.2. Here h_{Cx} and A_x are depth of immergence of center and area of the horizontal projection respectively.

2.6.2   Vertical force on curved surface

We can find the vertical force F_z by considering a volume of liquid enclosed by the curved surface and vertical elements extending to the free surface (Fig.2.6.1c). This volume of liquid is also in static equilibrium, the only vertical forces on this volume of liquid are the gravity force W downward, and F_z' the upward vertical force by the surface MN. The force F_z' acts on the curved fluid area is the counterforce of F_z, which is the z-component force acts on the cured solid surface (Fig.2.6.1a). Therefore

$$F_z = -F_z' = W = \rho g V \tag{2.6.3}$$

here W represents the weight of the enclosed liquid block and V as the volume of the imaginary liquid block.

If the pressure on the free surface is p_0, we can simply add the pressure force $p_0 A_z$ to the vertical force F_z, here A_z is the area of vertical projection.

2.6.3   Pressure prism

Hence the vertical force acting on any area is equal to the weight of the volume of liquid above it, is the product of the specific weight and the volume of the liquid block, as in Eq.2.6.3. Here V represents geometric volume of the imaginary liquid block which is called *pressure prism*. Actually, the calculation of vertical component force is the calculation of the volume of pressure prism. So the concept of pressure prism is very important to the calculation of hydrostatic force on curved surface.

A pressure prism is surrounded by the compressed curved surface, vertical elements along the edge of the curved surface and free surface or the extended surface of free surface.

The pressure prism from the integral of infinitesimal vertical elements, is a mathematical volume, there may or may not be fluid inside it.

When the liquid and pressure prism are on the same side of the curved surface as shown in Fig.2.6.2a, the vertical force component F_z is equal to the liquid weight of the pressure prism in magnitude and the direction is downward.

When the liquid and pressure prism are on the different sides of the curved surface as shown in Fig.2.6.2b, F_z is also equal to the liquid weight of the pressure prism in magnitude but the direction is upward, and there is no real fluid inside the pressure prism.

流体对空间封闭体的压力 F' 作用于竖直表面 $M'N$，这个表面正是曲面在水平方向的投影面，其面积可以用 A_x 表示，则曲面水平方向分力的大小为

$$F' = F_x = \rho g h_{C_x} A_x \qquad （2.6.2）$$

所以，任一曲面所受流体静压力的水平方向分量等于作用于曲面的水平投影面上的静压力。压力的大小和 2.5 节讨论的平面壁受力情况相同，等于投影面形心处的压强乘以投影面的面积，如式 2.6.2。式中 h_{C_x} 和 A_x 分别表示投影面形心的淹深和投影面面积。

2.6.2 曲面壁上竖直方向分力

曲面上竖直方向的分力 F_z 同样可以在曲面及其延伸至自由液面的投影封闭而成的区域中得出（图 2.6.1c）。这个"流体块"同样静力平衡，其所受竖直方向外力仅包括体积内流体的重力 W（竖直向下）和曲面 MN 对流体的竖直方向作用力 F_z'（竖直向上）。曲面对流体的竖直方向作用力 F_z' 是曲面所受流体静压力的竖直分力 F_z（见图 2.6.1a）的反作用力，所以有

$$F_z = -F_z' = W = \rho g V \qquad （2.6.3）$$

式中，W 是封闭空间内流体的重量；V 是这一假想封闭空间的体积。

如果需要考虑自由液面的压强 p_0，只需竖直分力 F_z 与 $p_0 A_z$ 相加，A_z 为曲面在竖直方向投影面的面积。

2.6.3 压力体

由式 2.6.3 可知，任意曲面在竖直方向所受的静压力等于曲面上方一定体积内流体的重量，即流体的重度和该空间体积的乘积。式中 V 代表假想的空间几何体的体积，称为压力体。所以计算竖直方向分力就是计算压力体的体积。可见压力体这一概念对曲面壁流体静压力的计算是十分重要的。

压力体由曲面、沿曲面边缘延伸至自由液面的竖直微元，以及自由液面（或自由液面的延伸部分）封闭而成。

压力体是对竖直体积微元（微压力体）积分而得到的数学意义的体积，其内部不一定存在流体。

如图 2.6.2a 所示的情况，流体与压力体在曲面的同侧，则竖直分力 F_z 与压力体内假想液体重量相等方向均为向下。

如果如图 2.6.2b 所示的情况，压力体与液体分别处于曲面的两侧，则 F_z 与压力体内假想液体重量相等但方向相反，且此时压力体内没有真实液体。

如果曲面在水平方向有重叠（见图 2.6.2c），可分段定义压力体，同侧加、异侧减。

For the curved surfaces whose horizontal projections overlap (Fig.2.6.2c), separate the defined pressure prisms and superpose them, i.e. add magnitudes if both act in the same direction and subtract if they act in opposite directions.

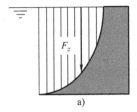 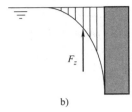 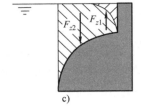

Fig.2.6.2   Pressure prisms

2.6.4   Resultant force on curved surface

The resultant hydrostatic force acting on the curved surface is a addition $F_x + F_z$, the magnitude of this resultant force is given in Eq.2.6.4

$$F = \sqrt{F_x^2 + F_z^2} \tag{2.6.4}$$

And the tangent of the angle it makes with the horizontal is

$$\tan \alpha = \frac{F_z}{F_x} \tag{2.6.5}$$

The exact location of the line of action of the resultant force (e.g., its distance from one of the end points of the curved surface) can be determined by taking a moment about an appropriate point. When the curved surface is a circular arc (full circle or any part of it), the resultant hydrostatic force acting on the surface always passes through the center of the circle. This is because the pressure forces are normal to the surface, and all lines normal to the surface of a circle pass through the center of the circle.

Example 2.6.1   A long solid quadrant cylinder of radius 1m and length 2m hinged at point A is used as an automatic gate, as shown in Fig.X2.6.1. The height of a water is controlled by the gate hinged to the reservoir. When the water level reaches 5m, the gate opens by turning about the hinge at point A. Determine the hydrostatic force acting on the cylinder and its line of action when the gate opens.

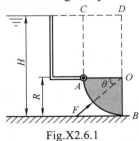

Fig.X2.6.1

Solution   Horizontal force on vertical projective plane:

$$F_x = \rho g h_{Cx} A_x = \rho g \left(H - \frac{R}{2} \right) RL = 88.2\text{kN}$$

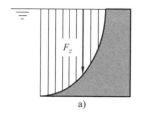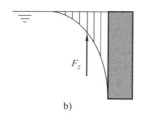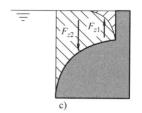

图 2.6.2　压力体

2.6.4　曲面壁上的合力

曲面壁上的合力是两个分力的和 $F_x + F_z$，其值由式 2.6.4 得出

$$F = \sqrt{F_x^2 + F_z^2} \qquad (2.6.4)$$

合力与水平方向夹角的正切值为

$$\tan \alpha = \frac{F_z}{F_x} \qquad (2.6.5)$$

合力的准确作用位置可通过力矩平衡求解。当曲面为圆或圆的一部分时，则合力的作用线一定通过圆心。这是因为曲面上的流体静压力一定垂直于作用面，而垂直于圆表面的线一定通过圆心。

例 2.6.1　一半径 1m 长 2m 的四分之一圆形水闸，如图 X2.6.1 所示，可自动绕 A 点转动以控制水位。当上游水位达到 5m 时，闸门绕 A 点转动开启。计算闸门开启时其所受的水静压力及压力作用方向。

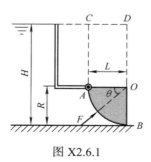

图 X2.6.1

解　竖直投影面上的水平分力大小：

$$F_x = \rho g h_{Cx} A_x = \rho g \left(H - \frac{R}{2} \right) RL = 88.2 \text{kN}$$

Vertical force on horizontal projective plane (upward):

$$F_z = \rho g V = \rho g V_{ABDCA} = \rho g \left(V_{ABO} + V_{AODC} \right)$$

$$= \rho g \left(\frac{\pi R^2}{4} + \left(H - R \right) R \right) L = 93.8 \text{kN}$$

Then the magnitude and direction of the hydrostatic force acting on the cylindrical surface become

$$F = \sqrt{F_x^2 + F_z^2} = 128.8 \text{kN}$$

$$\theta = \arctan \frac{F_z}{F_x} = 46.8°$$

Therefore, the magnitude of the hydrostatic force acting on the gate is 128.8kN, and its line of action passes through the center of the cylinder making an angle 46.8° with the horizontal.

Example 2.6.2   Find the horizontal and vertical components of the force exerted by the water on the horizontal cylinder of radius 2m and length 1m in Fig.X2.6.2a, the water to the left of the cylinder is with a free surface at an elevation coincident with the uppermost part of the cylinder.

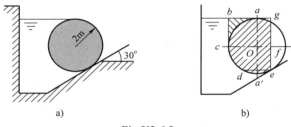

a)   b)

Fig.X2.6.2

Solution   The net projection on a vertical plane of the portion of the cylindrical surface under consideration is, from Fig.X2.6.2b

Horizontal force on vertical projective plane:

$$F_x = \rho g h_{Cx} A_x = \rho g 2 R^2 L = 78.48 \text{kN}$$

Vertical force on horizontal projective plane: the vertical component F_z, can be calculated through plotting the pressure prism of curved surface $bceg$. The curved surface $bceg$ was divided into two sections (cde and ac), for the pressure prism of $bcdeg$, the direction of F_{z1} is upward, and for the pressure prism of abc, the direction of F_{z2} is downward.

Therefore, the vertical component force F_z is determined by

$$F_z = F_{z1} - F_{z2} = \rho g \left(V_{bcdeg} - V_{abc} \right)$$

$$= \rho g \left[\frac{\pi R^2}{2} + 2R \cdot R \sin \theta + \left(R \cdot R \sin \theta - \frac{\pi R^2}{12} - \frac{1}{2} R \sin \theta R \cos \theta \right) \right]$$

$$= 99.89 \text{kN}$$

竖直分力大小（方向向上）：

$$F_z = \rho g V = \rho g V_{ABDCA} = \rho g \left(V_{ABO} + V_{AODC} \right)$$

$$= \rho g \left(\frac{\pi R^2}{4} + (H - R) R \right) L = 93.8 \text{kN}$$

静压力合力大小和方向

$$F = \sqrt{F_x^2 + F_z^2} = 128.8 \text{kN}$$

$$\theta = \arctan \frac{F_z}{F_x} = 46.8°$$

所以，水静压力的大小为 128.8kN，过 O 点且与水平方向夹角 46.8°。

例 2.6.2　水平放置的圆柱形挡水坝半径 2m 长 1m（见图 X2.6.2a），左侧水位与挡水坝最高点平齐。计算挡水坝水平方向和竖直方向的流体静压力。

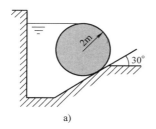 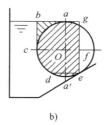

a)　　　　　　　　　　　　　b)

图 X2.6.2

解　竖直方向压力体截面形状如图 X2.6.2b 所示

水平分力大小：

$$F_x = \rho g h_{Cx} A_x = \rho g 2 R^2 L = 78.48 \text{kN}$$

竖直分力：竖直分力 F_z 可以通过求解压力体 $bceg$ 的体积（面积）计算得出。曲面可分成两部分（cde 和 ac），压力体 $bcdeg$ 的压力 F_{z1} 向上，而压力体 abc 的压力 F_{z2} 向下。

所以，竖直方向静压力 F_z 等于

$$F_z = F_{z1} - F_{z2} = \rho g \left(V_{bcdeg} - V_{abc} \right)$$

$$= \rho g \left[\frac{\pi R^2}{2} + 2R \cdot R \sin\theta + \left(R \cdot R \sin\theta - \frac{\pi R^2}{12} - \frac{1}{2} R \sin\theta R \cos\theta \right) \right]$$

$$= 99.89 \text{kN}$$

The resultant force of hydrostatic force is

$$F = \sqrt{F_x^2 + F_z^2} = 127.03\text{kN}$$

The angle the resultant force makes with the horizontal is

$$\alpha = \arctan\frac{F_z}{F_x} = 51.84°, \text{ and through the center of the cylinder.}$$

2.7   Fluid at Relative Rest

Many fluids need to be transported in tankers. In an accelerating tanker, each fluid particle assumes the same acceleration, no shear stresses exist within the fluid body since there is no deformation and thus no change in shape and the entire fluid moves like a rigid body. Rigid-body motion of a fluid also occurs when the fluid is contained in a tank that rotates about an axis.

2.7.1   Linearly accelerating container

In this section we obtain relations for the variation of pressure in a liquid mass in an open tank moving horizontally with a linear acceleration a, as in Fig.2.7.1a.

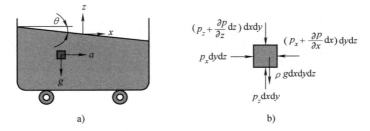

Fig.2.7.1   Linearly accelerating tank

a) The accelerating tant b) Force diagram of a fluid element

Consider a differential rectangular fluid element of side lengths dx, dy, and dz in the x-, y-, and z-directions, respectively, with the z-axis is being upward in the vertical direction (Fig.2.7.1b). Noting that the differential fluid element behaves like a rigid body, Newton's second law of motion for this element can be expressed as
in x-direction:

$$p_x\text{d}y\text{d}z - \left(p_x + \frac{\partial p}{\partial x}\text{d}x\right)\text{d}y\text{d}z = ma_x = \rho a\text{d}x\text{d}y\text{d}z \qquad (2.7.1)$$

in z-direction:

$$p_z\text{d}x\text{d}y - \left(p_z + \frac{\partial p}{\partial z}\text{d}z\right)\text{d}x\text{d}y - \rho g\text{d}x\text{d}y\text{d}z = ma_z = 0 \qquad (2.7.2)$$

合力的大小

$$F = \sqrt{F_x^{\,2} + F_z^{\,2}} = 127.03\text{kN}$$

合力与水平方向夹角

$$\alpha = \arctan \frac{F_z}{F_x} = 51.84° 且通过圆柱形挡水坝中心。$$

2.7　流体的相对静止

很多时候需要用容器运输流体。如果加速运动容器中的流体没有形状上的改变或变形，流体间不存在剪切应力，那么这部分流体就如同刚性物体一样在运动。绕某一轴线做旋转运动的容器中也可能发生这种流体刚性运动的情况。

2.7.1　直线加速运动的容器

本小节首先讨论以匀加速度 a 水平运动的开放容器内（图 2.7.1a）液体压强的分布规律。

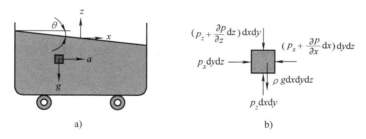

图 2.7.1　直线加速运动的容器

a）运动容器　b）流体微元受力

取一个正六面体形状流体微元，微元 x、y 和 z 方向的边长分别为 $\mathrm{d}x$、$\mathrm{d}y$ 和 $\mathrm{d}z$，z 轴正方向向上（见图 2.7.1b）。因为流体微元如刚体般运动，可以对其应用牛顿第二定律。

在 x 方向：

$$p_x\mathrm{d}y\mathrm{d}z - \left(p_x + \frac{\partial p}{\partial x}\mathrm{d}x \right)\mathrm{d}y\mathrm{d}z = ma_x = \rho a\mathrm{d}x\mathrm{d}y\mathrm{d}z \tag{2.7.1}$$

在 z 方向：

$$p_z\mathrm{d}x\mathrm{d}y - \left(p_z + \frac{\partial p}{\partial z}\mathrm{d}z \right)\mathrm{d}x\mathrm{d}y - \rho g\mathrm{d}x\mathrm{d}y\mathrm{d}z = ma_z = 0 \tag{2.7.2}$$

Eq.2.7.1 and Eq.2.7.2 are reduce to

$$\frac{\partial p}{\partial x}dx = -\rho a dx$$

$$\frac{\partial p}{\partial z}dz = -\rho g dz$$

$$(2.7.3)$$

Similarly, we can obtain a general result for a liquid mass that is accelerating in the x-, y- and z-directions, called *equation of motion*

$$\nabla p + \rho g \boldsymbol{k} = -\rho \boldsymbol{a} \qquad (2.7.4)$$

or, in scalar form in the three orthogonal directions as

$$\frac{\partial p}{\partial x} = -\rho a_x$$

$$\frac{\partial p}{\partial y} = -\rho a_y \qquad (2.7.5)$$

$$\frac{\partial p}{\partial z} = -\rho (a_z + g)$$

Substitute Eq.2.7.3 into Eq.2.3.1, the differential pressure equation is represented by

$$dp = -\rho a dx - \rho g dz \qquad (2.7.6)$$

Therefore, the pressure difference between any two points in the fluid can be calculated from

$$p_2 - p_1 = -\rho a (x_2 - x_1) - \rho g (z_2 - z_1) \qquad (2.7.7)$$

If points 1 and 2 lie on a surface of constant pressure, such as the free surface in Fig.2.7.1a, then dp=0 and we have

$$dz = -\frac{a}{g}dx \qquad (2.7.8)$$

integrating Eq.2.7.8 results in the equation for surfaces of constant pressure called *isobaric plane* (or isobars),

$$z = -\frac{a}{g}x + C \qquad (2.7.9)$$

Obviously, with different value of constant C in Equ.2.7.9, equipressure planes can be determined, so we can conclude that the isobars (including the free surface) in an incompressible fluid with constant acceleration in linear motion are parallel surfaces whose slope in the xz-plane is

$$\tan \theta = \frac{a}{g} \qquad (2.7.10)$$

式 2.7.1 和式 2.7.2 可化简为

$$\frac{\partial p}{\partial x}dx = -\rho a dx$$

$$\frac{\partial p}{\partial z}dz = -\rho g dz$$

（2.7.3）

对在 x、y 和 z 方向加速运动的流体列出与上式相同的等式，称为 **运动方程**

$$\nabla \boldsymbol{p} + \rho g \boldsymbol{k} = -\rho \boldsymbol{a}$$

（2.7.4）

也可以写作三个坐标方向的标量形式

$$\frac{\partial p}{\partial x} = -\rho a_x$$

$$\frac{\partial p}{\partial y} = -\rho a_y$$

$$\frac{\partial p}{\partial z} = -\rho \left(a_z + g \right)$$

（2.7.5）

将式 2.7.3 代入式 2.3.1，得到容器内液体的压强差

$$dp = -\rho a dx - \rho g dz$$

（2.7.6）

根据上式可以计算流体内任意两点间的压强差

$$p_2 - p_1 = -\rho a \left(x_2 - x_1 \right) - \rho g \left(z_2 - z_1 \right)$$

（2.7.7）

若点 1 和点 2 位于同一等压面上，例如图 2.7.1a 中的自由液面，则 dp=0，式 2.7.6 写为

$$dz = -\frac{a}{g}dx$$

（2.7.8）

对式 2.7.8 积分，可得出各点压强相等的面，即 **等压面方程**

$$z = -\frac{a}{g}x + C$$

（2.7.9）

显然式 2.7.9 中的常数 C 可以取不同值，也就对应着不同的等压面。由此可知，做匀加速运动的不可压缩流体，其等压面（包括自由液面）是一系列平行平面，且等压面在 xz 平面内的斜率为

$$\tan \theta = \frac{a}{g}$$

（2.7.10）

If we locate the center of the free surface at (0,0), then equation of the free surface is

$$z_0 = -\frac{a}{g}x \tag{2.7.11}$$

note that when the free surface is atmospheric, we have $p_0=0$.

The pressure distribution can be obtained by integration of Eq.2.7.6 and substitution of $p(0,0)=p_0$ (at the center of free surface)

$$p = -\rho ax - \rho gz + p_0 \tag{2.7.12}$$

Equ.2.7.12 is general expression for the law of pressure distribution when the liquid in uniformly accelerated linear motion along with the container is in a state of relative equilibrium.

Rewrite Eq.2.7.12 and substitute Eq.2.7.11 into it

$$p - p_0 = \rho g(z_0 - z) = \rho gh \tag{2.7.13}$$

where z_0-z is the submergence depth of the fluid particle (of the same x coordinate) below the free surface, denoted as h. Eq.2.7.13 has the same form with the pressure distribution equation of fluid at absolute rest.

Example 2.7.1   The tank shown in Fig.X2.7.1a is accelerated to the right. Calculate the acceleration a needed to cause the free surface, shown in F.X2.7.1b, to touch point A. Also, find p_B if the tank width is 1m.

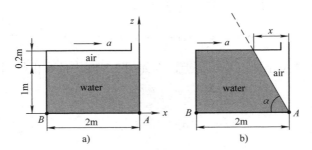

Fig.X2.7.1

Solution   The angle the free surface takes, is found by equating the air volume (actually, areas since the width is constant) before and after since no water spills out:

$$2\times0.2=\frac{1}{2}1.2x, \text{ and, we have } x=\frac{2}{3}\text{m}$$

From the slope of isobaric plane Eq.2.7.10

$$\tan\alpha = \frac{a}{g} = \frac{1.2}{x}, \text{ the acceleration is given as } a = 17.64\text{m/s}^2.$$

如果取自由液面的中心为坐标系原点（0，0），则自由液面方程为

$$z_0 = -\frac{a}{g}x \qquad (2.7.11)$$

如果自由液面上为大气压，则 $p_0=0$。

对式 2.7.6 积分，并代入自由液面上原点压强 $p(0，0)=p_0$，则可得到压强分布式

$$p = -\rho ax - \rho gz + p_0 \qquad (2.7.12)$$

式 2.7.12 为在做匀加速运动的容器内相对静止的流体内部压强分布规律。

整理式 2.7.12 并代入式 2.7.11，得

$$p - p_0 = \rho g(z_0 - z) = \rho gh \qquad (2.7.13)$$

式中，z_0-z 是流体质点在自由液面下（x 坐标位置相同处）的淹深，用 h 表示。可以看出，式 2.7.13 与绝对静止流体的压强分布规律式具有相同的形式。

例 2.7.1 图 X2.7.1a 所示的容器加速向右运动。若如图 X2.7.1b 所示，容器内液体的自由液面经过 A 点，则计算此时容器的运动加速度 a 及此时容器底部 B 点处的压强 p_B（容器宽度为 1m）。

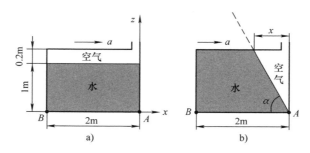

图 X2.7.1

解 容器内水没有溅出所以其体积不变，也就是容器内空气部分的体积（容器宽度一定，所以可取图中空气部分面积）不变：

$$2\times 0.2 = \frac{1}{2}1.2x, \text{ 得 } x = \frac{2}{3}\text{m}$$

由式 2.7.10 等压面斜率公式

$$\tan\alpha = \frac{a}{g} = \frac{1.2}{x}, \text{ 可得容器运动加速度 } a = 17.64\text{m/s}^2。$$

The free surface can be expressed as

$$z_0 = -\frac{17.64}{g}x$$

The point on free surface where $x'=x_B=-2$, we have $z_0'=3.6$, the submergence depth of point B where $z_B=0$, is determined as

$$h = z_0' - z_B = 3.6\text{m}$$

From Eq.2.7.13, pressure at point B is

$$p = \rho g(z_0 - z) = \rho gh = 35.28\text{kPa}$$

Or, from Eq.2.7.7, it is given that $p_A=p(0,0)=0$, then the pressure at point $B(-2,0)$ is

$$p_B - p_A = -\rho a(-2-0) - \rho g(0-0) = 35.28\text{kPa}$$

Example 2.7.2 A tank filled with water moving horizontally with a velocity of 10m/s, and will stop after 100m deceleration, D, L and h are given as 2m, 8m and 0.3m respectively. Find the pressure at center of front side A , where shown in Fig X2.7.2.

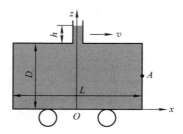

Fig.X2.7.2

Solution Deceleration of the tank: $0=10+at$ $\qquad s = -\frac{1}{2}at^2 = 100$ $\qquad a = -0.5\text{m/s}^2$

Take the center of the bottom of the tank as the origin of coordinate.

The pressure difference between point A and the center of free surface from Eq.2.7.7 is

$$p_2 - p_1 = -\rho a(x_2 - x_1) - \rho g(z_2 - z_1) = \left[0.5\times(4-0) - 9.8\times(1-2.3)\right]\text{kPa} = 14.74\text{kPa}$$

2.7.2 Rotating cylindrical containers

In this section we consider the situation of a liquid contained in a rotating container, such as that shown in Fig.2.7.2. Consider a vertical cylindrical container partially filled with a liquid. The container is now rotated about its axis at a constant angular velocity of ω, after a relatively short time the liquid reaches static equilibrium, i.e. the liquid move as a rigid body together with the container, there is no shear stress, and every fluid particle in the container moves with the same angular velocity.

自由液面方程为

$$z_0 = -\frac{17.64}{g}x$$

于是自由液面上 $x'=x_B=-2$ 处 $z_0'=3.6$，可计算得出 B 点（$z_B=0$）在自由液面下的淹深

$$h = z_0' - z_B = 3.6\text{m}$$

由式 2.7.13 计算得 B 点压强

$$p = \rho g(z_0 - z) = \rho g h = 35.28\text{kPa}$$

或由式 2.7.7，已知 $p_A = p(0,0) = 0$，则点 $B(-2,0)$ 处压强

$$p_B - p_A = -\rho a(-2-0) - \rho g(0-0) = 35.28\text{kPa}$$

例 2.7.2 一盛满水的容器以 10m/s 的速度水平方向前进，减速 100m 后停止。已知容器直径 D、长度 L 和罐口高度 h 分别为 2m、8m 和 0.3m。计算容器前侧面中心 A 点的压强。

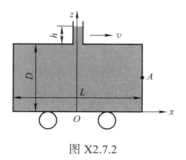

图 X2.7.2

解 容器的减速度：$0=10+at$，$s = -\frac{1}{2}at^2 = 100$ → $a = -0.5\text{m/s}^2$

取容器底部中心为坐标原点，如图 X2.7.2 所示

根据式 2.7.7，A 点与自由液面中心处的压强差为

$$p_2 - p_1 = -\rho a(x_2 - x_1) - \rho g(z_2 - z_1) = \left[0.5 \times (4-0) - 9.8 \times (1-2.3)\right]\text{kPa} = 14.74\text{kPa}$$

2.7.2 旋转容器

本小节讨论图 2.7.2 所示的圆柱形容器，容器内盛放部分液体并做旋转运动。现假设容器绕自身中心线以角速度 ω 转动且达到稳定，即容器内液体与容器一同做刚性旋转，液体内部无剪切，所有流体质点均以相同的角速度转动。

This problem is analyzed in cylindrical coordinates (r, θ, z), with z taken along the centerline of the container directed from the bottom toward the free surface. The centripetal acceleration of a fluid particle rotating with a constant angular velocity of ω at a distance r from the axis of rotation is $r\omega^2$ and is directed radially toward the axis of rotation (negative r-direction).

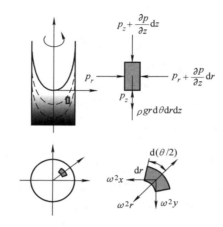

Fig.2.7.2   A container rotating with equal angular velocity

Applying Newton's second law ($\Sigma F_r = ma_r$) in the r-direction to the element, using $\sin\theta(d/2)=(d\theta)/2$, yields:
in the r-direction:

$$p_r r d\theta dz - \left(p_r + \frac{\partial p}{\partial r} dr\right) r d\theta dz = -\rho\omega^2 r \cdot r d\theta dr dz \qquad (2.7.14)$$

in the z-direction:

$$p_z r d\theta dr - \left(p_z + \frac{\partial p}{\partial z} dz\right) r d\theta dr - \rho g r d\theta dr dz = 0 \qquad (2.7.15)$$

Eq.2.7.14 and Eq.2.7.15 are reduce to

$$\frac{\partial p}{\partial r} dr = \rho\omega^2 r dr$$
$$\frac{\partial p}{\partial z} dz = -\rho g dz \qquad (2.7.16)$$

the differential pressure equation is represented by

$$dp = \frac{\partial p}{\partial r} dr + \frac{\partial p}{\partial z} dz = \rho\omega^2 r dr - \rho g dz \qquad (2.7.17)$$

therefore, the pressure difference between any two points in the fluid can be calculated from

对该运动系统建立柱坐标系（r，θ，z），z 轴通过圆柱容器中心线且向上通过自由液面。距中心线 r 处的流体质点以恒角速度 ω 转动，其向心加速度大小为 $r\omega^2$，沿 r 轴负方向指向转动中心。

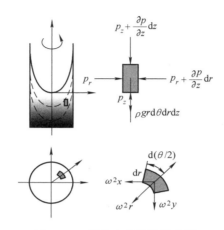

图 2.7.2　等角速度旋转的容器

对流体微元在 r 方向应用牛顿第二定律（$\Sigma F_r = ma_r$），并取 $\sin\theta(\mathrm{d}/2) = (\mathrm{d}\theta)/2$，则有

r 方向：

$$p_r r\mathrm{d}\theta\mathrm{d}z - \left(p_r + \frac{\partial p}{\partial r}\mathrm{d}r\right)r\mathrm{d}\theta\mathrm{d}z = -\rho\omega^2 r \cdot r\mathrm{d}\theta\mathrm{d}r\mathrm{d}z \qquad (2.7.14)$$

z 方向：

$$p_z r\mathrm{d}\theta\mathrm{d}r - \left(p_z + \frac{\partial p}{\partial z}\mathrm{d}z\right)r\mathrm{d}\theta\mathrm{d}r - \rho g r\mathrm{d}\theta\mathrm{d}r\mathrm{d}z = 0 \qquad (2.7.15)$$

整理并化简式 2.7.14 和式 2.7.15，得

$$\begin{aligned} \frac{\partial p}{\partial r}\mathrm{d}r &= \rho\omega^2 r\mathrm{d}r \\ \frac{\partial p}{\partial z}\mathrm{d}z &= -\rho g\mathrm{d}z \end{aligned} \qquad (2.7.16)$$

则压强差公式可表示为

$$\mathrm{d}p = \frac{\partial p}{\partial r}\mathrm{d}r + \frac{\partial p}{\partial z}\mathrm{d}z = \rho\omega^2 r\mathrm{d}r - \rho g\mathrm{d}z \qquad (2.7.17)$$

则容器内任意两点间的压强差可通过下式计算

$$p_2 - p_1 = \frac{\rho \omega^2}{2}\left(r_2^2 - r_1^2\right) - \rho g\left(z_2 - z_1\right) \tag{2.7.18}$$

The pressure distribution can be obtained by integration of Eq.2.7.17

$$p = \frac{\rho \omega^2}{2}r^2 - \rho g z + C \tag{2.7.19}$$

Locating the lowest point on the free surface at $(0,0)$, and $p(0,0)=0$, there results $C=0$, then we have the equation of pressure distribution:

$$p = \frac{\rho \omega^2}{2}r^2 - \rho g z \tag{2.7.20}$$

If points 1 and 2 lie on a surface of constant pressure, such as the free surface in Fig.2.7.2, then dp=0 and we have

$$\rho \omega^2 r \mathrm{d}r - \rho g \mathrm{d}z = 0 \tag{2.7.21}$$

Integrating Eq.2.7.21 results in the equation of isobaric planes

$$\omega^2 \frac{r^2}{2} - gz = C \tag{2.7.22}$$

Note that when the free surface is atmospheric, we have $p_0=0$, then the equation of free surface is represented by

$$z_0 = \frac{\omega^2 r^2}{2g} \tag{2.7.23}$$

Substitute Eq.2.7.23 into Eq.2.7.20, then the pressure at an arbitrary point can be given as

$$p = \rho g\left(z_0 - z\right) = \rho g h \tag{2.7.24}$$

where $z_0 - z$ is the submergence depth of the fluid particle below the free surface, as same as that of in equation of absolute rest and uniform linear accelerating motion, denoted as h. This indicates the variation of the pressure at any point with liquid depth is linear for the liquid in absolute or relative rest.

Example 2.7.3　The cylinder shown in Fig.X2.7.3 is rotated about its centerline. Calculate the rotational speed that is necessary for the water to just touch the origin O. Also, find the pressures at A and B.

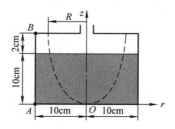

Fig.X2.7.3

$$p_2 - p_1 = \frac{\rho\omega^2}{2}\left(r_2^{\,2} - r_1^{\,2}\right) - \rho g\left(z_2 - z_1\right) \tag{2.7.18}$$

对压强差公式 2.7.17 积分，可得压强分布

$$p = \frac{\rho\omega^2}{2}r^2 - \rho gz + C \tag{2.7.19}$$

取自由液面的最低点为坐标系原点，由原点 $p(0,0)=0$ 可知 $C=0$，则压强分布规律为

$$p = \frac{\rho\omega^2}{2}r^2 - \rho gz \tag{2.7.20}$$

若在图 2.7.2 的自由液面上取点 1 和点 2，则该两点压强相等，压强差 $\mathrm{d}p=0$，于是有

$$\rho\omega^2 r\mathrm{d}r - \rho g\mathrm{d}z = 0 \tag{2.7.21}$$

对上式积分得出等压面方程

$$\omega^2 \frac{r^2}{2} - gz = C \tag{2.7.22}$$

若自由液面上为大气压，取表压强 $p_0=0$，则自由液面上各点位置可由下式确定

$$z_0 = \frac{\omega^2 r^2}{2g} \tag{2.7.23}$$

将式 2.7.23 代入式 2.7.20，得到容器内任意点压强

$$p = \rho g\left(z_0 - z\right) = \rho gh \tag{2.7.24}$$

式中，$z_0 - z$ 为流体质点在自由液面下的淹深，用 h 表示。则无论是绝对静止还是相对静止的流体，其压强分布规律相同，都与流体质点的淹深成线性关系。

例 2.7.3　图 X2.7.3 所示容器绕中心线旋转，若自由液面中心刚好处于容器底部中点 O，此时容器的转速是多少？A 点和 B 点的压强分别是多少？

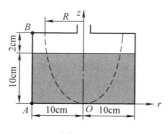

图 X2.7.3

Solution   Taking the center of the bottom surface of the rotating vertical cylinder as the origin ($r=0$, $z=0$), the equation of free surface is given as:

$$z_0 = \frac{\omega^2 r^2}{2g}$$

When $r=R$, $z_0=0.12$, we have $0.12 = \dfrac{\omega^2 R^2}{2g}$.

Since no water spills from the container, the air volume remains constant, and form Eq.2.7.21, that is

$$\int \pi r^2 dz = \int_0^R \pi r^2 \left(\frac{\omega^2 r}{g}\right) dr = 0.02 \times \pi \times 0.1^2$$

$$\frac{\omega^2 R^4}{4g} = 0.0002 = 0.12 \times \frac{R^2}{2}$$

$$R = 5.77\text{cm} \quad \rightarrow \quad \omega = 26.6\,\text{rad/s}$$

Since $p_0=0$, $A(0.1, 0)$, $B(0.1, 0.12)$, then

$$p_A - p_O = \frac{\rho\omega^2}{2}\left(r_A^2 - r_O^2\right) - \rho g\left(z_A - z_O\right) = 3.54\text{kPa}, \; p_A = 3.54\text{kPa}$$

$$p_B - p_O = \frac{\rho\omega^2}{2}\left(r_B^2 - r_O^2\right) - \rho g\left(z_B - z_O\right) = 2.36\text{kPa}, \; p_B = 2.36\text{kPa}$$

Exercises 2

2.1   The gage pressure in a liquid at a depth of 3m is read to be 25kPa. Determine the gage pressure in the same liquid at a depth of 8m.

2.2   The absolute pressure in water at a depth of 3m is read to be 130kPa. Determine the local atmospheric pressure.

2.3   Determine the pressure exerted on a diver at 6m below the free surface of the sea. Assume the atmospheric pressure is 101kPa and a specific gravity of 1.03 for seawater.

2.4   Under what conditions can a moving body of fluid be treated as a rigid body?

2.5   Estimate the pressure in the water pipe shown in Fig.E2.5 the manometer is open to the atmosphere.

A. 4kPa   B. 5kPa   C. 6kPa   D. 7kPa

2.6   Compare the pressure at three points shown in Fig.E2.6, the order from the smallest to the largest is

A. $p_1 > p_2 > p_3$ B. $p_1 = p_2 = p_3$ C. $p_1 < p_2 < p_3$ D. $p_2 < p_1 < p_3$

2.7   An inclined submerged rectangular plate, the relationship of the depth of centroid center h_C and center of hydrostatic pressure h_D is

A. $h_C > h_D$ B. $h_C = h_D$ C. $h_C < h_D$ D. incomparable

2.8   The unit of pressure head is

A. m   B. m/s   C. Pa   D. Pa, s

解　取容器底部中心为坐标系原点（$r=0$，$z=0$），自由液面方程为

$$z_0 = \frac{\omega^2 r^2}{2g}$$

当 $r=R$，$z_0=0.12$ 时，有 $0.12 = \frac{\omega^2 R^2}{2g}$。

因为容器内水无溢出，则空气体积不变，且根据式 2.7.21，于是

$$\int \pi r^2 \mathrm{d}z = \int_0^R \pi r^2 \left(\frac{\omega^2 r}{g}\right) \mathrm{d}r = 0.02 \times \pi \times 0.1^2$$

$$\frac{\omega^2 R^4}{4g} = 0.0002 = 0.12 \times \frac{R^2}{2}$$

$$R = 5.77\,\mathrm{cm} \quad \rightarrow \quad \omega = 26.6\,\mathrm{rad/s}$$

因为 $p_O=0$，且 A（0.1，0），B（0.1，0.12），所以

$$p_A - p_O = \frac{\rho\omega^2}{2}\left(r_A{}^2 - r_O{}^2\right) - \rho g(z_A - z_O) = 3.54\,\mathrm{kPa}, \quad p_A = 3.54\,\mathrm{kPa}$$

$$p_B - p_O = \frac{\rho\omega^2}{2}\left(r_B{}^2 - r_O{}^2\right) - \rho g(z_B - z_O) = 2.36\,\mathrm{kPa}, \quad p_B = 2.36\,\mathrm{kPa}$$

习题二

2.1　某液体 3m 深处表压强为 25kPa，则该液体 8m 深处的表压强是多少？

2.2　水下 3m 深处的绝对压强为 130kPa，则当地大气压是多？

2.3　海面下 6m 深处潜水者身体承受的压强是多少？当地大气压 101kPa，海水比重 1.03。

2.4　什么条件下流动的流体可以被视为刚性物体？

2.5　图 E2.5 中用一端开口的测压管测量水管内的压强，水压是多少？

A. 4kPa　　　　　B. 5kPa　　　　　C. 6kPa　　　　　D. 7kPa

2.6　图 E2.6 中三点压强由小到大的顺序是

A. $p_1>p_2>p_3$　　　B. $p_1=p_2=p_3$　　　C. $p_1<p_2<p_3$　　　D. $p_2<p_1<p_3$

2.7　一倾斜矩形平板淹没于水下，形心和静水压力中心的淹深关系为

A. $h_C>h_D$　　　B. $h_C=h_D$　　　C. $h_C<h_D$　　　D. 无法比较

2.8　压强水头的单位为

A. m　　　　　B. m/s　　　　　C. Pa　　　　　D. Pa·s

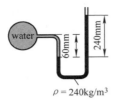

$\rho = 240\,kg/m^3$

Fig.E2.5

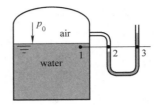

Fig.E2.6

2.9   Calculate the pressure at a depth of 10m in a liquid with specific gravity of (a) 0.68, (b) 1.2, (c) 13.6.

2.10   What depth is necessary in a liquid to produce a pressure of 50kPa if the specific gravity is (a) 0.68, (b) 1.2, (c) 13.6.

2.11   Plot the pressure distribution of an open tank if it contains layers of (a) 50cm of water and 10cm of mercury, (b) 50cm of oil (S_g=0.8) and 50cm of water, (c) 50cm of water and 30cm of carbon tetrachloride (S_g=1.6).

2.12   The specific gravity of a liquid varies linearly from 0.9 at the surface to 1.1 at a depth of 10m. Calculate the pressure at (a) the location of depth of 6m, (b) the bottom of the 10m depth of tank.

2.13   Several liquids are layered inside a tank with pressurized air at the top. If the air pressure is 3.2kPa, calculate the pressure at the bottom of the tank if the liquid layers include 20cm of S_g=0.68, 50cm of water, 25cm of S_g=1.24, and 30cm of S_g=1.6.

2.14   Neglecting the pressure on the surface and the compressibility of water, what is the pressure at a depth of Jiaolong 5000m below the surface of the sea? The specific weight of sea water under ordinary conditions is 1.03.

2.15   An open tank contains 3m of water covered with 2m of oil (S_g=0.85). Find the gage pressure (a) at the interface between the liquids and (b) at the bottom of the tank.

2.16   Calculate the pressure in a pipe transporting air if a U-tube manometer measures 360mm Hg, if the weight of air in the manometer is negligible. Include the weight of the air column, assuming that ρ_{air}=1.2kg/m³, and calculate the percent error of the former calculation.

2.17   For the setup shown in Fig.E2.17, calculate the pressure difference between container A and B if H=30cm is read.

2.18   For the setup shown in Fig.E2.17, calculate the manometer reading H if the pressure is measured of 32kPa at A and 14kPa at B.

2.19   What is the pressure in the pipe B shown in Fig.E2.19 if the pressure in the water pipe A is 16kPa?

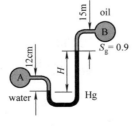

Fig.E2.17

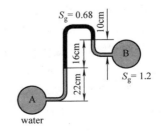

Fig.E2.19

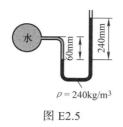

图 E2.5

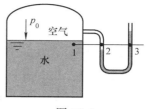

图 E2.6

2.9　计算以下比重液体 10m 深处的压强（1）0.68，（2）1.2，（3）13.6。

2.10　以下比重的液体深度为多少时其表压强可达 50kPa？（1）0.68，（2）1.2，（3）13.6。

2.11　绘制以下容器内各层液体的压强分布图。（1）50cm 水和 10cm 水银；（2）50cm 油（S_g=0.8）和 50cm 水；（3）50cm 水 30cm 四氯化碳（S_g=1.6）。

2.12　某种液体的比重在 10m 深度内由 0.9 线性增加到 11。计算（1）深度为 6m 处的压强；（2）10m 深容器底部的压强。

2.13　某容器内盛放多层液体且上部压缩空气的压强为 32kPa，计算容器底部的压强。容器内液体为：比重 S_g=0.68 液体 20cm，水 50cm，比重 S_g=1.24 的液体 25cm 及比重 S_g=1.6 的液体 30cm。

2.14　忽略大气压和水的压缩性，计算当蛟龙号深潜器下潜至海底 5000m 深处时所承受的压强，海水比重通常取 1.03。

2.15　开口容器内盛有 2m 深的油液（S_g=0.85）和 3m 深的水。计算以下位置的表压强：（1）液体间分界面；（2）容器底部。

2.16　用 U 形管测量某通气管道内的压强，测压管水银液面高度差为 360mm，忽略空气重量计算管道内压强。若考虑空气重量，取空气密度为 $\rho_{空气}$=1.2kg/m³，则计算以上测量的误差百分比。

2.17　如图 E2.17 所示，测压管读数 H=30cm，计算 A、B 容器压强差。

2.18　习题 2.17 中的情况，若测得 A 容器内压强为 32kPa，B 容器内压强为 14kPa，则测压管读数 H 的值应为多少？

2.19　若图 E2.19 中 A 管内水压为 16kPa，则 B 管内压强是多少？

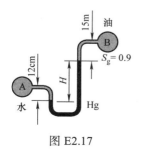

图 E2.17

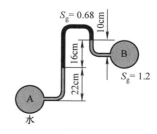

图 E2.19

2.20   For the tank shown in Fig.E2.20, determine the pressure of air.

2.21   For the tank shown in Fig.E2.21, what will the pressure gage read?

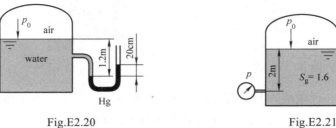

Fig.E2.20   Fig.E2.21

2.22   Find the pressure in the water pipe shown in Fig.E2.22.

2.23   For the setup shown in Fig.E2.23, calculate the pressure difference between water pipe A and B.

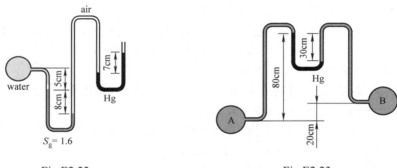

Fig.E2.22   Fig.E2.23

2.24   Determine the location of oil surface H in the tank as shown in Fig.E2.24, if $h_1=20$mm, $h_2=240$mm and $h_3=220$mm.

2.25   Find the pressure of air confined above water in the tank (Fig.E2.25) by a manometer which measuring fluid is mercury reads $z_0=2080$mm, $z_1=600$mm, $z_2=960$mm, $z_3=720$mm and $z_4=940$mm.

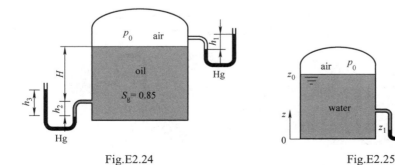

Fig.E2.24   Fig.E2.25

2.20　计算图 E2.20 中空气部分的压强。

2.21　图 E2.21 中压力表的读数应该是多少?

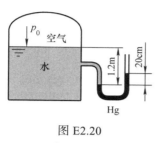

图 E2.20

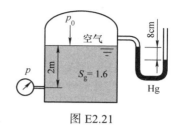

图 E2.21

2.22　计算图 E2.22 中水管内的压强。

2.23　计算图 E2.23 中管道 A、B 间的压差。

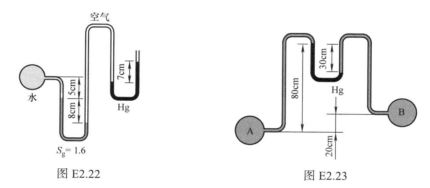

图 E2.22

图 E2.23

2.24　若图 E2.24 中测压管各段液柱高度分别为 h_1=20mm、h_2=240mm 和 h_3=220mm，计算容器内液位高度 H。

2.25　用水银测压管测量容器（见图 E2.25）上方空气压强，液面和各段水银柱高度分别为：z_0=2080mm、z_1=600mm、z_2=960mm、z_3=720mm 和 z_4=940mm。

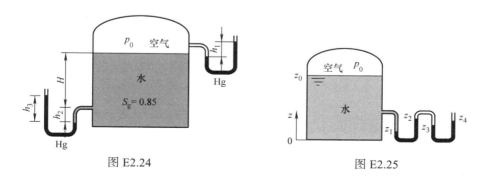

图 E2.24

图 E2.25

2.26 A rectangular plate 0.5m×0.6m is at an angle of 30° with the horizontal, and the 0.5m side is horizontal. Find the magnitude of the force on one side of the plate and the depth of its center of pressure when the top edge is (a) at the water surface; (b) 1m below the water surface.

2.27 A rectangular area is 4m×6m, with the 4m side horizontal. It is placed with its centroid 4m below a water surface and is 60° with the horizontal. Find the magnitude of the force on one side and the distance between the center of pressure and the centroid of the plane.

2.28 The cubic tank shown in Fig.E2.28 is half full of water. (a) Plot the pressure distribution along the wall of the tank. (b) Find the pressure on the bottom of the tank.

2.29 Find the magnitude and depth of center of the force on the rectangular gate on the wall of a hydraulic oil tank shown in Fig.E2.29 which top edge at the oil surface (S_{goil}=0.86).

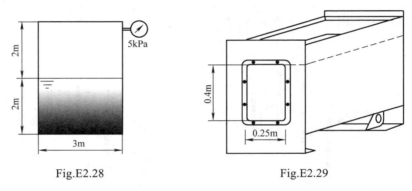

Fig.E2.28   Fig.E2.29

2.30 A water pool shown in Fig.E2.30 is filled with 3m of water. Its bottom is square and measures 6m on each side. One end makes an angle of 60° with the horizontal and the other is vertical. Calculate the force of the water on (a) the bottom of the water pool; (b) the vertical end; (c) the 60° end.

2.31 The plate shown in Fig.E2.31 is submerged in water and lies in a vertical plane. Find the magnitude and location of the hydrostatic force acting on one side of the plate.

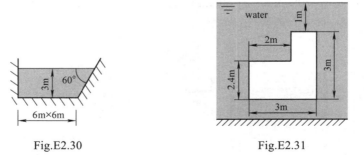

Fig.E2.30   Fig.E2.31

2.32 The rectangular gate shown in Fig.E2.32 is 2m wide (into the paper). Find the force F needed to hold the gate in the position.

2.26　边长为 0.5m×0.6m（0.5m 为水平边）与水平方向夹角 30° 的矩形平板，计算以下情况平板一侧所受流体静压力大小及压力中心的淹深：（a）平板顶部与液面平齐；（b）平板顶部位于水下 1m 深。

2.27　边长 4m×6m 的矩形平板（4m 为水平边），其形心位于水面下 4m 深，且与水平方向夹角 60°。计算平板一侧所受流体静压力以及形心和力心间的距离。

2.28　如 E2.28 图所示半满水的方形容器，（1）绘制沿容器壁面的压强分布图；（2）计算容器底部流体静压强。

2.29　图 E2.29 中的液压油箱，计算侧面清洁口盖板所受流体静压力的大小及压力中心的深度。油箱内油液（$S_{g油}$=0.86）的液面与窗顶部平齐。

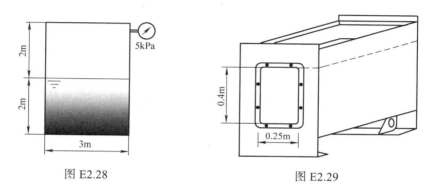

图 E2.28　　　　　　　　　　　图 E2.29

2.30　图 E2.30 中水池水深 3m，底部为边长 6m 的正方形，一侧面与水平方向成 60° 角，另一侧面竖直。计算以下表面水压力：（a）水池底部；（b）竖直侧面；（c）60° 侧面。

2.31　图 E2.31 为淹没于水下的竖直平板，计算其一侧的流体静压力大小和作用点位置。

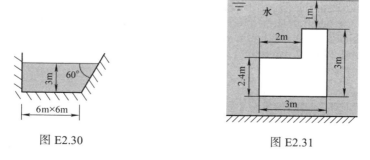

图 E2.30　　　　　　　　　　　图 E2.31

2.32　图 E2.32 所示 2m 宽（垂直页面）矩形水闸，计算保持闸门关闭力 F 的最小值。

2.33 Find the force F needed to hold the 5m-long, 3m-wide rectangular gate as shown in Fig.E2.33, if: (a) the force F is applied in the horizontal direction; (b) normal to the gate.

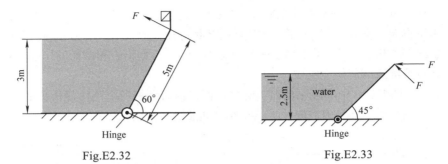

Fig.E2.32   Fig.E2.33

2.34 The gate of 2m wide (perpendicular to the plane of the figure) rotates about the hinge in Fig.E2.34. What torque applied to the shaft through the hinge is required to hold the gate closed?

2.35 The rigid gate will open automatically when water surface H is 2m above the top of the 1.8m-high and 3m-wide rectangular gate, determine the height of the hinge on the centerline as shown in Fig.E2.35.

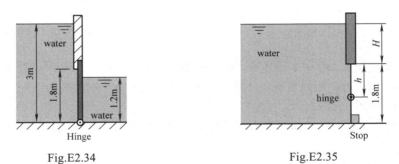

Fig.E2.34   Fig.E2.35

2.36 Calculate the force F necessary to hold the 5m-long and 4m-wide (into the paper) gate in the position shown in Fig.E2.36. Neglect the weight of the gate.

2.37 Calculate the maximum water depth H that the gate of 5m-long 4m-wide (into the paper) and 200×10^3 kg weight can keep in the position shown in Fig.E2.37.

Fig.E2.36   Fig.E2.37

2.33　图 E2.33 中矩形闸门长 5m、宽 3m，计算以下两种施力方式下使闸门保持关闭所需力 F 的最小值：（1）力 F 为水平方向；（2）力 F 垂直于闸门。

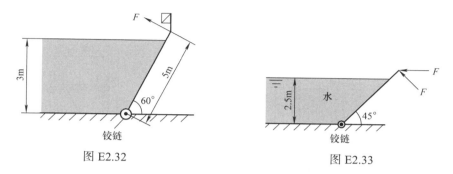

图 E2.32　　　　　　　　　图 E2.33

2.34　图 E2.34 中闸门宽 2m 且可绕铰链转动，计算保持闸门关闭需加载于铰链的力矩。

2.35　刚性矩形闸门如图 E2.35 所示，高 1.8m 宽 3m。请设计闸门铰链的安装位置尺寸 h，以保证水位达到闸门顶部距离 H 为 2m 时闸门自动开启。

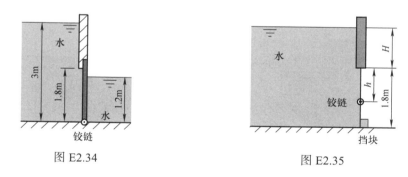

图 E2.34　　　　　　　　　图 E2.35

2.36　闸门长 5m 宽 4m，忽略其重量，计算图 E2.36 所示保持闸门关闭所需的力 F 大小。

2.37　图 E2.37 中闸门长 5m 宽 4m 自重 200×10^3kg，计算闸门自动开启时的水位 H。

图 E2.36　　　　　　　　　图 E2.37

2.38 Determine the force F needed to hold the 5m-long and 4m-wide (into the paper) gate in the position shown in Fig.E2.38.

2.39 The flow of water from a reservoir is controlled by a 4m-wide (into the paper) gate hinged at point O, as shown in Fig.E2.39. If it is desired that the gate open when the water height is 3m, determine the mass of the required weight W.

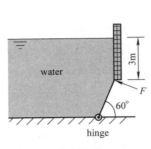

Fig.E2.38

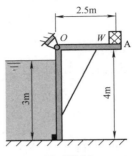

Fig.E2.39

2.40 For the gate shown in Fig.E2.40, calculate the height H that will result in the gate opening automatically. Neglect the weight of the gate.

2.41 Calculate the maximum water depth H that the gate of 5m-long 4m-wide (into the paper) and $W = 200 \times 10^3$ kg weight can keep in the position shown in Fig.E2.41.

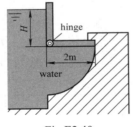

Fig.E2.40

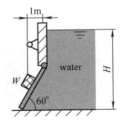

Fig.E2.41

2.42 Calculate the force F necessary to hold the 2.4m-wide gate (into the paper) in the position shown in Fig.E2.42. Neglect the weight of the gate.

2.43 The cross section of a tank is as shown in Fig.E2.43. BC is a cylindrical surface with $r=2$m, and wide of 4m. If the tank contains liquid of $S_g=0.9$, determine the magnitude of the horizontal- and vertical-force components acting on the tank wall ABC.

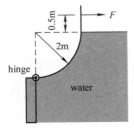

Fig.E2.42

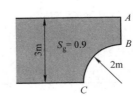

Fig.E2.43

2.38　计算图 E2.38 中 5m 长 4m 宽闸门保持关闭所需的力 F 大小。

2.39　图 E2.39 所示 4m 宽闸门可绕 O 点转动，用于控制水库的水位。若希望水位达到 3m 时开启闸门放水，则闸门上方配重 W 的质量应为多少？

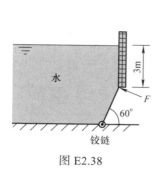

图 E2.38

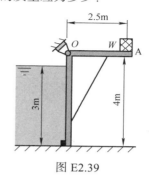

图 E2.39

2.40　计算图 E2.40 中闸门自动开启时的水位高度 H。忽略闸门自重。

2.41　计算图 E2.41 中 5m 长 4m 宽（垂直页面方向）配重 $W = 200 \times 10^3$kg 的闸门可承受的最大水位高度 H。

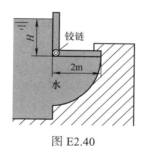

图 E2.40

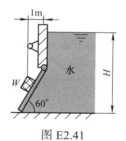

图 E2.41

2.42　计算保持图 E2.42 中 2.4m 宽闸门关闭所需的最小力 F。忽略闸门自重。

2.43　某容器截面为如图 E2.43 所示的形状，BC 段为圆柱面，其半径 r=2m 且宽 4m。若容器内液体比重 S_g=0.9，计算容器壁 ABC 表面水平和竖直方向压力的大小。

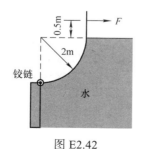

图 E2.42

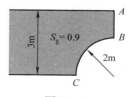

图 E2.43

2.44   Determine the magnitude, direction, and line of action of the hydrostatic force acting on Tainter gate surface shown in Fig.E2.44, with 2m-radius and 4m-width.

2.45   Determine the magnitude, direction, and line of action of the hydrostatic force acting on the submerged Tainter gate shown in Fig.E2.45, which has a radius of 2m and a width of 1m.

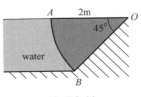

Fig.E2.44

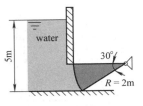

Fig.E2.45

2.46   Find the horizontal and vertical forces acting on the 2m wide Tainter gate in Fig.E2.46. Indicate the line of action of the vertical force without actually computing its location.

2.47   The hemispherical body shown in Fig.E2.47 (r=2m) projects into a tank. Find the horizontal and vertical forces acting on the hemispherical projection for the tank.

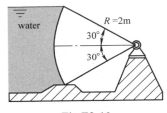

Fig.E2.46

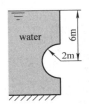

Fig.E2.47

2.48   Find the force F needed to hold the 8m-long cylindrical object in position as shown in Fig.E2.48. Neglect the weight of the cylinder.

2.49   A 3m-long quarter-circular gate of radius 2m and of negligible weight is hinged about its upper edge A, as shown in Fig.E2.49. Determine the minimum force F required to keep the gate closed when the water level rises to the upper edge of the gate.

2.50   The tank, with an initial pressure of p=25kPa, is accelerated as shown in Fig. E2.50 at the rate of a=10m/s^2. Determine the force on the 100mm-diameter plug.

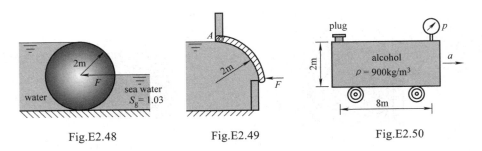

Fig.E2.48   Fig.E2.49   Fig.E2.50

2.44　如图 E2.44 所示圆弧闸门半径 2m 宽 4m，计算其所受静压力的大小、方向和作用点位置。

2.45　如图 E2.45 所示淹没于水下的圆弧闸门半径 2m 宽 1m，计算其所受静压力的大小、作用方向和位置。

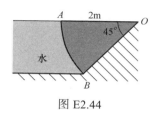

图 E2.44

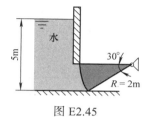

图 E2.45

2.46　计算图 E2.46 中圆弧闸门所受水平方向和竖直方向静压力的大小，标出竖直压力的作用方向（无须计算作用点位置）。

2.47　如图 E2.47 所示容器内部凸入一半球体（$r = 2m$），计算该半球表面所受的水平和竖直方向静压力。

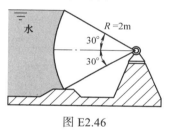

图 E2.46

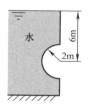

图 E2.47

2.48　如图 E2.48 所示 8m 长圆柱体，忽略圆柱体自重，计算保持其固定所需施加的力 F。

2.49　如图 E2.49 所示长度 3m 半径 2m 的四分之一圆柱形水闸可绕顶点 A 转动。计算当水位到达闸门顶部时，依然可保持闸门关闭的最小力 F。

2.50　图 E2.50 中储液罐初始加压 $p=25kPa$，以 $a=10m/s^2$ 加速度加速前进，计算罐顶部直径 100mm 的罐盖所受压力。

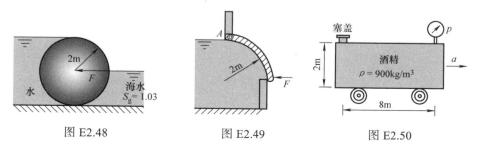

图 E2.48　　　　　图 E2.49　　　　　图 E2.50

2.51   The tank shown in Fig.E2.51 is completely filled with water and accelerated. Calculate the maximum pressure in the tank.

2.52   The tank of 2m high has 1.5 water level shown in Fig.E2.52 is accelerated to the right at 3m/s². Find p_A, p_B and p_C.

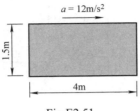

Fig.E2.51

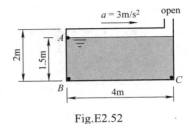

Fig.E2.52

2.53   Suppose the rectangular tank of Fig.E2.53 is completely open at the top. If it is initially filled to the top, how much liquid will be spilled if it is given a horizontal acceleration $a_x=0.5g$ in the direction of its length?

2.54   For the cylinder shown in Fig.E2.54, determine the pressure at point A and the center of the bottom of the tank.

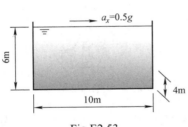

Fig.E2.53

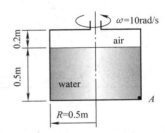

Fig.E2.54

2.51　图 E2.51 中所示容器盛满水且加速运动，计算容器内压强最大处的压强值。

2.52　图 E2.52 所示容器高 2m，静止时液面高 1.5m，以 3m/s² 的加速度向右运动，计算图中 A、B、C 三点处的压强 p_A、p_B、p_C。

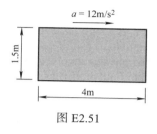

图 E2.51

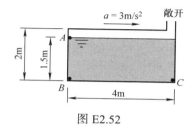

图 E2.52

2.53　如图 E2.53 所示的长方形容器，顶部完全敞开，初始时刻盛满水。当容器以 $a_x=0.5g$ 的加速度沿长度方向水平运动时，会有多少水溢出容器？

2.54　图 E2.54 中的圆柱形容器，计算其旋转时容器底部 A 点和底部中心处的压强。

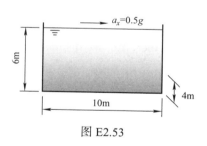

图 E2.53

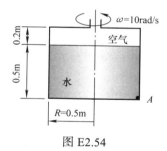

图 E2.54

CHAPTER 3
BASICS OF FLUID KINEMATIC

Τhis chapter serves for an introduction to all the following chapters that deal with fluid motions. Fluid motions indicate themselves in many different ways. In engineering applications, it is important to describe the fluid motions as simply as can be justified, it is the engineer's responsibility to make simplifying assumptions. This, of course, requires experience and, more importantly, an understanding of the physics involved. Some common assumptions used to simplify a flow situation are related to fluid properties. For example, we derive the basic equations in chapter 4 by neglecting the effects of viscosity to the flow, in others, the viscosity affects the internal flow significantly in chapter 6.

This chapter has three sections. In the first section we introduce the approaches used to describe fluid motions. In the second section we give a brief overview of different types of flow. The third section introduces some basic conceptions that helps to represent fluid motions mathematically.

3.1　Description of Fluid Motion

In engineering applications, the first important thing is to describe the fluid flow as simply and accurately as can be justified. In this section, we discuss the mathematical description of fluid motion. In fluid dynamics, the study of how to describe fluid motion, that deals with velocities and flow paths without considering forces or energy is known as *fluid kinematics*. From a fundamental point of view, there are two distinct ways to describe motion.

3.1.1　Lagrangian and Eulerian descriptions

The *Lagrangian description* after the Italian mathematician Joseph Louis Lagrange (1736—1813), same with the most familiar method you learned in high school physics, follows the path of individual object, describes the motion of individual particles which are observed as a function of time. Each individual particle is considered to be a small mass of fluid, consisting of a large number of molecules that occupies a small volume $\Delta\Omega$ that moves with the flow. The position, velocity, and acceleration of each particle are listed as $s(x, y, z, t)$, $v(x, y, z, t)$, and $a(x, y, z, t)$, and quantities of interest can be calculated. Here (x, y, z) represent the position of fluid particle.

The Lagrangian description requires us to track the position and velocity of each individual fluid particle as functions of time. The point (x_0, y_0, z_0) locates the starting point when $t = 0$ as the name of each particle, and the position of each fluid particle is the function of start position (x_0, y_0, z_0) and time t, that is

$$\begin{cases} x = x\left(x_0, y_0, z_0, t\right) \\ y = y\left(x_0, y_0, z_0, t\right) \\ z = z\left(x_0, y_0, z_0, t\right) \end{cases} \tag{3.1.1}$$

When (x_0, y_0, z_0) is given and t changes, Eq.3.1.1 follows the track of a certain fluid particle. If a time t is chosen, positions of different particles at this time are coordinated as

第 3 章
流体运动基础

本章介绍描述流体运动的方法，是后续章节分析的基础。描述流体运动的方式有很多，工程应用中需要工程师依据自己的经验和掌握的知识对流动问题做必要的假设加以简化。这些简化通常是基于流体的特性。例如在第 4 章推导基本方程时，我们会忽略流体黏性对流动的影响，而第 6 章的内流问题，则必须考虑黏性的显著影响。

本章主要包括三部分内容。第一部分介绍描述流体运动的方法；第二部分简要介绍流动的分类；第三部分介绍与流动相关的基本概念，以帮助我们建立数学模型。

3.1 描述流体运动的方法

本节我们介绍描述流体运动的数学方法，但是工程应用中首要的工作是尽可能在保证足够精度的前提下简化流动问题。在流体动力学中只描述流体运动的几何性质而不涉及力和能量的内容，称为**流体运动学**。

总的来说描述流体运动的方式有两种。

3.1.1 拉格朗日法和欧拉法

拉格朗日法以意大利数学家 Joseph Louis Lagrange（1736—1813）命名，其基本思想与我们高中物理学中的方法一致：跟踪独立的运动目标，把每个质点的运动轨迹都作为时间的函数来描述。组成流动的每个流体质点都包含无数多个流体分子，具有很小的质量和体积 $\Delta\Omega$。每个与流动相关的物理量如位置、速度以及加速度等都可以用如 $s(x,y,z,t)$，$v(x,y,z,t)$ 和 $a(x,y,z,t)$ 等的函数形式表达和计算。

拉格朗日法要求对大量流体质点位置和速度随时间的变化情况做跟踪记录，如以各质点在 $t=0$ 初始时刻的位置 (x_0, y_0, z_0) 为其命名，则每个质点的位置均由其初始位置 (x_0, y_0, z_0) 和时间 t 的函数来表示，即

$$\begin{cases} x = x(x_0, y_0, z_0, t) \\ y = y(x_0, y_0, z_0, t) \\ z = z(x_0, y_0, z_0, t) \end{cases} \quad (3.1.1)$$

当 (x_0, y_0, z_0) 不变而 t 变化时，式 3.1.1 描述的就是某个质点的运动轨迹；若选定某一时刻 t，则表示该时刻各流体质点所处的位置，如

$$\begin{cases} x' = x(x_0', y_0', z_0', t) \\ y' = y(x_0', y_0', z_0', t), \\ z' = z(x_0', y_0', z_0', t) \end{cases} \begin{cases} x'' = x(x_0'', y_0'', z_0'', t) \\ y'' = y(x_0'', y_0'', z_0'', t) \dots \\ z'' = z(x_0'', y_0'', z_0'', t) \end{cases} \qquad (3.1.2)$$

The starting point location (x_0, y_0, z_0), the name of each particle, do not change with time. By Lagrangian description, the velocity of each particle is defined as that following a fluid particle, the rate of change of the particle's position with respect to time, expressed as v_x, v_y, v_z

$$\begin{cases} v_x = \dfrac{\partial x}{\partial t} \\[2mm] v_y = \dfrac{\partial y}{\partial t} \\[2mm] v_z = \dfrac{\partial z}{\partial t} \end{cases} \qquad (3.1.3)$$

In the same way, acceleration of a fluid particle represented by a_x, a_y and a_z, defined as the second derivative of the coordinate of track versus time, such as

$$\begin{cases} a_x = \dfrac{\partial^2 x}{\partial t^2} \\[2mm] a_y = \dfrac{\partial^2 y}{\partial t^2} \\[2mm] a_z = \dfrac{\partial^2 z}{\partial t^2} \end{cases} \qquad (3.1.4)$$

Lagrangian analysis follows a mass of fixed identity, in the description many particles need to be followed and their influence on one another noted. This method of describing motion is much a difficult task as the number of particles becomes extremely large in even the simplest fluid flow.

A more common method of describing fluid motion is the *Eulerian description*, named after the Swiss mathematician Leonhard Euler (1707—1783), which identify points in space and then observe the velocity of particles passing each point.

Eulerian description of motion thinks of flow properties as functions of both space and time. In Cartesian coordinates the velocity is expressed as $V = V(x, y, z, t)$. The region of flow that is being considered is called a *flow field*. For example, the pressure field is a scalar field variable; for general unsteady three-dimensional fluid flow in Cartesian coordinates

$$p = p(x, y, z, t) \qquad (3.1.5)$$

We define the velocity field as a vector field variable in similar fashion

$$V = V(x, y, z, t) \qquad (3.1.6)$$

Likewise, the acceleration field is also a vector field variable

$$a = a(x, y, z, t) \qquad (3.1.7)$$

$$\begin{cases} x' = x(x_0', y_0', z_0', t) \\ y' = y(x_0', y_0', z_0', t), \\ z' = z(x_0', y_0', z_0', t) \end{cases} \begin{cases} x'' = x(x_0'', y_0'', z_0'', t) \\ y'' = y(x_0'', y_0'', z_0'', t)\dots \\ z'' = z(x_0'', y_0'', z_0'', t) \end{cases} \quad (3.1.2)$$

各流体质点的初始位置（x_0, y_0, z_0），即该质点的名称，不随时间改变。拉格朗日法定义流体质点的运动速度为其位置对时间的变化率，用 v_x、v_y 和 v_z 表示

$$\begin{cases} v_x = \dfrac{\partial x}{\partial t} \\[2mm] v_y = \dfrac{\partial y}{\partial t} \\[2mm] v_z = \dfrac{\partial z}{\partial t} \end{cases} \quad (3.1.3)$$

同样，流体质点的加速度定义为位置坐标对时间的二阶导数，用 a_x、a_y 和 a_z 表示

$$\begin{cases} a_x = \dfrac{\partial^2 x}{\partial t^2} \\[2mm] a_y = \dfrac{\partial^2 y}{\partial t^2} \\[2mm] a_z = \dfrac{\partial^2 z}{\partial t^2} \end{cases} \quad (3.1.4)$$

拉格朗日法以一定质量的流体为研究对象，跟踪且记录大量流体质点的运动以及相互间的作用，即使对于最简单的流动，大量数据的记录和计算也是一项艰巨的工作。

另一种广泛应用的方法是以瑞士数学家 Leonhard Euler（1707—1783）命名的**欧拉法**，该方法着眼于空间的固定点，观察流体质点经过这些位置时的速度（或其他物理量）变化情况。

欧拉法把流动相关的物理量视为空间和时间的函数，在直角坐标系下流速可表示为 $V = V(x, y, z, t)$，待研究的流动区域称为**流场**。例如，一般三维非稳定流动的压强是一个标量场变量，在直角坐标系下表示为

$$p = p(x, y, z, t) \quad (3.1.5)$$

类似地，速度为向量场变量

$$V = V(x, y, z, t) \quad (3.1.6)$$

同理，加速度场也同样是向量场

$$a = a(x, y, z, t) \quad (3.1.7)$$

The field variable at a particular location at a particular time is the value of the variable for whichever fluid particle happens to occupy that location at that time.

We can observe the rate of change of velocity as the particles pass each point, that is, $\partial V/\partial x$, $\partial V/\partial y$, and $\partial V/\partial z$, and we can observe if the velocity is changing with time at each particular point, that is, $\partial V/\partial t$.

Collectively, these field variables define the flow field, the velocity field of Eq.3.1.6 is expanded in Cartesian coordinates

$$V = (u,v,w) = u(x,y,z,t)\boldsymbol{i} + v(x,y,z,t)\boldsymbol{j} + w(x,y,z,t)\boldsymbol{k} \tag{3.1.8}$$

where u is the x-component of the velocity vector (Eulerian description), equal to the rate of change of the particle's x-position with respect to time by Eq.3.1.3 (Lagrangian description), is $u = dx/dt$. Similarly, $v = dy/dt$, and $w = dz/dt$.

A similar expansion can be performed for the acceleration field of Eq.3.1.7. In the Eulerian description, we don't really care what happens to individual fluid particles, rather we are concerned with the pressure, velocity, acceleration, etc., of whichever fluid particle happens to be at the location of interest at the time of interest, all such field variables are defined at any location (x, y, z) and at any instant in time t.

An example may clarify these two ways of describing motion. Certain measures need to be taken to improve the traffic flow on a busy road. The researcher has two alternatives: Request college students to travel in each automobile throughout the road recording the appropriate observations (the Lagrangian approach), or ask help to traffic police officer who has been watching each section and intersection and recording the required information in control-room (the Eulerian approach). In this example it may not be obvious which approach would be preferred, a correct interpretation of each set of data would lead to the same solution. At any instant in time under consideration, the position vector of the fluid particle in the Lagrangian frame is equal to the position vector in the Eulerian frame.

There are many occasions in which the Lagrangian description is useful while the Eulerian description is often more convenient for fluid mechanics applications. Furthermore, experimental measurements are generally more suited to the Eulerian description. In a wind tunnel, for example, velocity or pressure probes are usually placed at a fixed location in the flow, measuring $V(x, y, z, t)$ or $p(x, y, z, t)$. However, whereas the equations of motion in the Lagrangian description following individual fluid particles are well known as Newton's second law, examples such as drifting buoys used to study ocean currents.

3.1.2 Acceleration

In a flow field, the velocities may be everywhere different in magnitude and direction, and also change with time, as gives in Eq.3.1.6. The acceleration of a fluid particle in the Lagrangian frame can be expressed as

$$\boldsymbol{a} = \frac{dV}{dt} = \frac{\partial V}{\partial t} + \frac{\partial V}{\partial x}\frac{dx}{dt} + \frac{\partial V}{\partial y}\frac{dy}{dt} + \frac{\partial V}{\partial z}\frac{dz}{dt} \tag{3.1.9}$$

We replace the dx/dt, dy/dt and dz/dt with the components of velocity vector u, v and w, the Eq.3.1.9 can be written as

$$\boldsymbol{a} = \frac{dV}{dt} = \frac{\partial V}{\partial t} + u\frac{\partial V}{\partial x} + v\frac{\partial V}{\partial y} + w\frac{\partial V}{\partial z} \tag{3.1.10}$$

场变量在某一时间某一位置的取值，就是流体质点刚好在此时间流经此位置时物理量的值。

我们可以观察到流体质点经过各个位置时速度的变化率，记为 $\partial V/\partial x$、$\partial V/\partial y$ 和 $\partial V/\partial z$，同样可以记录各位置处运动速度随时间的变化率，记为 $\partial V/\partial t$。

各种场变量定义出流场特征，式 3.1.6 的速度场可以进一步展开为

$$V = (u,v,w) = u(x,y,z,t)\boldsymbol{i} + v(x,y,z,t)\boldsymbol{j} + w(x,y,z,t)\boldsymbol{k} \tag{3.1.8}$$

式中，u 是速度向量在 x 方向的分量（欧拉法），等于式 3.1.3 中 x 坐标位置随时间的变化率（拉格朗日法），即 $u = \mathrm{d}x/\mathrm{d}t$，同样，$v = \mathrm{d}y/\mathrm{d}t$ 及 $w = \mathrm{d}z/\mathrm{d}t$。

式 3.1.7 中的加速度也可以同样展开。欧拉描述法不需要去关注各个流体质点的具体运动情况，而是关注流体质点刚好在我们需要研究的时间处于待研究位置时的压强、速度、加速度等变量情况，并可以用任意位置（x,y,z）和时间 t 来描述这些变量。

我们用一个形象的例子来比较一下这两种方法的区别。假设我们要调查和改善某一繁忙路段的交通情况，研究人员可以采用两种不同的调研方式：请大学生志愿者乘坐该路上的每一辆交通工具，全程观察并记录相关情况（拉格朗日法）；或者求助交通控制中心的警官，从监控系统中调取各路口和路段中监控视频记录（欧拉法）。本例中的两种方法都可以达成目标，无所谓优劣之分。同一时刻，用拉格朗日法或欧拉法观察到的流体质点其位置向量都是相同的。

流体力学中欧拉法更加便于应用，也更方便与实验法配合使用，拉格朗日法在很多情况下则更有助问题的描述。例如风洞实验中在流场固定位置安装速度、压强等传感器，测量并记录 $V(x,y,z,t)$ 和 $p(x,y,z,t)$ 属于欧拉的方法。为研究洋流问题而设置的浮标则是基于拉格朗日法跟踪单个流体质点并描述其运动的方法，与牛顿第二定律相契合。

3.1.2　加速度

流场中各点速度的大小和方向都是随时间变化的，见式 3.1.6。拉格朗日法对流体质点加速度的定义可以表示为

$$\boldsymbol{a} = \frac{\mathrm{d}V}{\mathrm{d}t} = \frac{\partial V}{\partial t} + \frac{\partial V}{\partial x}\frac{\mathrm{d}x}{\mathrm{d}t} + \frac{\partial V}{\partial y}\frac{\mathrm{d}y}{\mathrm{d}t} + \frac{\partial V}{\partial z}\frac{\mathrm{d}z}{\mathrm{d}t} \tag{3.1.9}$$

用速度分量 u、v 和 w 代替 $\mathrm{d}x/\mathrm{d}t$、$\mathrm{d}y/\mathrm{d}t$ 和 $\mathrm{d}z/\mathrm{d}t$，式 3.1.9 可写为

$$\boldsymbol{a} = \frac{\mathrm{d}V}{\mathrm{d}t} = \frac{\partial V}{\partial t} + u\frac{\partial V}{\partial x} + v\frac{\partial V}{\partial y} + w\frac{\partial V}{\partial z} \tag{3.1.10}$$

Thus we may express the acceleration of a fluid particle as a field variable to transform from the Lagrangian to the Eulerian frame of reference. That is, at any instant in time t, the acceleration field of Eq.3.1.10 must equal the acceleration of the fluid particle that happens to occupy the location (x, y, z) at that time t.

In vector form Eq.3.1.10 is written as

$$a = \frac{dV}{dt} = \frac{\partial V}{\partial t} + (V \cdot \nabla)V \tag{3.1.11}$$

Where ∇ is the gradient operator.

$$\nabla = \left(\frac{\partial}{\partial x}, \frac{\partial}{\partial y}, \frac{\partial}{\partial z}\right) = \frac{\partial}{\partial x}i + \frac{\partial}{\partial y}j + \frac{\partial}{\partial z}k \tag{3.1.12}$$

The scalar component equations of the above vector equation for Cartesian coordinates are written as

$$a_x = \frac{\partial u}{\partial t} + u\frac{\partial u}{\partial x} + v\frac{\partial u}{\partial y} + w\frac{\partial u}{\partial z}$$
$$a_y = \frac{\partial v}{\partial t} + u\frac{\partial v}{\partial x} + v\frac{\partial v}{\partial y} + w\frac{\partial v}{\partial z} \tag{3.1.13}$$
$$a_z = \frac{\partial w}{\partial t} + u\frac{\partial w}{\partial x} + v\frac{\partial w}{\partial y} + w\frac{\partial w}{\partial z}$$

The time derivative term on the right-hand side of Eq.3.1.10 and 3.1.11, dV/dt, is called the local acceleration, and the remaining terms on right side of each equation form the convective acceleration. Hence the acceleration of a fluid particle is the sum of the local acceleration and convective acceleration.

The acceleration is a scalar variable, there is only one velocity or acceleration component at each point which moves along the streamline S.

$$a_S = \frac{\partial u}{\partial t} + u\frac{\partial u}{\partial s} \tag{3.1.14}$$

Example 3.1 A velocity (m/s) field in a particular flow is given by $V = 16xyi - 20y^2j$. Calculate the acceleration at the point $(1, -1)$.

Solution Here we have $u = 16xy$ and $v = -20y^2$.

Use Eq. 3.1.13 and find each component of the acceleration

$$a_x = u\frac{\partial u}{\partial x} + v\frac{\partial u}{\partial y} = 16xy \cdot 16y - 20y^2 \cdot 16x = -64xy^2$$

$$a_y = u\frac{\partial v}{\partial x} + v\frac{\partial v}{\partial y} = 16xy \cdot 0 + 20y^2 \cdot 40y = 800y^3$$

All particles passing through the point $(1, -1)$ have the acceleration(m/s^2)

这样流体质点加速度的表达从拉格朗日方法转换至欧拉方法。可以理解为，在任意瞬时 t，式 3.1.10 所描述的加速度场在（x, y, z）点的值，一定等于在这一瞬时刚好处于（x, y, z）位置的流体质点的加速度。

式 3.1.10 的向量形式为

$$a = \frac{\mathrm{d}V}{\mathrm{d}t} = \frac{\partial V}{\partial t} + (V \cdot \nabla)V \qquad (3.1.11)$$

上式 ∇ 代表梯度算子，且

$$\nabla = \left(\frac{\partial}{\partial x}, \frac{\partial}{\partial y}, \frac{\partial}{\partial z} \right) = \frac{\partial}{\partial x} i + \frac{\partial}{\partial y} j + \frac{\partial}{\partial z} k \qquad (3.1.12)$$

加速度在直角坐标系中的标量形式可写为

$$a_x = \frac{\partial u}{\partial t} + u \frac{\partial u}{\partial x} + v \frac{\partial u}{\partial y} + w \frac{\partial u}{\partial z}$$

$$a_y = \frac{\partial v}{\partial t} + u \frac{\partial v}{\partial x} + v \frac{\partial v}{\partial y} + w \frac{\partial v}{\partial z} \qquad (3.1.13)$$

$$a_z = \frac{\partial w}{\partial t} + u \frac{\partial w}{\partial x} + v \frac{\partial w}{\partial y} + w \frac{\partial w}{\partial z}$$

式 3.1.10 和式 3.1.11 中等式右侧的时间导数项，$\mathrm{d}V/\mathrm{d}t$，称为局部加速度，其他导数项称为对流加速度。所以，流体质点加速度由局部加速度和对流加速度组成。

当流体质点仅沿流线 S 运动时，其流速和加速度都仅有一个分量，所以流体质点的加速度为标量

$$a_S = \frac{\partial u}{\partial t} + u \frac{\partial u}{\partial s} \qquad (3.1.14)$$

例 3.1　已知某流场内速度（m/s）分布规律 $V = 16xy i - 20y^2 j$，计算点（1，−1）处的加速度值。

解　可知两坐标轴方向的速度分量 $u = 16xy$ 和 $v = -20y^2$。
根据式 3.1.13 可得出加速度的各个分量

$$a_x = u \frac{\partial u}{\partial x} + v \frac{\partial u}{\partial y} = 16xy \cdot 16y - 20y^2 \cdot 16x = -64xy^2$$

$$a_y = u \frac{\partial v}{\partial x} + v \frac{\partial v}{\partial y} = 16xy \cdot 0 + 20y^2 \cdot 40y = 800y^3$$

任意经过点（1，−1）的流体质点加速度（m/s²）为

$$a = -64i - 800j$$

3.1.3 Substantial derivative

The total derivative operator d/dt in Eq.3.1.11 is given a special name, the *substantial derivative*, and a special symbol that is written as

$$\frac{D}{Dt} = \frac{d}{dt} = \frac{\partial}{\partial t} + (V \cdot \nabla) \tag{3.1.15}$$

We use the special symbol D/Dt instead of d/dt, in order to emphasize that it is formed by following a fluid particle as it moves through the flow field, that is, we followed the substance (or material). Substantial derivative can also be called material derivative, particle derivative, or Lagrangian derivative.

When we apply the substantial derivative of Eq.3.1.15 to the velocity field, the result is the acceleration field as expressed by Eq.3.1.11.

Equation 3.1.15 can also be applied to other fluid properties besides velocity, both scalars and vectors. For example, the substantial derivative of pressure is written as

$$\frac{Dp}{Dt} = \frac{dp}{dt} = \frac{\partial p}{\partial t} + (V \cdot \nabla) p \tag{3.1.16}$$

Equation 3.1.15 represents the time rate of change of pressure following a fluid particle as it moves through the flow and contains both local and convective components.

Similarly, we can write the substantial derivative of density or any other variable (vetor) N which following a fluid particle as it moves through the flow field as Eq.3.1.17 and Eq. 3.1.18.

$$\frac{D\rho}{Dt} = \frac{d\rho}{dt} = \frac{\partial \rho}{\partial t} + (V \cdot \nabla) \rho \tag{3.1.17}$$

and

$$\frac{DN}{Dt} = \frac{\partial N}{\partial t} + (V \cdot \nabla) N \tag{3.1.18}$$

3.2 Classification of Fluid Flows

In this section we introduce the general classification of fluid flows. To solve the problems encountered in practice and make it feasible to study them in groups, we should first classify the fluid flows on the basis of some common characteristics. There are many ways to classify fluid flow problems, and here we present some general categories.

3.2.1 Viscous and inviscid flows

Viscosity is one of the inherent properties of fluid, when we deal with problem related to real fluids, the effects of viscosity are introduced into the problem. This results in the development of shear stresses between neighboring fluid particles when they are moving at different velocities. Flows in which the effects of viscosity are significant are called *viscous flows*. Viscous flows, such as flows in pipes and in open channels, where viscous effects cause substantial"losses" and account for the huge amounts of energy that must be used to transport oil and gas in pipelines. Viscous flow in pipes, for example, the velocity adjacent to the wall will be zero which resulted by the no-slip condition, it will increase rapidly within a short distance from the wall and produce a velocity profile such

$$a = -64i - 800j$$

3.1.3　物质导数

式 3.1.11 中的全微分算子 d/dt 有一个特别的名称，称为**物质导数**，表示形式为

$$\frac{D}{Dt} = \frac{d}{dt} = \frac{\partial}{\partial t} + (V \cdot \nabla) \tag{3.1.15}$$

用大写微分符号 D/Dt 代替 d/dt，是为了特别强调该变量为跟随流场中的流体质点，即跟随物质，所以称为物质导数，或质点导数、拉格朗日导数等。

把速度代入式 3.1.15 中，就可以得出式 3.1.11 形式的加速度表达式。

不仅仅速度，与流动相关的其他向量和标量也都可以用式 3.1.15 的形式表示，如用物质导数形式表示压强

$$\frac{Dp}{Dt} = \frac{dp}{dt} = \frac{\partial p}{\partial t} + (V \cdot \nabla) p \tag{3.1.16}$$

上式为运动流体质点所受压强随时间的变化率，同样包含局部导数项和对流导数项。

同样，还可以用物质导数表示密度或在流场内随流体质点移动的任意物理量 **N**（向量），如下两式

$$\frac{D\rho}{Dt} = \frac{d\rho}{dt} = \frac{\partial \rho}{\partial t} + (V \cdot \nabla) \rho \tag{3.1.17}$$

和

$$\frac{D\boldsymbol{N}}{Dt} = \frac{\partial \boldsymbol{N}}{\partial t} + (V \cdot \nabla) \boldsymbol{N} \tag{3.1.18}$$

3.2　流动的分类

本节介绍一些常见的流动类型。为了更有效地分析和归类解决各种实际问题，我们首先需要根据基础特征对流动进行分类。对流动分类的方法很多，本书介绍一些常用的分类形式。

3.2.1　黏性流动和无黏流动

黏性是流体的固有性质，实际流体的黏性都会对流动产生影响。流体质点以不同速度运动时，黏性使相邻质点间产生剪切力。黏性影响明显的流动称为**黏性流动**。管道及明渠等黏性流动会导致明显的能量损失，所以在设计油、气输送管道时必须考虑黏性损失的问题。例如，管道内的黏性流动，由于壁面不滑移效应的影

as shown in Fig.3.2.1a.

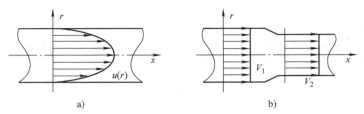

Fig.3.2.1　Flows of viscous an inviscid in pipe
a) viscous flow　b) inviscid flow

An *inviscid flow* is one in which viscous effects do not significantly influence the flow and are thus neglected. Consideration of viscosity complicates analysis and dynamical model, increases the computation load of algorithm. We may presume an idealized situation that an ideal fluid has no viscosity. However, in many flows of practical interest, there are regions where viscous forces are negligibly small compared to inertial or pressure forces, instances in engineering problems where the assumption of an ideal fluid is helpful. Neglecting the viscous terms in such inviscid flow greatly simplifies the analysis without much loss in accuracy. Nonetheless, it is very difficult to create an inviscid flow experimentally, because all fluids of interest, even as water and air, have viscosity. To model an inviscid flow in a straight conduit analytically, we can obviously make all viscous effects zero, all particles move in parallel lines with equal velocity such as shown in Fig.3.2.1b.

3.2.2　Uniform flow and non-uniform flow

The flow that convective components of Eq.3.1.18 are not zero is called *non-uniform flow*. Contrarily a flow that the convective components are all zero is called uniform flow.

There are many engineering problems in fluid mechanics in which a flow field is simplified to a *uniform flow*: the velocity, and other fluid properties, are constant over the area, as shown in Fig.3.2.1b. This simplification is made when the velocity is essentially constant over the area, a rather common occurrence. The average velocity may change from one section to another, the flow conditions depend only on the space variable in the flow direction. For large conduits, however, it may be necessary to consider hydrostatic variation in the pressure normal to the streamlines, this will be discussed further in section 4.3.

3.2.3　One-, two- and three-dimensional flows

In the Eulerian description of motion the velocity vector, in general, depends on the three space variables and time. A flow field is characterized by its velocity distribution, and thus a flow is said to be *one-, two-, or three-dimensional flow* if the flow velocity varies in one, two, or three space variables respectively.

A fluid flow involves a three-dimensional geometry, and the velocity may vary in all three dimensions, for example, rendering the flow three-dimensional $V(x, y, z)$ in rectangular or $V(r, \theta, z)$ in cylindrical coordinates. However, the variation of velocity in certain direction may be very small relative to the variation in other directions and can be ignored with negligible error. In such cases, the flow can be modeled conveniently as being one- or two-dimensional, which is easier to analyze. Often a three-dimensional flow can be approximated as a two-dimensional flow.

响，管道壁处的流体流速为零，那么在距离管道壁很近的范围内就会出现流速激增的情况，如图 3.2.1a 所示的速度曲线。

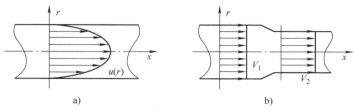

图 3.2.1　管道内的黏性和无黏流动

a）黏性流动　b）无黏流动

无黏流动是指黏性影响可以忽略的流动。考虑黏性的动力学模型增加了其复杂性且计算量随之增大。许多场合黏性力的影响相较惯性力或压力等小很多，甚至可以忽略，我们可以假想存在一种理想化的无黏性流体，这种理想流体的假设为解决工程实践问题带来了很大的帮助。无黏流动中忽略黏性的影响，可以在不明显降低精度的同时大大简化问题。即便如此，若想在实验中制造无黏流动还是非常困难的，因为实际流体都是有黏性的，即使是流动性很好的水和空气也一样具有黏性。在平直管道内建立无黏流动模型仅需假设所有的黏性影响均为零，流体质点都如图 3.2.1b 所示沿平行流线等速运动。

3.2.2　均匀流动和非均匀流动

式 3.1.18 中位置导数项不为零的流动称为**非均匀流动**。反之，位置导数项全部为零的流动称为**均匀流动**。

工程中许多流动可以简化为**均匀流动**，即，流速或其他流动相关物理量在某截面上均匀分布，如图 3.2.1b 所示。最常用的均匀流动是假设流速在截面上恒定不变，不同截面上的平均流速随截面所处的空间位置沿流动方向发生改变。但对于某些大型管道，可能需要讨论与流线垂直截面上的静压强分布，这一问题在 4.3 节分析。

3.2.3　一维、二维和三维流动

欧拉法把流速定义为三个空间坐标位置和时间的函数。当我们用流速描述一个流场的特征时，根据流速随一个、两个还是三个空间坐标变化，可以把流动分为**一维、二维或三维流动**。

一个三维流动的流速随三个坐标值而变化，如在直角坐标系中 $V(x, y, z)$ 或柱坐标系中 $V(r, \theta, z)$。如果某方向上流速的变化较其他方向小很多，那么在不引起明显误差的情况下可以将其忽略，如此一来，三维流动就可以被化简为二维、甚至一维流动，其分析也将大大简化。实践中，通常可以把三维流动近似为二维流动。

如图 3.2.2 所示的管道流动，在入口附近流速随半径 r 和 x 坐标位置改变，为

Considering the flow shown in Fig.3.2.2, it is two-dimensional in the entrance region of the pipe since the velocity changes in both the r- and x-directions. The flow in the region of fully developed is one-dimensional since the velocity varies in the radial r-direction but not in the axial x-directions.

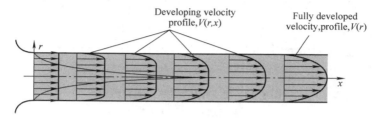

Fig.3.2.2   dimensional developing flow and 1-dimemsional fully developed flow

Note that the dimensionality of the flow also depends on the choice of coordinate system and its orientation. The pipe flow discussed, such as shown in Fig.3.2.1a, $u(r)$ is one-dimensional in cylindrical coordinates, but two-dimensional represented by $u(y, z)$ in rectangular coordinate, illustrating the importance of choosing the most appropriate coordinate system.

3.2.4   Incompressible and compressible flows

Flow can also be classified as that of an incompressible or compressible fluid depending on the level of variation of density during flow. An *incompressible flow* exists if the density of each fluid parcel remains relatively constant as it moves through the flow field, that is

$$\frac{\mathrm{D}\rho}{\mathrm{D}t} = \frac{\mathrm{d}\rho}{\mathrm{d}t} = 0 \tag{3.2.1}$$

Or

$$\rho = \text{Constant} \tag{3.2.2}$$

Since liquids are relatively incompressible, we generally treat them as wholly *incompressible fluids*. Under particular conditions where there is little pressure variation, we may also consider the flow of gases to be incompressible, though generally we should consider the effects of the compressibility of the gas. Gas flows can be approximated as incompressible when $Ma<0.3$ which usually the case that the density changes under about 5 percent. If $Ma>0.3$, the density variations influence the flow and compressibility effects should be accounted for, such flows are compressible flows. Therefore, the compressibility effects of air at room temperature can be neglected at speeds under about 100m/s.

In this book, we restrict the incompressible flow of liquid that are studied.

3.2.5   Steady and unsteady flows

Steady flow means steady with respect to time. Thus all properties of the flow, velocity, density, pressure, etc., at every point remain constant with respect to time, that is

$$\frac{\mathrm{D}(\ )}{\mathrm{D}t} = 0 \quad \text{or} \quad \frac{\partial(\ )}{\partial t} = 0 \tag{3.2.3}$$

二维流动；但是进入充分发展流区域后，流速仅随半径 r 变化，不再受 x 坐标位置影响，因此转变为一维流动。

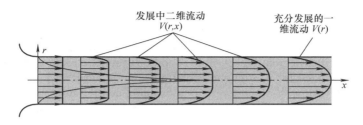

图 3.2.2　二维流动和一维流动的转变

　　流动的维数与坐标系的建立和原点的选取有关。图 3.2.1a 中的管道流，在柱坐标系内 $u(r)$ 是一维流动，但是在直角坐标系内 $u(y,z)$ 则是二维流动，可见，合理建立坐标系是十分重要的。

3.2.4　不可压缩流动和可压缩流动

　　根据流体在流动过程中密度是否发生改变，还可以把流动类型分为不可压缩流动和可压缩流动。若流场内流体微团的密度均恒定不变，则该流动为**不可压缩流动**，有

$$\frac{\mathrm{D}\rho}{\mathrm{D}t} = \frac{\mathrm{d}\rho}{\mathrm{d}t} = 0 \qquad (3.2.1)$$

或

$$\rho = \text{Constant} \qquad (3.2.2)$$

　　由于液体几乎都是不可压缩的，所以液体的流动均视为**不可压缩流动**。气体虽然具有明显的压缩性，但当气体所受压强变化不大时，也可以将其视为不可压缩流体。通常气体流动的马赫数 $Ma<0.3$ 时，其体积的变化率不会超过 5%，可当作不可压缩流体；但当马赫数 $Ma>0.3$ 时，气体密度变化的影响不可忽视，需要作为可压缩流体来分析。室温下，只要空气流速不超过 100m/s，其压缩性都可以忽略。

　　本书中仅讨论不可压缩液体的流动。

3.2.5　稳定流动和非稳定流动

　　稳定流动是指相对时间稳定。流动的各项参数，如流速、密度、压强等，在各点相对时间均保持恒定，即

$$\frac{\mathrm{D}(\ )}{\mathrm{D}t} = 0 \quad 或 \quad \frac{\partial (\ )}{\partial t} = 0 \qquad (3.2.3)$$

In *unsteady flow*, the flow properties at a point change with time, then

$$\frac{D(\ )}{Dt} \neq 0 \text{ , or } \frac{\partial(\ )}{\partial t} \neq 0 \tag{3.2.4}$$

The terms steady and uniform are used frequently in engineering. The term steady implies no change at a point with time, and the term uniform implies no change with location over a specified region. Steady-flow conditions can be closely approximated by devices that are intended for continuous operation such as turbines, pumps. However, the fluid properties vary with time in a periodic manner, and the flow through these devices can still be analyzed as a steady-flow process by using time-averaged values for the properties. When a gear pump is used as supplier for a hydraulic system, for example, the instantaneous flow pulsation is neglected and the flow rate is considered to be steady.

Most of the discussion in this textbook deal with steady flows.

3.2.6 Laminar and turbulent flows

A viscous flow can be classified as either a laminar flow or a turbulent flow.

The highly ordered fluid motion characterized by smooth layers of fluid is called *laminar*. In a laminar flow the fluid flows with no significant mixing of neighboring fluid particles. Viscous shear stresses always influence a laminar flow. The highly disordered fluid motion characterized by velocity fluctuations is called *turbulent*. A distinguishing characteristic of turbulence is its irregularity. In a turbulent flow fluid motions vary irregularly so that quantities such as velocity and pressure show a random variation with time and space coordinates.

Whether a flow is laminar or turbulent depends on the dimensionless Reynolds number, determination of the flow regime is discussed in Sec6.2.

3.3 Related Concepts of Fluid Flow

The topic in this section introduces two different flow lines that are useful in our objective of describing a fluid flow. Some other concepts of fluid motion are presented.

3.3.1 Streamline, pathline and stream tube

A *streamline* is a line (or curve) in the flow field possessing the following property: the velocity vector of each particle occupying a point on the streamline is tangent to the streamline. Streamlines are useful as indicators of the instantaneous direction of fluid motion throughout the flow field, obviously an Eulerian concept. This is shown graphically in Fig.3.3.1.

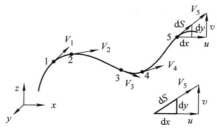

Fig.3.3.1   Streamline in a flow field

非稳定流动的各项参数随时间变化，即

$$\frac{D(\ )}{Dt} \neq 0 \ , \ \text{或} \ \ \frac{\partial(\ )}{\partial t} \neq 0 \tag{3.2.4}$$

稳定和均匀是工程中最常用到的假设，稳定的含义是参数不随时间改变；而均匀则是指在一定范围内参数不随位置改变。一些持续工作的流体设备，如涡轮、泵等，都可以视为稳定流动。还有一些流动其参数呈周期性变化但时间平均值是稳定的，例如液压系统采用齿轮泵供油，如果忽略输出流量的瞬时脉动，也可以被看作稳定流动。

本书中所讨论均为稳定流动。

3.2.6　层流和紊流

黏性流动还可以分为层流和紊流。

层流是指高度有序的，层次分明的流动。层流中的流体没有明显的掺混，黏性力主导流动。**紊流**，顾名思义是指高度无序、速度波动剧烈的流动，其显著特征就是无规律和随机性。紊流中流体质点的运动没有规律和秩序，所以如速度、压强等物理量也呈现随空间和时间无规律变化的现象。

应用无量纲数雷诺数可以判定流体的流态为层流或紊流，这部分内容将在 6.2 节详细讨论。

3.3　流体运动相关概念

本节介绍与流动有关的两种曲线以及描述流体运动的其他相关概念。

3.3.1　流线、迹线和流管

流线是指流场中具有以下特征的曲线（或直线）：流经流线上各点处的流体质点，其运动速度方向均在该点与流线相切。流线可以指示流场内各位置流体质点的运动趋势，属于欧拉表示法，如图 3.3.1 所示。

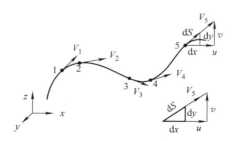

图 3.3.1　流场中的流线

We can mathematically write a simple expression for a streamline based on its definition. Consider an infinitesimal arc length $\mathrm{d}S = \mathrm{d}x\boldsymbol{i} + \mathrm{d}y\boldsymbol{j} + \mathrm{d}z\boldsymbol{k}$ start from point 5, in Fig.3.3.1, along the streamline, $\mathrm{d}\boldsymbol{S}$ must be parallel to the local velocity vector at point 5 $\boldsymbol{V} = u\boldsymbol{i} + v\boldsymbol{j} + w\boldsymbol{k}$ by definition of the streamline. Using similar triangle rules, we know that the components of $\mathrm{d}\boldsymbol{S}$ must be proportional to those of $\boldsymbol{V}$ (Fig.3.3.1). Then, here is the equation for a streamline:

$$\frac{\mathrm{d}S}{V} = \frac{\mathrm{d}x}{u} = \frac{\mathrm{d}y}{v} = \frac{\mathrm{d}z}{w} \tag{3.3.1}$$

where $\mathrm{d}S$ is the magnitude of $\mathrm{d}\boldsymbol{S}$ and V is speed, the magnitude of velocity vector $\boldsymbol{V}$. For a known velocity field, we integrate Eq.3.3.1 to obtain equations for the streamlines. Note that, on normal circumstance streamlines can't intersect.

A *pathline* is the actual path traveled by an individual fluid particle over some time period, it is a Lagrangian concept in that we simply follow the path of an individual fluid particle as it moves around in the flow field. In a physical experiment, you can imagine a tracer fluid particle that is marked by color or brightness. Streamlines cannot be directly observed experimentally except in steady flow fields, in which they are coincident with pathlines. In a steady flow, the velocity vector of a particle at a given point will be tangent to the line that the particle is moving along, the pathline, here the line is also a streamline. In an unsteady flow, the streamline pattern may change significantly with time.

Example 3.3.1 For a steady, incompressible, two-dimensional velocity field, the velocity vector is given by $\boldsymbol{V} = (u,v) = (x+2)\boldsymbol{i} - (y+1)\boldsymbol{j}$. Find the equation of streamline through point (1, 1).

Solution The flow is two-dimensional, implying no z-component of velocity and no variation of u or v with z, and the velocity components in x and y directions are:

$u = x+2, \quad v = -(y+1)$

Equation 3.3.1 is applicable here, thus, along a streamline

$$\frac{\mathrm{d}x}{u} = \frac{\mathrm{d}y}{v} = \frac{\mathrm{d}x}{x+2} = -\frac{\mathrm{d}y}{y+1}$$

We solve this differential equation,

$$\frac{\mathrm{d}x}{x+2} = -\frac{\mathrm{d}y}{y+1} \quad \rightarrow \quad \frac{\mathrm{d}(x+2)}{x+2} + \frac{\mathrm{d}(y+1)}{y+1} = 0$$

$$\rightarrow \ln(x+2) + \ln(y+1) = \ln c$$

$$\rightarrow (x+2)(y+1) = C$$

This is the equation of streamlines in this flow field, where C is a constant of integration that can be set to various values by passing through different location in the flow field.

At (1, 1), C = 6, so that the streamline passing through (1, 1) has the equation

$$(x+2)(y+1) = 6$$

　　根据流线定义可以写出其数学表达式。在图 3.3.1 中的点 5 处取一段微元，其长度 $\mathrm{d}S = \mathrm{d}x\boldsymbol{i} + \mathrm{d}y\boldsymbol{j} + \mathrm{d}z\boldsymbol{k}$，根据流线的定义 $\mathrm{d}\boldsymbol{S}$ 在此点一定与速度向量 $\boldsymbol{V} = u\boldsymbol{i} + v\boldsymbol{j} + w\boldsymbol{k}$ 平行。根据三角形相似规则，图 3.3.1 中 $\mathrm{d}\boldsymbol{S}$ 的各分量一定与 $\boldsymbol{V}$ 的分量具有相同比例。因此，可以写成流线方程：

$$\frac{\mathrm{d}S}{V} = \frac{\mathrm{d}x}{u} = \frac{\mathrm{d}y}{v} = \frac{\mathrm{d}z}{w} \tag{3.3.1}$$

式中，$\mathrm{d}S$ 是微弧 $\mathrm{d}\boldsymbol{S}$ 的长度；V 是速度向量 $\boldsymbol{V}$ 的值，已知速度场的情况下，对式 3.3.1 积分可以得到流线方程。特别指出，流线通常不可能相交。

　　迹线是流体质点在一段时间内经过的轨迹线，属于拉格朗日的方法，只需跟随流场中的一个流体质点运动就可以得到其迹线。实验中使用颜料就可以标记出一个质点的迹线，但却无法直接标记流线，除非在稳定流动中，流线和迹线重合。流体质点的速度向量与其所沿的运动曲线，亦即其迹线，相切，而在稳定流动中，此曲线也是流线。非稳定流动的流线形状会随时间变化。

　　例 3.3.1　已知某稳定、不可压缩、二维流动的速度场为 $\boldsymbol{V} = (u,v) = (x+2)\boldsymbol{i} - (y+1)\boldsymbol{j}$。求经过流场区域内点（1，1）处的流线方程。

　　解　二维流动意味着没有 z 方向速度分量，速度分量 u 和 v 也不会随 z 坐标变化，仅有 x 方向和 y 方向的速度分量，分别是：$u = x+2$，$v = -(y+1)$

沿流线应用式 3.3.1

$$\frac{\mathrm{d}x}{u} = \frac{\mathrm{d}y}{v} = \frac{\mathrm{d}x}{x+2} = -\frac{\mathrm{d}y}{y+1}$$

求解该微分方程

$$\frac{\mathrm{d}x}{x+2} = -\frac{\mathrm{d}y}{y+1} \quad \rightarrow \quad \frac{\mathrm{d}(x+2)}{x+2} + \frac{\mathrm{d}(y+1)}{y+1} = 0$$

$$\rightarrow \ln(x+2) + \ln(y+1) = \ln c$$

$$\rightarrow (x+2)(y+1) = C$$

　　上式为该流场内流线的方程，式中 C 为积分常数，经过流场内不同的点对应不同的常数值。

　　在点（1，1）处 C = 6，所以经过点（1，1）处的流线方程为

$$(x+2)(y+1) = 6$$

A *streamtube* is a tube whose walls are streamlines, as it is shown in Fig.3.3.2. Since the velocity is tangent to a streamline, and streamlines can't intersect, no fluid particle can cross the walls of a stream tube. In other words, fluid within a streamtube must remain there and cannot cross the boundary of the streamtube. A pipe is a streamtube since its walls are streamlines; an open channel is a streamtube since no fluid crosses the walls of the channel. The streamtube is of particular interest in fluid mechanics, we often sketch a streamtube with a small cross section in the interior of a flow for demonstration purposes.

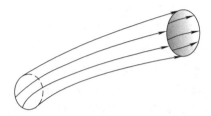

Fig.3.3.2　A streamtube consists of a bundle of individual streamlines

3.3.2　Flow rate and average velocity

We call the quantity of fluid flowing per unit time through a cross section of a whole stream the *flow rate*. We may express it in terms of volume flow rate or discharge, or in terms of mass flow rate, or in terms of weight flow rate. m^3/s, kg/s, and N/s are SI units for expressing volume, mass, and weight flow rates, respectively. When dealing with incompressible fluids, we commonly use volume flow rate, whereas weight flow rate or mass flow rate is more convenient with compressible fluids.

Figure 3.3.3 depicts a streamline in steady flow lying in the xz plane. Element of area dA normal to the streamline at point P. The average velocity at point P is u. The volume flow rate passing through the element of area dA is

$$dq = \boldsymbol{u}dA = \boldsymbol{u} \cdot \boldsymbol{n}dA = (u\cos\theta)dA = u_n dA \qquad (3.3.2)$$

here the component of the flow velocity normal to dA, which we denoted as u_n.

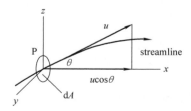

Fig.3.3.3　The volume flow rate at a point on the streamline

This indicates that the volume flow rate is equal to the magnitude of the average velocity multiplied by the flow area at right angles to the direction of the average velocity.

　　流管是由流线围成的封闭小管，如图 3.3.2 所示。因为流线与流速相切，且流线不能相交，因此没有流体质点能够穿过流管壁流入或流出。也就是说，流管内流体的量保持不变。管道和明渠都可以看作是流管，同样没有流体能够穿过管壁或渠壁。在流体力学中我们常常会在流场内取一部分截面并画出一段流管，对其中的流体进行分析和证明。

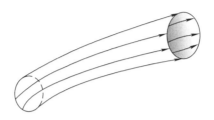

图 3.3.2　由流线封闭而成的流管

3.3.2　流量和平均流速

　　单位时间内通过某一截面的总流量称为**流量**。流量可以表示为体积流量、质量流量或重量流量，体积流量、质量流量和重量流量的单位分别表示为 m^3/s、kg/s 和 N/s。对于不可压缩流体，我们通常使用体积流量，可压缩流体使用质量流量或重量流量更加方便。

　　图 3.3.3 所示为 xz 平面内的一条流线。微元面积 dA 在 P 点与流线垂直，且 P 点处平均流速为 u。流过微元面积 dA 的体积流量为

$$dq = \boldsymbol{u}dA = \boldsymbol{u} \cdot \boldsymbol{n}dA = (u\cos\theta)dA = u_n dA \qquad (3.3.2)$$

式中，u_n 表示与微元面积 dA 垂直的速度分量，即正交速度。

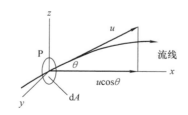

图 3.3.3　流线上一点处的体积流量

　　式 3.3.2 证明，体积流量等于垂直于流线的截面面积与平均流速之积。

In a real fluid the local time average velocity u will vary across the section in some manner, such as the sections that shown in Fig.3.2.1, and so we can express the flow rate as

$$q_V = \int_A u \mathrm{d}A = AV \qquad (3.3.3)$$

where q is the volume flow rate, u is the time average velocity through an infinitesimal area $\mathrm{d}A$, while V is the average velocity $\bar{V}$ (or, mean velocity) over the entire sectional area A, in examples and problems the overbar is often omitted. The section A that perpendicular to all streamlines is called *cross section* of flow, such as cross section 1 and 2 in Fig.3.3.4.

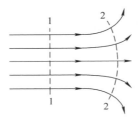

Fig.3.3.4   Across section in fluid flow

We can compute the mass flow rate and the weight flow rate by multiplying the volume flow rate by the density and specific weight of the fluid respectively. For incompressible fluid

$$q_m = \int_A u\rho \mathrm{d}A = \rho \int_A u \mathrm{d}A = \rho q_V \qquad (3.3.4)$$

$$q_g = \rho g \int_A u \mathrm{d}A = \gamma AV = \gamma q_V \qquad (3.3.5)$$

If we have determined the flow rate directly by some method, by flowmeter for example, then from Eq.3.3.3 we can find the average velocity

$$V = \frac{q}{A} \qquad (3.3.6)$$

For the very common occurrence of flow through a circular pipe, we can substitute $A = \pi D^2/4$ into Eq.3.3.6, to yield the average velocity as

$$V = \frac{4q}{\pi D^2} \qquad (3.3.7)$$

here D is diameter of the pipe.

Example 3.3.2   The velocity of a liquid ($S_g = 1.4$) in a 150mm-diameter pipeline is 0.8m/s. Calculate the rate of flow in terms of volume and mass.

Solution   From Eq.3.3.3, the volume flow rate:

　　实际流场中各点流速 u 可能随位置变化，如图 3.3.2 所示，那么，我们可由下式计算体积流量

$$q_V = \int_A u \mathrm{d}A = AV \tag{3.3.3}$$

式中，q 是体积流量；u 是通过微元面积 $\mathrm{d}A$ 的速度的时间平均值；V 是整个截面上速度时间平均值的平均值 $\overline{V}$，为书写方便本书中省略字母上方的横线。与所有流线都垂直的截面 A 称为**有效截面**，如图 3.3.4 中的截面 1 和 2。

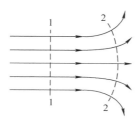

图 3.3.4　有效截面

体积流量乘以流体的密度或重度就是质量流量或重量流量。对于不可压缩流体

$$q_m = \int_A u\rho \mathrm{d}A = \rho \int_A u \mathrm{d}A = \rho q_V \tag{3.3.4}$$

$$q_g = \rho g \int_A u \mathrm{d}A = \gamma AV = \gamma q_V \tag{3.3.5}$$

如果使用流量计测量等方法已知了体积流量，则由式 3.3.3 可得出平均流速

$$V = \frac{q}{A} \tag{3.3.6}$$

　　对于最常见的管道流动，把面积 $A = \pi D^2/4$ 代入式 3.3.6，可得出管道流动的平均流速

$$V = \frac{4q}{\pi D^2} \tag{3.3.7}$$

式中，D 是管道直径。

　　例 3.3.2　直径为 150mm 的管道内某种液体（相对密度 $S_g = 1.4$）的平均流速为 0.8m/s。计算其体积流量和质量流量。

　　解　根据式 3.3.3，体积流量为

$$q_V = VA = 0.8 \times \frac{\pi \times 0.15^2}{4} \, \text{m}^3/\text{s} = 0.02826 \text{m}^3/\text{s}$$

Relative density of the liquid is 1.4: $\rho = 1400 \text{kg/m}^3$

From Eq.3.3.4, the mass flow rate:

$$q_m = \rho q = 1400 \times 0.02826 \text{kg/s} = 39.564 \text{kg/s}$$

Example 3.3.3 The outlet flow of a pump is 60L/min, diameter of the discharge pipe is 60mm. Calculate the average velocity in the pipeline.

Solution From Eq.3.3.7, the average velocity at the cross section of the pipe

$$V = \frac{4q_V}{\pi D^2} = \frac{4 \times 60 \times 10^{-3}}{3.14 \times 0.06^2 \times 60} \, \text{m/s} = 0.354 \text{m/s}$$

3.3.3 System and control volume

We mentioned in section 3.1 that, the Lagrangian analysis follows a mass of fixed identity, here we define this fixed collection of material particles as system. A fluid *system* refers to a specific mass of fluid within the boundaries defined by a closed surface. The shape of the system, and so the boundaries, may change with time, as when liquid flows through a constriction or when gas is compressed, as a fluid moves and deforms, so the system containing it moves and deforms. The size and shape of a system is entirely optional, but no mass crosses its boundaries.

For example, if we consider flow through a pipe, we could identify a fixed quantity of fluid at time t as the system (Fig.3.3.5), this system would then move due to velocity to a downstream location at time $t+\Delta t$.

In fluid dynamics, it is more common to work with a *control volume*, defined as a region of space chosen for study, into which fluid enters and/or from which fluid leaves, of course, it is of Eulerian approach. A control volume, always fixed and nondeformable, allows mass to flow in or out across its boundaries, which are called the *control surface*.

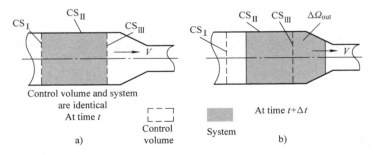

Fig.3.3.5 The fixed control volume and a system at different time

a) time t b) time $t+\Delta t$

The difference between a control volume and a system is illustrated in Fig.3.3.5. The

$$q_V = VA = 0.8 \times \frac{\pi \times 0.15^2}{4} \, \mathrm{m^3/s} = 0.02826 \mathrm{m^3/s}$$

液体的相对密度为 1.4，所以 $\rho = 1400 \mathrm{kg/m^3}$

由式 3.3.4 可得，质量流量为

$$q_m = \rho q = 1400 \times 0.02826 \mathrm{kg/s} = 39.564 \mathrm{kg/s}$$

例 3.3.3 水泵的输出流量为 60L/min，出流管直径为 60mm。计算出流管内的平均流速。

解 由式 3.3.7 可得，管道截面上的平均流速为

$$V = \frac{4q_V}{\pi D^2} = \frac{4 \times 60 \times 10^{-3}}{3.14 \times 0.06^2 \times 60} \, \mathrm{m/s} = 0.354 \mathrm{m/s}$$

3.3.3 系统和控制体

如 3.1 节所述，拉格朗日法跟随一定质量的流体，这里我们把这部分流体质点的集合称为系统。所以，**系统**是指封闭在某闭合边界面内、质量一定的流体。系统随流体运动或变形，例如液体流过狭缝或气体被压缩等，因此系统及其边界面的形状甚至大小都会随时间变化，但在边界面上没有质量交换。

例如我们分析管道流动，在时刻 t 选取一定质量的流体作为系统（见图 3.3.5），那么一段时间后的 $t+\Delta t$ 时刻，这一系统会移动到下游的某个位置。

流体力学中我们会更加经常用到**控制体**（CV），是指为研究方便而选取的一个空间区域，流体会流入、流出该区域，所以这一概念属于欧拉的方法。控制体通常具有固定的位置和形状，具有质量的流体通过其边界面，也称**控制面**（CS），流入或流出控制体。

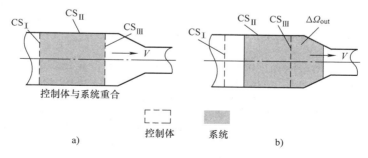

图 3.3.5 控制体和系统在不同时刻

a）t 时刻 b）$t+\Delta t$ 时刻

图 3.3.5 说明了系统和控制体的区别，图中系统和控制体在 t 时刻重合，而在

figure indicates that the system occupies the control volume at time t and has partially moved out of it at time $t+\Delta t$.

3.3.4　System to control volume transformation

Since it is often more convenient to focus on a control volume rather than on a system, the most important is to find a transformation that will allow us to express the substantial derivative of a system (a Lagrangian description) in terms of quantities associated with a control volume (an Eulerian description) so that the basic laws can be applied directly to a control volume.

Most principles of fluid mechanics are adopted from solid mechanics, where the physical laws dealing with the time rates of change of extensive properties are expressed for systems.The mass of fluid system can be expressed in integral form as

$$m_{sys} = \int_{sys} \rho \, d\Omega \tag{3.3.8}$$

where $d\Omega$ is the volume element occupied by the fluid particle, and ρ is the density of fluid which may change from point to point in the system.

We will use the symbol N_{sys} to denote the extensive property such as the integral quantities at right side of Eq.3.3.8. N_{sys} could represent the mass, the momentum, or the energy of the system, either a scalar or a vector quantity. It is helpful to introduce the variable η for the intensive property, the property of the system per unit mass. Any extensive property of fluid can be given by

$$N_{sys} = \int_{sys} \eta \rho \, d\Omega \tag{3.3.9}$$

Then the momentum of a fluid particle is a vector quantity given by

$$M_{sys} = \int_{sys} V \rho \, d\Omega \tag{3.3.10}$$

Let us think of N as the mass of a fluid system, the specified fluid parcel of the system is indicated by subscript sys, and either that of control volume indicated by subscript CV, or the boundary surfaces of control volume indicated by subscript CS. In this case simply assumed that there is only one inlet control surface CS_I and one outlet CS_{III}, and the velocities are approximately normal to the surfaces CS_I and CS_{III}, no mass exchange on surface CS_{II}.

Consider the general flow situation of Fig.3.3.6, system flow into the control volume through control surface CS_I and out from CS_{III}, there is no mass exchange on control surface CS_{II}.The system and control volume at time t are shown in Fig3.3.6a. At instant Δt later,the system has changed its shape and position; a small amount of system has entered the control volume through the control surface CS_I, and another small amount of system has left the control volume through CS_{III} as shown in Fig.3.3.6b. These small volumes carry small amounts of property N (mass, moment, etc.) with them, so that $N_{CS_{in}}$ enters and $N_{CS_{out}}$ leaves the control volume. Defining the outward normal direction to the control volume is positive while the inward normal is negative. Comparing N in the various volumes, we see that $(N_{sys})_t = N_I + (N_{CV})_t + N_{III}$ and $(N_{sys})_{t+\Delta t} = N_I' + (N_{CV})_{t+\Delta t} + N_{III}'$. Subtracting the first equation from the second one, we obtain

$t+\Delta t$ 时刻，部分系统已经流出控制体。

3.3.4　系统和控制体间转换关系

　　物理量间的关系在系统上可以明确的体现，但是在系统上建模远不如在控制体上方便，所以我们非常有必要寻求一个转换方法把系统上的微分关系（拉格朗日法）通过控制体（欧拉法）上的变量表达出来，即，把明确的物理关系直接应用于控制体。

　　流体力学中的大部分定理都是从固体力学的规律推演而来的，而固体力学中关于广延性质时间变化率的规律都是建立在系统之上的。如通过积分形式表示流体的质量

$$m_{sys} = \int_{sys} \rho d\Omega \qquad (3.3.8)$$

式中，$d\Omega$ 是流体质点占有的微元体积；ρ 是流体的密度，其在系统中的分布可能不均匀。

　　我们用符号 N_{sys} 表示系统上可积的一些物理量（向量或标量），如式 3.3.8 右侧所示的质量、动量或能量等。变量 η 为这类可积物理量的密度属性，是单位质量系统所具有的物理量 N。那么，流体的任何广延属性都可以表示为

$$N_{sys} = \int_{sys} \eta \rho d\Omega \qquad (3.3.9)$$

　　例如流体质点的动量可以表示为

$$M_{sys} = \int_{sys} V \rho d\Omega \qquad (3.3.10)$$

　　假设 N 为流体的质量，下标 sys 表示限定为特定系统内流体微团的质量，同样的方式，我们用下标 CV 表示控制体，CS 表示控制面。图 3.3.5 例的控制体中，我们假设流体仅从一个控制面 CS_I 流入，一个控制面 CS_{III} 流出，且流入、流出的方向与控制面垂直，控制面 CS_{II} 上无质量交换。

　　考虑图 3.3.6 的一般情况，系统仅由控制面 CS_I 流入控制体，从 CS_{III} 流出控制体，在控制面 CS_{II} 上无质量交换，在 t 时刻系统和控制体如图 3.3.6a 所示。而经过 Δt 时间间隔后，系统的形状和位置发生了改变；一部分流体经控制面 CS_I 流入控制体，一部分流体经控制面 CS_{III} 流出了控制体，如图 3.3.6b 所示。流入、流出的流体含有一定的物理量 N（质量、动量等），也就是说 $N_{cs_{in}}$ 物理量流入、$N_{cs_{out}}$ 流出了控制体。这里规定控制体的外法向为正，内法向为负，把几部分的物理量 N 相加、减，可得 $(N_{sys})_t = N_I + (N_{CV})_t + N_{III}$ 和 $(N_{sys})_{t+\Delta t} = N_I' + (N_{CV})_{t+\Delta t} + N_{III}'$。计算该时间间隔 Δt 内物理量 N 的变化量，有

$$\Delta\left(N_{\text{sys}}\right)=\left(N_{\text{sys}}\right)_{t+\Delta t}-\left(N_{\text{sys}}\right)_{t}$$
$$=\left(N_{\text{CV}}\right)_{t+\Delta t}-\left(N_{\text{CV}}\right)_{t}+N'_{\text{III}}-N_{\text{III}}+N'_{\text{I}}-N_{\text{I}}$$
$$=\Delta N_{\text{CV}}+N_{\text{CS}_{\text{out}}}+N_{\text{CS}_{\text{in}}} \qquad (3.3.11)$$

dividing by Δt and letting $\Delta t \to 0$, we get

$$\frac{dN_{\text{sys}}}{dt}=\frac{dN_{\text{CV}}}{dt}+\frac{dN_{\text{CS}_{\text{out}}}}{dt}-\frac{dN_{\text{CS}_{\text{in}}}}{dt} \qquad (3.3.12)$$

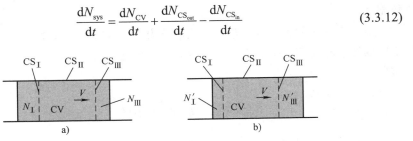

Fig.3.3.6　The relationship of system and control volume
a) System and control volume at time t　b) System and control volume at time $t+\Delta t$

The left-hand side of Eq.3.3.12 is the rate of change of the total amount of any extensive property N within the moving system. The first term at right hand, dN_{CV}/dt, is the rate of change of that property contained within the fixed control volume. The last two terms are the net rate of outflow of N passing through the control surface.Eq.3.3.12 states that the difference between the rate of change of N within the system and that within the control volume is equal to the net rate of outflow through the control surface.

Substituting Eq.3.3.9 into the Eq.3.3.12, and by Eq.3.3.2, rewrite the infinitesimal flow volume dq as the infinitesimal element of control volume dV, where the strikethrough is used to distinguish with the symbol of average velocity, that is $d\cancel{V}=V\cdot n\,dA$, then we get

$$\frac{DN_{\text{sys}}}{Dt}=\frac{d}{dt}\int_{\text{CV}}\rho\eta\,d\cancel{V}+\int_{\text{CS}}\rho\eta V\cdot n\,dA \qquad (3.3.13)$$

where the substantial derivative D/Dt is used since we are following a specified group of substantial particles.

Since the control volume is not changing with time, the time derivative on the right-hand side can be moved inside the integral, it is irrelevant whether we differentiate or integrate first, we have the alternative form of Eq.3.3.13

$$\frac{DN_{\text{sys}}}{Dt}=\int_{\text{CV}}\frac{\partial}{\partial t}(\rho\eta)\,d\cancel{V}+\int_{\text{CS}}\rho\eta V\cdot n\,dA \qquad (3.3.14)$$

These two equations state that, the time rate of change of the property N of the system is equal to the time rate of change of N of the control volume plus the net flux of N crossing the control surface. Addressing the motion of fluid as it moves through the control volume, in contrast to the motion of system, this relationship more formally known as

$$\Delta\left(N_{\text{sys}}\right) = \left(N_{\text{sys}}\right)_{t+\Delta t} - \left(N_{\text{sys}}\right)_{t}$$

$$= \left(N_{\text{CV}}\right)_{t+\Delta t} - \left(N_{\text{CV}}\right)_{t} + N'_{\text{III}} - N_{\text{III}} + N'_{\text{I}} - N_{\text{I}}$$

$$= \Delta N_{\text{CV}} + N_{\text{CS}_{\text{out}}} + N_{\text{CS}_{\text{in}}} \tag{3.3.11}$$

除以时间间隔 Δt 并令 $\Delta t \to 0$，可以得到

$$\frac{\mathrm{d}N_{\text{sys}}}{\mathrm{d}t} = \frac{\mathrm{d}N_{\text{CV}}}{\mathrm{d}t} + \frac{\mathrm{d}N_{\text{CS}_{\text{out}}}}{\mathrm{d}t} - \frac{\mathrm{d}N_{\text{CS}_{\text{in}}}}{\mathrm{d}t} \tag{3.3.12}$$

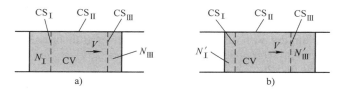

图 3.3.6　系统 - 控制体关系

a）t 时刻的系统和控制体　b）$t+\Delta t$ 时刻的系统和控制体

式 3.3.12 左边代表任一随系统运动的广延物理量 N 的时间变化率；右边第一项，$\mathrm{d}N_{\text{CV}}/\mathrm{d}t$，为固定的控制体内该物理量随时间变化率；另外两项代表物理量 N 经控制面的净出流量。该式说明，物理量 N 在系统和控制体内的变化率之差，等于其在控制面上的净流量。

将式 3.3.9 代入式 3.3.12，并重写式 3.3.2，将式中流体微元体积 $\mathrm{d}q$ 替换为控制体内的微元体积 $\mathrm{d}V$，字母 V 中间的横线是为了将该体积符号与平均流速的符号 V 相区别，因此有 $\mathrm{d}V = V \cdot \boldsymbol{n}\mathrm{d}A$，则系统内物理量 N 变化率为

$$\frac{\mathrm{D}N_{\text{sys}}}{\mathrm{D}t} = \frac{\mathrm{d}}{\mathrm{d}t}\int_{\text{CV}}\rho\eta\mathrm{d}V + \int_{\text{CS}}\rho\eta V \cdot \boldsymbol{n}\mathrm{d}A \tag{3.3.13}$$

由于系统是一定的流体质点，所以式中使用了物质导数 $\mathrm{D}/\mathrm{D}t$ 的表达形式。

因为控制体的体积不随时间变化，所以我们将上式右侧的时间导数移至积分符号内，积分与微分孰先孰后并不会对结果产生影响，这样，我们可得到式 3.3.13 的另一种形式：

$$\frac{\mathrm{D}N_{\text{sys}}}{\mathrm{D}t} = \int_{\text{CV}}\frac{\partial}{\partial t}\left(\rho\eta\right)\mathrm{d}V + \int_{\text{CS}}\rho\eta V \cdot \boldsymbol{n}\mathrm{d}A \tag{3.3.14}$$

以上两式说明，物理量 N 在系统内的时间变化率等于其在控制体内的时间变化率与其通过控制体表面的净通量之和。这就是著名的**雷诺输运定理**，以英国工程师

the *Reynolds transport theorem*, named after the English engineer, Osborne Reynolds (1842—1912), provides the link between the Lagrangian and Eulerian approaches, i.e. a system-to-control volume transformation.

Exercises 3

3.1 What is the definition of a streamline? What do streamlines indicate?

3.2 What is the definition of a pathline? How do pathlines differ from streamlines?

3.3 A fluid particle moves along a pathline or a streamline?

3.4 A stationary probe is placed at one location in a fluid flow and measures pressure and temperature as functions of time. Is this a Lagrangian or Eulerian measurement?

3.5 A buoyant electronic probe is released into the drain pipe and transmits 20 position readings per minute as it flow in the pipe. Is this a Lagrangian or Eulerian measurement?

3.6 Briefly explain the purpose of the Reynolds transport theorem.

3.7 The flow velocity components are defined by $u = 2$, $v = 3$, and $w = 4$. What is the velocity vector of the flow?

3.8 The velocity along a streamline is given by $u(s) = 3+s^{0.6}$. What are the accelerations at $s = 2$ and $s = 5$?

3.9 A velocity field is given by $V = 2xy i - y^2 j$. (a) Find the acceleration vector field. (b) Calculate the magnitude of the acceleration at $(-1, 2)$.

3.10 A flow field is defined by $u = 3x$ and $v = -y$. Derive expressions for the x and y components of acceleration. Find the magnitude of the velocity and acceleration at the point $(1, 2)$.

3.11 A flow field is defined by $u = 3x$ and $v = -xy$. Derive expressions for the x and y components of acceleration. Find the magnitude of the velocity and acceleration at the point $(1, 2)$.

3.12 The velocity of water in a converging pipe is given by $V(x) = 12/(5-x)^2$, calculate the acceleration at $x = 3$.

3.13 A velocity field is given by $V = (2+x)i - y^2 t j - 3z k$. (a) Find the acceleration vector field. Determine the vector of (b) velocity and (c) acceleration of a fluid particle at the origin and at the point $(1, -1, 0)$ for the velocity fields when $t = 2$.

3.14 A flow is defined by $u = 3(1+t)$, $v = 5(2+t)$, $w = 6(3+t)$. What is the velocity of flow at the point $(3, 2, 1)$ at $t = 2$? What is the acceleration at that point at $t = 2$? Whether the flow in this case is a steady flow or a uniform flow?

3.15 A steady, two-dimensional velocity field is given by $V = (3+2x)i + (4-5y)j$. Is there a stagnation point (a point where the local velocity is zero) in this flow field? If so, where is it?

3.16 Consider the steady, incompressible, two-dimensional velocity field of Exercise 3-15, (a) Calculate the acceleration at the point $(1, -3)$. (b) Find the streamline through the point $(1, -3)$.

3.17 The location of the stagnation point in a steady, two-dimensional velocity field which given by $V = (bx - ay)i + (3-by)j$ is at point $(2, -2)$, determine the value of constants a and b.

Osborne Reynolds（1842—1912）命名。雷诺输运定理把拉格朗日法和欧拉法联系起来了，建立了系统到控制体的转换方法。

习题三

3.1　流线的定义是什么？流线代表什么？

3.2　迹线的定义是什么？迹线与流线的区别是什么？

3.3　流体质点沿迹线还是流线运动？

3.4　在流体流动区域内放置一个固定的传感器测量压强和温度随时间变化的数据，这属于拉格朗日法还是欧拉法？

3.5　在排污管内投放浮球随污水流动并每分钟发出 20 次其位置坐标，这属于拉格朗日法还是欧拉法？

3.6　简单解释雷诺输运定理的意义。

3.7　流动速度分量 $u = 2$，$v = 3$ 和 $w = 4$。写出速度向量。

3.8　沿流线的速度为 $u(s) = 3 + s^{0.6}$。计算在点 $s = 2$ 和 $s = 5$ 的加速度。

3.9　已知流动的速度场为 $V = 2xy\boldsymbol{i} - y^2\boldsymbol{j}$。（1）写出加速度场；（2）计算点（-1, 2）处加速度的值。

3.10　某速度场定义为 $u = 3x$，$v = -y$。写出 x 和 y 方向的加速度分量，并计算点（1, 2）处速度和加速度的值。

3.11　某速度场定义为 $u = 3x$，$v = -xy$。写出 x 和 y 方向的加速度分量，并计算点（1, 2）处速度和加速度的值。

3.12　渐缩管内水的流速为 $V(x) = 12/(5-x)^2$，计算 $x = 3$ 处的加速度。

3.13　已知速度场 $V = (2+x)\boldsymbol{i} - y^2\boldsymbol{j} - 3z\boldsymbol{k}$，（1）写出加速度场；（2）分别写出在 $t = 2$ 时刻位于原点和点（1, -1, 0）处的流体质点的速度向量；（3）分别写出在 $t = 2$ 时刻位于原点和点（1, -1, 0）处的流体质点的加速度向量。

3.14　已知流场 $u = 3(1+t)$，$v = 5(2+t)$，$w = 6(3+t)$。（1）计算 $t = 2$ 时点（3, 2, 1）的速度；（2）同时刻该点的加速度；（3）判断该流动为稳定流动还是均匀流动？

3.15　已知二维稳定流动的速度场 $V = (3+2x)\boldsymbol{i} + (4-5y)\boldsymbol{j}$。判断该流场内是否存在驻点（此处流速为零的点）？如果有，在哪里？

3.16　习题 3.15 中的流场，（1）计算点（1, -3）处的加速度；（2）写出经过该点的流线方程。

3.17　已知流场 $V = (bx - ay)\boldsymbol{i} + (3 - by)\boldsymbol{j}$ 的驻点位于（2, -2）处，确定常数 a 和 b 的值。

3.18　The velocity field of a steady, incompressible, two-dimensional flow through a converging duct is given by $V = (u_0 + mx)\mathbf{i} + ny\mathbf{j}$, write (a) the acceleration components and (b) the acceleration vector.

3.19　A steady, incompressible, two-dimensional velocity field is given by components in the xy-plane: $u = 2.5 + 1.732x + 0.707y$, $v = 0.88 - 4.47x - 2.45y$. Find the acceleration field and calculate the acceleration at the point $(1, -1)$.

3.20　Consider the velocity field of a steady, incompressible, plane flow: $V = (5.22 + 3.26x)\mathbf{i} + (4.09 - 8.1y)\mathbf{j}$. Find the expression for the flow streamline which through point $(2, 2)$.

3.21　The velocity of a liquid ($S_g = 1.6$) in a 120-mm-diameter pipeline is 0.8m/s. Calculate the flow rate of (a) volume and (b) mass.

3.22　Water flows through a 10mm-diameter hose in the bottom of a large tank. A boy tries to collect water in a 20Liter bucket. Neglecting frictions, the average velocity on the section of hose is given by $V = \sqrt{2gh}$, here h is the elevation of water surface to the center of the hose. Determine how long it will takes the boy to fulfill the bucket when the water surface is at 2, 1, and 0.5m from the hose?

3.23　We know that for a system, Newton's second law is

$$\sum F = ma = m\frac{\mathrm{d}V}{\mathrm{d}t} = \frac{\mathrm{d}}{\mathrm{d}t}(mV)_{\mathrm{sys}},$$ Use the Reynolds transport theorem and the above equation to derive the linear momentum equation for a control volume.

3.18   已知渐缩管内二维、不可压缩、稳定流动的速度分布 $V = (u_0 + mx)\boldsymbol{i} + ny\boldsymbol{j}$，（1）写出加速度分量；（2）写出加速度向量。

3.19   已知稳定、不可压缩、平面流的速度分布：$u = 2.5 + 1.732x + 0.707y$，$v = 0.88 - 4.47x - 2.45y$。求加速度场和点（1，−1）处的加速度值。

3.20   已 知 稳 定、不 可 压 缩 平 面 流 的 速 度 分 布： $V = (5.22 + 3.26x)\boldsymbol{i} +$ $(4.09 - 8.1y)\boldsymbol{j}$。求过点（2，2）的流线方程。

3.21   比重为 $S_g = 1.6$ 的液体在直径 120mm 的管道内流动，其流速为 0.8m/s，计算其体积流量和质量流量。

3.22   大水罐下方有一个直径 10mm 的出水口，一个小男孩用容积为 20L 的水桶接水，忽略摩擦，出水管截面上的平均流速为 $V = \sqrt{2gh}$，式中 h 为水罐内液面至出水口中心的距离。分别计算当罐内水面位于出水口上方 2m、1m 和 0.5m 时，小男孩装满一桶水所需的时间。

3.23   根据牛顿第二定律可知，系统的动量变化率 $\sum \boldsymbol{F} = m\boldsymbol{a} = m\dfrac{\mathrm{d}V}{\mathrm{d}t} = \dfrac{\mathrm{d}}{\mathrm{d}t}(mV)_{\text{sys}}$。应用上式和雷诺输运定理，写出控制体内的动量变化率。

CHAPTER 4
BASICS OF FLUID DYNAMICS

In this chapter we will introduce the three basic laws for fluid in motion and basic equations commonly used in fluid mechanics: the continuum, the Bernoulli and the momentum. The continuum equation is an expression of the conservation of mass principle. The Bernoulli equation is concerned with the conservation of kinetic, potential, and pressure energies of a fluid stream and their conversion to each other in regions of flow where restrictive conditions apply. The momentum equation for control volumes is used to determine the forces and torques associated with fluid flow.

4.1 Basic Conservation Laws

In this section three basic laws of conservation are introduced, and it is followed by three basic equations for the fluid mechanics. We are already familiar with numerous conservation laws such as the laws of conservation of mass, conservation of energy, and conservation of momentum. The conservation laws are applied to a system, and extended to control volumes. Balance equations are derived from conservation relations since any conserved quantity must balance during a process. We now give a brief description of the conservation of mass and energy relations, and the momentum equation.

4.1.1 Conservation of Mass

The conservation of mass, which states that matter is indestructible, or, the mass of a system remains constant. In fact, mass is conserved even during chemical reactions.

The conservation of mass relation for a closed moving system is expressed as $m_{sys} =$ Constant or

$$\frac{dm_{sys}}{dt} = 0 \tag{4.1.1}$$

In fluid mechanics, the conservation of mass relation involving the rate of change of mass within the system and that within the control volume as while as to the net rate of outflow through the control surface, written for a differential control volume is usually called the continuity equation, is discussed in section 4.2.

4.1.2 Conservation of Energy

The second fundamental law is the conservation of energy, which is also known as the first law of thermodynamics: The total energy of an isolated system remains constant. It is noted that the total energy consists of potential, kinetic, and internal energy, the latter being the energy content due to the temperature of the system. In this book we limit our consideration to mechanical form of energy only.

Energy can be transferred to or from a closed system, and the conservation of energy principle requires that the net energy transfer to or from a system during a process be equal to the change in the energy content of the system. Control volumes involve energy transfer via mass flow also, and the conservation of energy principle is presented in section 4.3.

第 4 章
流体动力学基础

$\mathbf{本}$ 章介绍流体动力学三大基本守恒定律和三个基本方程：连续性方程、伯努利方程和动量方程。连续性方程是质量守恒的体现；伯努利方程描述了严格限制流动条件下流体的动能、位势能和压力能的守恒及相互转化关系；动量方程则用于控制体求解力和力矩。

4.1 基本守恒定律

本节介绍三个基本守恒定律，它们对应着流体力学的三个基本方程。这是我们已经熟悉的几个守恒定律，分别是质量守恒、能量守恒和动量守恒。这些守恒定律是基于系统的，同样也延伸到控制体。基于物理量在整个流动过程中的守恒，我们从中导出了几个平衡方程。我们首先简单介绍一下质量、能量和动量的守恒关系。

4.1.1 质量守恒

质量守恒的内容是指物质是不会消失的，或者说，系统的质量是恒定不变的。事实上，物质的质量甚至在化学反应的过程中也依然是守恒的。

一个封闭的运动系统，其质量守恒关系可以描述为 m_{sys} = 常数，或

$$\frac{\mathrm{d}m_{sys}}{\mathrm{d}t} = 0 \qquad (4.1.1)$$

流体力学中的质量守恒关系可以用系统质量的变化率表示，也可以用微控制体内质量的变化量和控制体表面的质量通量来表示，称为连续性方程，我们将在 4.2 节中详细学习。

4.1.2 能量守恒

第二个基本守恒关系是能量守恒，也就是通常所说的热力学第一定律：一个孤立系统的总能量是恒定不变的。这里的总能量是指包括势能、动能和内能，后者取决于系统的温度。本书中，我们仅讨论机械能的守恒。

能量可以传入或传出封闭系统，能量守恒定律就是指系统运动过程中，传入或传出系统的净能量等于系统内能量的变化量。系统的能量随具有质量的流体一起流经控制体，关于能量守恒方程的内容见 4.3 节。

4.1.3 Conservation of Momentum

In fluid mechanics, Newton's second law is usually referred to as the momentum equation. Newton's second law states that: The momentum of a system remains constant only when the net force acting on it is zero, and thus the momentum of such a system is conserved. This is known as the conservation of momentum principle. It has another statement: The acceleration of a body is proportional to the net force acting on it and is inversely proportional to its mass (Eq.4.1.2).

$$F = ma \tag{4.1.2}$$

In fluid mechanics, Newton's second law is usually referred to as the linear momentum equation. The conservation of momentum principle states that: The momentum of a system remains constant only when the net force acting on it is zero, and thus the momentum of such a system is conserved. The product of the mass and the velocity of a body is called the momentum of the body. The momentum of a rigid body of mass m moving with a velocity V is mV.

In fluid mechanics, however, the net force acting on a system is typically not zero, and we prefer to work with the linear momentum equation rather than the conservation of momentum principle. And then the Newton's second law can also be stated as: The sum of all external forces acting on a system is equal to the time rate of change of momentum of the system, this can be expressed as Eq.4.1.3.

$$F = \frac{\mathrm{d}(mV)}{\mathrm{d}t} \tag{4.1.3}$$

Newton's second law for rotating rigid bodies can be stated as: The flow rate of change of the angular momentum of a body is equal to the net torque acting on it, and is expressed as

$$M = I\alpha = \frac{\mathrm{d}H}{\mathrm{d}t} = I\frac{\mathrm{d}\omega}{\mathrm{d}t} = \frac{\mathrm{d}(I\omega)}{\mathrm{d}t} \tag{4.1.4}$$

where M is the net moment or torque applied on the body, I is the moment of inertia of the body about the axis of rotation, α is the angular acceleration, H is the angular momentum, and ω is the angular velocity.

Therefore, the resultant force or moment acting on a system equals the rate at which the momentum of the system is changing, the equation of momentum is further discussed in section 4.5.

4.2 Equation of Continuity

The conservation of mass principle for closed systems is implicitly used by requiring that the mass of the system remain constant during a process, tracking the substantial fluid particles, however, is not an easy task.

4.2.1 Continuity equation in integral form

For control volumes, instead, mass can cross the boundaries, and so we may keep track of the amount of mass entering and leaving the control volume. The conservation of mass principle for a control volume can be expressed as: The net mass transform to or from a control volume during a time interval Δt is equal to the net change (increase or decrease) of the total mass within the control volume during Δt, that is referred to as mass balance expressed by:

4.1.3　动量守恒

　　流体力学中的动量方程对应牛顿第二定律。牛顿第二定律的内容是：当系统所受的合外力为零时其动量保持不变，也就是系统的动量守恒。这也被称作动量守恒定律。它的另一种方式是：运动物体的加速度与其所受的合外力成正比，与物体的质量成反比（见式 4.1.2）

$$F = ma \tag{4.1.2}$$

　　牛顿第二定律的思想在流体力学中更经常通过动量方程的形式体现。动量守恒定律的内容是：系统所受的合外力为零时其动量保持不变，即系统的动量守恒。物体质量和运动速度的乘积称为动量。一个质量为 m、运动速度为 V 的刚性物体，其动量等于 mV。

　　但是一个流体系统所受的外力往往并不为零，所以在具体的分析中，更适用的是动量方程而不是动量守恒定律。这时，牛顿第二定律的解读为：一个系统所受的外力之和等于系统动量随时间的变化率，见式 4.1.3

$$F = \frac{\mathrm{d}(mV)}{\mathrm{d}t} \tag{4.1.3}$$

　　对于旋转的刚性物体，牛顿第二定律可以解释为：物体角动量的变化率等于物体所受的合力矩，可以表示为

$$M = I\alpha = \frac{\mathrm{d}H}{\mathrm{d}t} = I\frac{\mathrm{d}\omega}{\mathrm{d}t} = \frac{\mathrm{d}(I\omega)}{\mathrm{d}t} \tag{4.1.4}$$

式中，M 是物体所受的合力矩；I 是物体绕某一轴转动的惯性矩；α 是角加速度；H 是角动量；ω 是角速度。

　　所以，系统所受的合外力或合力矩等于系统动量或动量矩的变化率，我们将在 4.5 节对动量方程做深入讨论。

4.2　连续性方程

　　封闭系统的质量守恒定律描述了系统在运动过程中质量的守恒，这也就意味着我们要跟踪流体质点的移动来记录其质量的变化，但这是非常困难的。

4.2.1　积分形式的连续性方程

　　而对于控制体，由于流体质量可以通过其表面，所以我们只需跟踪记录流入和流出控制体的流体质量。这样，我们可以把质量守恒关系通过控制体来描述：在时间间隔 Δt 内流入和流出控制体的质量净通量等于控制体内在同样时间间隔 Δt 内流体质量的变化量，这种质量守恒关系式如下：

$$\dot{m}_{CV_{out}} - \dot{m}_{CV_{in}} = \frac{dm_{CV}}{dt} \tag{4.2.1}$$

where $\dot{m}$ is the mass flow rate, the amount of mass flowing through a cross section per unit time, the dot over a symbol is used to indicate time rate of change.

Consider a control volume of arbitrary shape, the mass of a differential volume $d\mathcal{V}$ within the control volume is $dm = \rho d\mathcal{V}$. The total mass within the control volume at any instant in time t is determined by integration to be

$$m_{CV} = \int_{CV} \rho d\mathcal{V} \tag{4.2.2}$$

Then the time rate of change of mass within the control volume is expressed as

$$\frac{dm_{CV}}{dt} = \frac{d}{dt} \int_{CV} \rho d\mathcal{V} \tag{4.2.3}$$

Consider mass flow into or out of the control volume through a differential area dA on the control surface of a fixed control volume, rearranging the volume flow rate passing through the element of area dA shown in Eq.3.3.2, the differential mass flow rate is given as

$$\delta \dot{m} = \rho (V \cdot n) dA = \rho V_n dA \tag{4.2.4}$$

The net flow rate into or out of the control volume through the entire control surface is obtained by integrating $\delta \dot{m}$ over the entire control surface

$$\dot{m}_{net} = \int_{CS} \delta \dot{m} = \int_{CS} \rho V_n dA \tag{4.2.5}$$

Note that V_n is positive for outward normal direction (outflow) and negative for inward normal direction (inflow). Therefore, the direction of flow is automatically accounted for, a positive value for $\dot{m}_{net}$ indicates a net outflow of mass and a negative value indicates a net inflow of mass.

By the conservation of mass principle, the general conservation of mass relation for a fixed control volume is then expressed as

$$\frac{d}{dt} \int_{CV} \rho d\mathcal{V} + \int_{CS} \rho V_n dA = 0 \tag{4.2.6}$$

It states that the time rate of change of mass within the control volume equal to net mass flow rate through the control surface.

The general conservation of mass relation for a control volume can also be derived using the Reynolds transport theorem (RTT) by taking the property N to be the mass m (section 3.3.4). Then we have $\eta = 1$ since dividing mass by mass to get the property per unit mass gives unity. Therefore Eq.3.3.13 is rewriten as

$$\frac{dm_{sys}}{dt} = \frac{d}{dt} \int_{CV} \rho d\mathcal{V} + \int_{CS} \rho V_n dA \tag{4.2.7}$$

Also the conservation of mass relation for a closed moving system by Eq.4.1.1 is $dm_{sys}/dt = 0$, then the Reynolds transport equation reduces immediately to Eq.4.2.6, and it illustrates that the Reynolds transport theorem is a very useful tool.

$$\dot{m}_{\text{CV}_{\text{out}}} - \dot{m}_{\text{CV}_{\text{in}}} = \frac{\mathrm{d}m_{\text{CV}}}{\mathrm{d}t} \qquad (4.2.1)$$

式中，$\dot{m}$ 是流体质量的变化量，是单位时间内通过某有效截面的流体质量，质量符号 m 上方的点表示该变量是随时间的变化率。

对于任意形状的控制体，体积为 $\mathrm{d}V$ 的微小控制体内流体的质量等于 $\mathrm{d}m = \rho\mathrm{d}V$。在任意时刻 t 控制体内流体的总质量可由积分得到

$$m_{\text{CV}} = \int_{\text{CV}} \rho\mathrm{d}V \qquad (4.2.2)$$

那么，控制体内流体质量随时间的变化率可以表示为

$$\frac{\mathrm{d}m_{\text{CV}}}{\mathrm{d}t} = \frac{\mathrm{d}}{\mathrm{d}t}\int_{\text{CV}} \rho\mathrm{d}V \qquad (4.2.3)$$

在固定控制体的控制面上取一面积微元 $\mathrm{d}A$，根据式 3.3.2，通过 $\mathrm{d}A$ 面的体积流量可改写为质量流量

$$\delta\dot{m} = \rho(V \cdot n)\mathrm{d}A = \rho V_{\text{n}}\mathrm{d}A \qquad (4.2.4)$$

对 $\delta\dot{m}$ 在控制面上积分，可以得到通过全部控制面流入和流出控制体的质量净通量

$$\dot{m}_{\text{net}} = \int_{\text{CS}} \delta\dot{m} = \int_{\text{CS}} \rho V_{\text{n}}\mathrm{d}A \qquad (4.2.5)$$

这里规定 V_{n} 沿外法线方向流出控制体为正，沿内法线方向流入为负。所以计算时流体质点的流动方向会自动计入，如果 $\dot{m}_{\text{net}}$ 为正值，表示有净质量的流体流出控制体，$\dot{m}_{\text{net}}$ 为负则表示有净质量的流体流入控制体。

根据质量守恒定律，对于固定控制体的总质量守恒关系可以表示为

$$\frac{\mathrm{d}}{\mathrm{d}t}\int_{\text{CV}} \rho\mathrm{d}V + \int_{\text{CS}} \rho V_{\text{n}}\mathrm{d}A = 0 \qquad (4.2.6)$$

上式的意义为：控制体内流体质量随时间的变化率等于通过控制面的质量净通量。

控制体内流体质量守恒的关系式也可以通过雷诺输运定理导出。把定理中的物理量 N 用质量 m 代替，密度属性 $\eta = 1$，因为质量与其自身相除就是单位值。式 3.3.13 可改写为

$$\frac{\mathrm{d}m_{\text{sys}}}{\mathrm{d}t} = \frac{\mathrm{d}}{\mathrm{d}t}\int_{\text{CV}} \rho\mathrm{d}V + \int_{\text{CS}} \rho V_{\text{n}}\mathrm{d}A \qquad (4.2.7)$$

同样应用式 4.1.1 封闭的运动系统中 $\mathrm{d}m_{\text{sys}}/\mathrm{d}t = 0$，雷诺输运定理可以化简为式 4.2.6，这也再次证明了雷诺输运定理的应用优势。

Many engineering devices such as nozzles, diffusers, turbines, compressors, and pumps involve only one inlet and one outlet. For these cases, we typically denote the inlet state by area A_1 across which fluid enters the control volume and the outlet state by area A_2 across which fluid leaves the control volume (a single stream), and the velocity vector is normal to the area, V_n simply becomes the magnitude of the average velocity V. Then the general conservation of mass relation Eq.4.2.6 reduces to

$$\frac{d}{dt}\int_{CV}\rho dV + \int_{A_2}\rho_2 V_2 dA - \int_{A_1}\rho_1 V_1 dA = 0 \tag{4.2.8}$$

There is considerable flexibility in the selection of a control volume, however a control volume should not introduce any unnecessary complications. A simple rule in selecting a control volume is to make the control surface normal to the flow at all locations where it crosses the fluid flow (Fig.4.2.1), whenever possible, this can make the solution of problem rather easy.

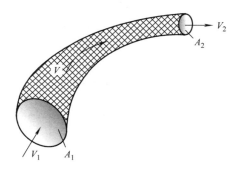

Fig.4.2.1   Portion of single stream tube as control volume

In engineering many flows of interest are steady flows, so that $\frac{d}{dt}\int_{CV}\rho dV = 0$, Eq.4.2.8 then takes the form

$$\int_{A_1}\rho_1 V_1 dA = \int_{A_2}\rho_2 V_2 dA \tag{4.2.9}$$

For a uniform flow with one entrance and one exit, there results

$$\rho_1 V_1 A_1 = \rho_2 V_2 A_2 \tag{4.2.10}$$

During a steady-flow process where density is constant, that is $\rho_1 = \rho_2 = \rho$, the continuity equation (4.2.6) then reduces to

$$V_1 A_1 = V_2 A_2 \quad \text{or} \quad \frac{V_1}{V_2} = \frac{A_2}{A_1} \tag{4.2.11}$$

This integral form of the continuity equation is used quite often, particularly with liquids and low-speed gas flows. Where V_1 and V_2 are the average velocities over the areas at section 1 and 2, it should be kept in mind, however, that actual velocity profiles are usually not uniform.

很多工程应用如喷头、导流装置、涡轮、压缩机和泵等，仅涉及一个入口和一个出口。我们通常用 A_1 表示流体流入控制体的入口通流面积，用 A_2 表示流出控制体的出口截面积。而此时垂直于通流面的平均速度 V 的法向分量 V_n 就等于速度的值。这样，式 4.2.6 的质量守恒关系式就可以化简为

$$\frac{\mathrm{d}}{\mathrm{d}t}\int_{\mathrm{CV}}\rho\mathrm{d}\mathcal{V}+\int_{A_2}\rho_2 V_2\mathrm{d}A-\int_{A_1}\rho_1 V_1\mathrm{d}A=0 \qquad (4.2.8)$$

选择控制体的方式有很多，但是应该以不增加问题的复杂性为准。选择控制体的一个最简单的规则是选取的控制面尽可能垂直于流经此处的流体的流动方向，如图 4.2.1 所示，这样可以极大地简化问题。

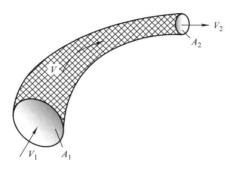

图 4.2.1　选取部分流管作为控制体

工程应用中很多问题涉及的是稳定流动，那么 $\dfrac{\mathrm{d}}{\mathrm{d}t}\displaystyle\int_{\mathrm{CV}}\rho\mathrm{d}\mathcal{V}=0$，式 4.2.8 可以写为

$$\int_{A_1}\rho_1 V_1\mathrm{d}A=\int_{A_2}\rho_2 V_2\mathrm{d}A \qquad (4.2.9)$$

对于只有一个入口和一个出口的均匀流动，有

$$\rho_1 V_1 A_1 = \rho_2 V_2 A_2 \qquad (4.2.10)$$

稳定流动的密度为常数，即 $\rho_1 = \rho_2 = \rho$。式 4.2.6 的连续性方程可进一步化简为

$$V_1 A_1 = V_2 A_2 \quad \text{或} \quad \frac{V_1}{V_2}=\frac{A_2}{A_1} \qquad (4.2.11)$$

上式是连续性方程的积分形式，也是最常用的形式，尤其对于液体和低速流动的气体。式中 V_1 和 V_2 是流体流经截面 1 和截面 2 时的平均流速，这一点一定要注意，因为实际流体的速度分布往往是不均匀的。

Example 4.2.1 Water flows at a uniform velocity of 3m/s into a reducing pipe that changes the diameter from 200 mm to 100 mm (Fig.X4.2.1). Calculate the water's (a) average velocity leaving the smaller area, and (b) the flow rate.

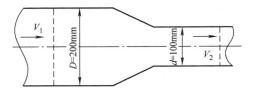

Fig.X4.2.1

Solution The control volume is selected inside of the pipe as shown.
(a) Eq.4.2.11 is used

$$\frac{V_1}{V_2} = \frac{A_2}{A_1} = \frac{d^2}{D^2}$$

It is given that $V_1 = 3$m/s, then $V_2 = 12$m/s.
(b) The flow rate is found to be

$$q = VA = \frac{\pi D^2}{4} V_1 = 0.0942 \text{m}^3/\text{s}$$

This first example represents the primary use of the continuity equation. It allows us to calculate the velocity at one section if it is known at another section. It is solved by first selecting a control volume. There often is only one proper choice for the control volume. We must position the inlet and exit areas at locations where the integrands are either known or where they can be approximated; also, the quantity is often included at an inlet or exit area.

Example 4.2.2 A 1-m-diameter cylindrical water tank whose top is open to the atmosphere is initially filled with water of 1.8-m-high. Now the discharge plug near the bottom of the tank is pulled out, and a water jet whose diameter is 20mm streams out (Fig.X.4.2.2). The average velocity of the jet is approximated as $V = \sqrt{2gh}$, where h is the height of water in the tank measured from the center of the hole (a variable) and g is the gravitational acceleration. Determine how long it takes for the water level in the tank to drop to half the initial height. Note that the distance between the bottom of the tank and the center of the hole is negligible compared to the total water height.

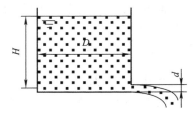

Fig.X4.2.2

例 4.2.1 渐缩管道大入口端直径为 200mm，小出口端直径为 100mm，水以 3m/s 的均匀流速流入（见图 X4.2.1）。计算（1）从小端流出时的平均流速，（2）管道内流体的流量。

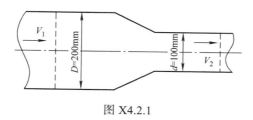

图 X4.2.1

解 取控制体在变径部分两端，如图 X4.2.1 所示。
（1）应用式 4.2.11 得

$$\frac{V_1}{V_2} = \frac{A_2}{A_1} = \frac{d^2}{D^2}$$

已知 $V_1 = 3$m/s，则有 $V_2 = 12$m/s。
（2）计算体积流量

$$q = VA = \frac{\pi D^2}{4} V_1 = 0.0942 \text{m}^3/\text{s}$$

例 4.2.1 是连续性方程最基础的应用。用于已知一截面流速的情况下计算另一截面上的流速。求解时首先需要选取控制体，通常最合适的选取方案是，流入、流出控制面上的变量都是已知或可以近似确定的。当然，待求解的变量也要在控制体范围内。

例 4.2.2 直径为 1m 的圆柱形水罐，顶部连通大气，罐内最初水位 1.8m 高。现取下罐底处直径为 20mm 的塞子使水柱喷出（见图 X4.2.2）。射流的平均流速 $V = \sqrt{2gh}$，式中 h 为罐内水位至出口中心的高度（变化值），g 是重力加速度。试计算何时水位可下降至最初高度的一半。罐底至出口中心的高度可忽略不计。

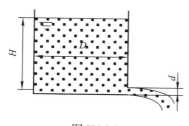

图 X4.2.2

Solution   It is given that:the initial water level height in the tank is $H = 1.8$m, diameter of water tank $D = 1$m, diameter of discharge hole $d = 20$mm.

The volume occupied by water is taken as the control volume. The control volume is variable since it decreases in the case as the water level drops.

The conservation of mass for a control volume is given in rate form as Eq.4.2.1

$$\dot{m}_{CV_{in}} - \dot{m}_{CV_{out}} = \frac{dm_{CV}}{dt}$$

During the discharging process no mass enters the control volume ($\dot{m}_{CV_{in}} = 0$), and the mass outflow rate is

$$\dot{m}_{CV_{out}} = \rho VA = \rho\sqrt{2gh}\,\frac{\pi d^2}{4}$$

The mass of water in the tank at any time is

$$m_{CV} = \rho A_T h = \rho h \frac{\pi D^2}{4}$$

Substituting the abovementioned two equations into the mass balance relation (Eq.4.2.1) gives

$$\left(-\rho\sqrt{2gh}\,\frac{\pi d^2}{4}\right)dt = \left(\rho\frac{\pi D^2}{4}\right)dh \;\rightarrow\; -dt = \frac{1}{\sqrt{2gh}}\frac{D^2}{d^2}dh$$

$$\int_0^t dt = \int_H^{\frac{H}{2}} -\frac{1}{\sqrt{2gh}}\frac{D^2}{d^2}dh \;\rightarrow\; t = \frac{2}{\sqrt{2\times9.8}}\left(\frac{1}{0.02}\right)^2\left(\sqrt{1.8}-\sqrt{0.9}\right)s = 443.8s = 7.4\text{min}$$

Therefore, it takes 7.4 min after the discharge hole is unplugged for half of water in the tank have been discharged.

Using the same relation with $h_2 = 0$ gives $t = 25.25$min for the discharge of the entire amount of water in the tank. Therefore, emptying the bottom half of the tank takes much longer than emptying the top half. This is due to the decrease in the average discharge velocity ($V = \sqrt{2gh}$) with decreasing h.

Example 4.2.3   We want to determine the rate at which the hydraulic oil level rises in an oil tank if the oil returning in through a 12-cm^2 pipe has a velocity of 0.5m/s and the flow rate pumping out is 15L/min (Fig.X4.2.3). The tank has a section A_3 of 0.5m^2.

解　已知，罐内水位初始高度 $H = 1.8\mathrm{m}$，水罐内径 $D = 1\mathrm{m}$，出水口直径 $d = 20\mathrm{mm}$。

取水所占据的体积为控制体。控制体的体积随水位的下降而减小。

应用式 4.2.2 控制体内水的质量变化率来描述其质量守恒关系

$$\dot{m}_{CV_{in}} - \dot{m}_{CV_{out}} = \frac{dm_{CV}}{dt}$$

排水过程中，没有质量流入控制体（$\dot{m}_{CV_{in}} = 0$），流出的质量流量为

$$\dot{m}_{CV_{out}} = \rho V A = \rho \sqrt{2gh}\frac{\pi d^2}{4}$$

罐内水的质量变化规律为

$$m_{CV} = \rho A_T h = \rho h \frac{\pi D^2}{4}$$

把前面两个变量代入质量平衡关系式 4.2.1，得到

$$\left(-\rho\sqrt{2gh}\frac{\pi d^2}{4}\right)dt = \left(\rho\frac{\pi D^2}{4}\right)dh \;\rightarrow\; -dt = \frac{1}{\sqrt{2gh}}\frac{D^2}{d^2}dh$$

$$\int_0^t dt = \int_H^{\frac{H}{2}} -\frac{1}{\sqrt{2gh}}\frac{D^2}{d^2}dh \;\rightarrow\; t = \frac{2}{\sqrt{2\times9.8}}\left(\frac{1}{0.02}\right)^2\left(\sqrt{1.8}-\sqrt{0.9}\right)s = 443.8s = 7.4\min$$

所以，取下塞子 7.4min 后，罐内一半的水被排放出去。

同样的方法，取积分上限为 0 可得出 $t = 25.25\min$ 可排出罐内全部水。可见排出下半部分水所需的时间长于排放上半部分水的耗时。这是由于排水口流速与剩余水位高度 h 有关（$V = \sqrt{2gh}$）。

例 4.2.3　如图 X4.2.3 所示的液压油箱，回油管截面积为 $12\mathrm{cm}^2$，管内油液平均流速为 0.5m/s。同时液压泵以 15L/min 的流量从油箱内吸油。试计算油箱内液压油液面上升的速度。已知油箱的底面积 A_3 为 $0.5\mathrm{m}^2$。

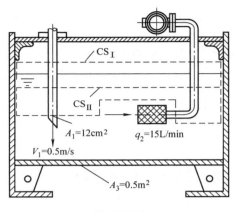

Fig.X4.2.3

Solution 1   First we select a control volume that extends above the oil surface as CS_I shown in Fig.X4.2.3. Apply the continuity equation Eq.4.2.8

$$\frac{d}{dt}\int_{CV}\rho d\mathcal{V} + \int_{A_2}\rho_2 V_2 dA - \int_{A_1}\rho_1 V_1 dA = 0 \;\rightarrow\; \frac{d}{dt}\int_{CV}\rho d\mathcal{V} + \rho V_1 A_1 - \rho q_2 = 0$$

In which the first term discribes the flow rate of change of mass in the control volume, neglecting the air mass above the oil.

Divided by the constant ρ, we have

$$A_3\frac{dh}{dt} + V_1 A_1 - q_2 = 0$$

The flow rate at which the oil level rises is then

$$\frac{dh}{dt} = \frac{q_2 - V_1 A_1}{A_3} = \frac{15\times10^{-3}/60 - 0.5\times12\times10^{-4}}{0.5}\,\text{m/s} = -0.7\text{mm/s}$$

The negative sign indicates that the oil level is actually decreasing.

Solution 2   We have another choice for the control volume, one with its top surface below the oil level as CS_{II} shown in Fig.X4.2.3.

The velocity at the top surface is then equal to the flow rate at which the surface rising, i.e., dh/dt. The flow condition inside the control volume is steady. Hence we can apply Eq.4.2.11, there are three areas across which fluid flows. On area CS_{II} oil flow out from the control volume, the velocity is dh/dt, hence the continuity equation takes the form

$$V_1 A_1 - q_2 - \frac{dh}{dt}A_3 = 0 \;\rightarrow\; \frac{dh}{dt} = \frac{q_2 - V_1 A_1}{A_3} = -0.7\text{mm/s}$$

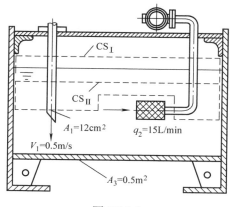

图 X4.2.3

解法 1　首先选取控制体的控制面 CS_I 高于油箱内油液的自由液面，如图 X4.2.3 所示。

应用连续性方程式 4.2.8

$$\frac{d}{dt}\int_{CV}\rho d\mkern-11mu\mathchar'26\mkern2mu V + \int_{A_2}\rho_2 V_2 dA - \int_{A_1}\rho_1 V_1 dA = 0 \quad \rightarrow \quad \frac{d}{dt}\int_{CV}\rho d\mkern-11mu\mathchar'26\mkern2mu V + \rho V_1 A_1 - \rho q_2 = 0$$

式中左侧第一项表示控制体内油液质量的变化率，忽略上部空气质量。

等式两侧除以常数密度 ρ，得到

$$A_3\frac{dh}{dt} + V_1 A_1 - q_2 = 0$$

那么，油箱内液压油液面上升速度为

$$\frac{dh}{dt} = \frac{q_2 - V_1 A_1}{A_3} = \frac{15\times10^{-3}/60 - 0.5\times12\times10^{-4}}{0.5}\text{m/s} = -0.7\text{mm/s}$$

负号说明油液的液面实际是下降的。

解法 2　选取控制体最上方的控制面 CS_{II} 始终低于油液的自由液面，如图 X4.2.3 所示。

此时上控制面处油液的流速等于油箱内液压油液面上升的速度，即 dh/dt。油箱内油液的流动是稳定流动，可以应用式 4.2.11。油液流过三个截面，从截面 CS_{II} 流出控制体，其流速为 dh/dt，连续性方程可改写为

$$V_1 A_1 - q_2 - \frac{dh}{dt} A_3 = 0 \quad \rightarrow \quad \frac{dh}{dt} = \frac{q_2 - V_1 A_1}{A_3} = -0.7\text{mm/s}$$

This is the same result as given in solution 1.

This example shows that there may be more than one practicable choice for a control volume.

4.2.2 Differential Continuity Equation

In previous section, the continuity equation was expressed in terms of a fixed control volume. Now we begin the partial differential equation by applying the conservation of mass to a small volume in a fluid flow. Consider the mass flux through each surface of the fixed infinitesimal control volume with length of each side dx, dy and dz, shown in Fig.4.2.2. We set the net flux of mass entering the volume equal to the flow rate of change of the mass inside the volume, that is

$$m_{\text{out}} - m_{\text{in}} = \dot{m}_{\text{CV}} \tag{4.2.12}$$

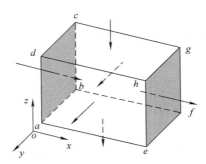

Fig.4.2.2   Mass flux through each surface of the fixed infinitesimal control volume.

Consider the mass flux through the surfaces a-b-c-d and e-f-g-h along x-direction in the interval time of Δt. We take the outflow direction is positive, then the mass flux in the x-direction is

$$\rho_{2x} u_2 \mathrm{d}t \mathrm{d}y \mathrm{d}z - \rho_{1x} u_1 \mathrm{d}t \mathrm{d}y \mathrm{d}z = \dot{m}_x \tag{4.2.13}$$

where ρ_{2x} and ρ_{1x} are the density of fluid at center of surface a-b-c-d and e-f-g-h respectively, and u_2, u_1 are velocity components at corresponding location.

The product of ρu is considered a variable, and put the Eq.4.2.13 in form

$$\frac{\partial(\rho u)}{\partial x} \mathrm{d}x \mathrm{d}y \mathrm{d}z \mathrm{d}t = \dot{m}_x \tag{4.2.14}$$

Similarly, we can express the total mass flux through the six surfaces of the infinitesimal element as

$$\frac{\partial(\rho u)}{\partial x} \mathrm{d}x \mathrm{d}y \mathrm{d}z \mathrm{d}t + \frac{\partial(\rho v)}{\partial y} \mathrm{d}x \mathrm{d}y \mathrm{d}z \mathrm{d}t + \frac{\partial(\rho w)}{\partial z} \mathrm{d}x \mathrm{d}y \mathrm{d}z \mathrm{d}t = \dot{m}_x + \dot{m}_y + \dot{m}_z \tag{4.2.15}$$

Refer to Eq.4.2.12, the flow rate of change of the mass inside the volume which due to the change of density with time can be written as

与解法 1 的结果相同。

此例说明，控制体可以有不止一种可行的选择方案。

4.2.2　微分形式的连续性方程

前一节，我们在控制体上表达了积分形式的连续性方程。接下来我们应用质量守恒定律导出体积微元内的微分方程。如图 4.2.2 所示位置固定的微控制体，边长分别为 dx、dy 和 dz。通过控制体表面的质量净通量应该等于控制体内质量的变化率

$$m_{\text{out}} - m_{\text{in}} = \dot{m}_{\text{CV}} \tag{4.2.12}$$

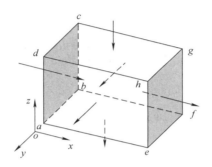

图 4.2.2　通过微控制体表面的质量通量

分析在 Δt 时间间隔内，沿 x 方向由 $abcd$ 面流入、从 $efgh$ 面流出的流体质量。取流出方向为正，则 x 方向的质量通量为

$$\rho_{2x} u_2 \mathrm{d}t\mathrm{d}y\mathrm{d}z - \rho_{1x} u_1 \mathrm{d}t\mathrm{d}y\mathrm{d}z = \dot{m}_x \tag{4.2.13}$$

式中，ρ_{2x} 和 ρ_{1x} 分别是 $abcd$ 面和 $efgh$ 面中心处的流体密度；u_2 和 u_1 为对应点的 x 方向速度分量。

把 ρu 当作一个变量，式 4.2.13 可以写为

$$\frac{\partial(\rho u)}{\partial x}\mathrm{d}x\mathrm{d}y\mathrm{d}z\mathrm{d}t = \dot{m}_x \tag{4.2.14}$$

同样地，可以写出流体微元六个表面上的质量净通量之和

$$\frac{\partial(\rho u)}{\partial x}\mathrm{d}x\mathrm{d}y\mathrm{d}z\mathrm{d}t + \frac{\partial(\rho v)}{\partial y}\mathrm{d}x\mathrm{d}y\mathrm{d}z\mathrm{d}t + \frac{\partial(\rho w)}{\partial z}\mathrm{d}x\mathrm{d}y\mathrm{d}z\mathrm{d}t = \dot{m}_x + \dot{m}_y + \dot{m}_z \quad (4.2.15)$$

根据式 4.2.12，控制体内由于密度随时间变化而引起的质量变化率可以表示为

$$\dot{m}_{CV} = \rho_{t+\Delta t}dxdydz - \rho_t dxdydz$$
$$= \frac{\partial \rho}{\partial t}dtdxdydz \tag{4.2.16}$$

Combining the Eq.4.2.16 and 4.2.15, the mass conservation of the fluid element takes the form

$$\frac{\partial(\rho u)}{\partial x}dxdydzdt + \frac{\partial(\rho v)}{\partial y}dxdydzdt + \frac{\partial(\rho w)}{\partial z}dxdydzdt = \frac{\partial \rho}{\partial t}dtdxdydz \tag{4.2.17}$$

Dividing by $dxdydzdt$, yields

$$\frac{\partial(\rho u)}{\partial x} + \frac{\partial(\rho v)}{\partial y} + \frac{\partial(\rho w)}{\partial z} = \frac{\partial \rho}{\partial t} \text{ or } \frac{\partial(\rho u)}{\partial x} + \frac{\partial(\rho v)}{\partial y} + \frac{\partial(\rho w)}{\partial z} - \frac{\partial \rho}{\partial t} = 0 \tag{4.2.18}$$

This is the most general form of the differential continuity equation expressed in Cartesian coordinates, it states that: The variation of fluid mass in an infinitesimal fluid element is zero.

For steady flow, the time differential term set to be zero, we have

$$\frac{\partial(\rho u)}{\partial x} + \frac{\partial(\rho v)}{\partial y} + \frac{\partial(\rho w)}{\partial z} = 0 \tag{4.2.19}$$

Furtherly, for incompressible fluid which ρ = constant, dividing by ρ, the differential continuity equation is reduce to

$$\frac{\partial u}{\partial x} + \frac{\partial v}{\partial y} + \frac{\partial w}{\partial z} = 0 \tag{4.2.20}$$

We can introduce the gradient operator in rectangular coordinates, is

$$\nabla = \frac{\partial}{\partial x}\boldsymbol{i} + \frac{\partial}{\partial y}\boldsymbol{j} + \frac{\partial}{\partial z}\boldsymbol{k} \tag{4.2.21}$$

Then the continuity equation for steady flow of incompressible fluid can be written in the form

$$\nabla \cdot V = 0 \tag{4.2.22}$$

where $V = ui + vj + wk$, $\nabla \cdot V$ is called the divergence of the velocity.

The divergence of the velocity vector is zero for an incompressible steady flow.

This form of the continuity equation does not refer to any particular coordinate system, and thus is used to express the differential continuity equation using various coordinate systems. In cylindrical coordinate the differential continuity equation for an incompressible steady flow is presented as

$$\frac{1}{r}\frac{\partial(rv_r)}{\partial r} + \frac{1}{r}\frac{\partial v_\theta}{\partial \theta} + \frac{\partial v_z}{\partial z} = 0 \tag{4.2.23}$$

Example 4.2.4 The velocity distribution in an incompressible steady flow is given

$$\dot{m}_{CV} = \rho_{t+\Delta t}dxdydz - \rho_t dxdydz$$
$$= \frac{\partial \rho}{\partial t}dtdxdydz$$

（4.2.16）

联立式 4.2.16 和式 4.2.15，流体微元的质量守恒可以表示为

$$\frac{\partial(\rho u)}{\partial x}dxdydzdt + \frac{\partial(\rho v)}{\partial y}dxdydzdt + \frac{\partial(\rho w)}{\partial z}dxdydzdt = \frac{\partial \rho}{\partial t}dtdxdydz \quad （4.2.17）$$

两边都除以 $dxdydzdt$，得

$$\frac{\partial(\rho u)}{\partial x} + \frac{\partial(\rho v)}{\partial y} + \frac{\partial(\rho w)}{\partial z} = \frac{\partial \rho}{\partial t}，或 \frac{\partial(\rho u)}{\partial x} + \frac{\partial(\rho v)}{\partial y} + \frac{\partial(\rho w)}{\partial z} - \frac{\partial \rho}{\partial t} = 0 \quad （4.2.18）$$

这是直角坐标系下微分形式连续性方程的一般式，其意义为：流体微元内流体质量的变化量为零。

对于稳定流动，时间微分项为零，则有

$$\frac{\partial(\rho u)}{\partial x} + \frac{\partial(\rho v)}{\partial y} + \frac{\partial(\rho w)}{\partial z} = 0$$

（4.2.19）

对于不可压缩流体，$\rho =$ 常数，等式两端除以 ρ，则微分方程化简为

$$\frac{\partial u}{\partial x} + \frac{\partial v}{\partial y} + \frac{\partial w}{\partial z} = 0$$

（4.2.20）

引用直角坐标系内的梯度算子，如

$$\nabla = \frac{\partial}{\partial x}\boldsymbol{i} + \frac{\partial}{\partial y}\boldsymbol{j} + \frac{\partial}{\partial z}\boldsymbol{k}$$

（4.2.21）

则不可压缩流体稳定流动的连续性方程可以表示为

$$\nabla \cdot \boldsymbol{V} = 0$$

（4.2.22）

式中，向量 $\boldsymbol{V} = u\boldsymbol{i} + v\boldsymbol{j} + w\boldsymbol{k}$，$\nabla \cdot \boldsymbol{V}$ 称为速度散度。

所以，对于不可压缩流体稳定流动，速度的散度为零。

连续性方程的这种微分形式可用于多种坐标系，如柱坐标系内不可压缩流体稳定流动的微分形式连续性方程可以表示为

$$\frac{1}{r}\frac{\partial(rv_r)}{\partial r} + \frac{1}{r}\frac{\partial v_\theta}{\partial \theta} + \frac{\partial v_z}{\partial z} = 0$$

（4.2.23）

例 4.2.4 已知不可压缩流体稳定流动的速度分布为：$u = ax^2 + x$，$v = (x+b)y$，

by $u = ax^2 + x$, $v = (x+b)y$, determine the value of constant a and b.

Solution The differential continuity equation for an incompressible steady, plane flow is

$$\frac{\partial u}{\partial x} + \frac{\partial v}{\partial y} = 0 \quad \rightarrow \quad (2ax+1)+(x+b) = 0 \quad \rightarrow \quad (2a+1)x+(1+b) = 0$$

Its solution is

$$a = -\frac{1}{2}, \quad b = -1$$

Example 4.2.5 The x-component of velocity at points A, B, and C, which are 10mm apart, is measured to be 3.14m/s, 4.06m/s, and 5.03m/s, respectively, in the plane steady, symmetrical, incompressible flow shown in Fig.X.4.2.5 in which $\omega = 0$. Approximate the y component of velocity 5mm above B.

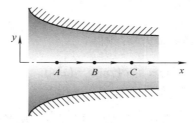

Fig.X4.2.5

Solution The y-component of velocity 5mm above B is found using the Eq.4.2.20 as follows:

$$\frac{\partial u}{\partial x} = -\frac{\partial v}{\partial y}$$

$$\frac{\partial u}{\partial x} \approx \frac{\Delta u}{\Delta x} = \frac{5.03-3.14}{0.02} \text{m/s} = 94.5 \text{m/s}$$

$$\frac{\partial v}{\partial y} \approx \frac{\Delta v}{\Delta y} = -\frac{\partial u}{\partial x} = -94.5 \text{m/s}$$

We know that $v = 0$ at location B, hence, at 5mm above B, the y-component velocity $\Delta v = -94.5\Delta y = -0.4725$m/s $\rightarrow$ $v = -0.4725$m/s, downward.

4.3 The Bernoulli Equation

The second basic equation of fluid dynamics approaches flow from the viewpoint of energy considerations.

求常数 a 和 b 的值。

解　题中不可压缩流体稳定平面流动的微分式连续性方程为

$$\frac{\partial u}{\partial x}+\frac{\partial v}{\partial y}=0 \;\rightarrow\; (2ax+1)+(x+b)=0 \;\rightarrow\; (2a+1)x+(1+b)=0$$

解得

$$a=-\frac{1}{2},\quad b=-1$$

例 4.2.5　如图 X.4.2.5 所示不可压缩流体稳定对称平面流动，相距均为 10mm 且 $\omega=0$ 的 A、B 和 C 三点，其 x 方向速度分量分别为 3.14m/s，4.06m/s 和 5.03m/s。试估算距 B 点上方 5mm 处沿 y 方向的速度分量。

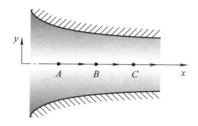

图 X4.2.5

解　距 B 点上方 5mm 处 y 方向的速度分量可由式 4.2.20 表示为

$$\frac{\partial u}{\partial x}=-\frac{\partial v}{\partial y}$$

$$\frac{\partial u}{\partial x}\approx\frac{\Delta u}{\Delta x}=\frac{5.03-3.14}{0.02}\,\text{m/s}=94.5\,\text{m/s}$$

$$\frac{\partial v}{\partial y}\approx\frac{\Delta v}{\Delta y}=-\frac{\partial u}{\partial x}=-94.5\,\text{m/s}$$

已知 B 点 $v=0$，所以位于其上方 5mm 处，y 方向的速度分量为

$$\Delta v=-94.5\Delta y=-0.4725\text{m/s}\;\rightarrow\; v=-0.4725\text{m/s}，方向向下。$$

4.3　伯努利方程

流体动力学的第二个基本方程是从能量的角度解读流体的流动。

4.3.1   The Kinetic Differential Equation along a Streamline

Consider the motion of a fluid particle in a frictionless steady flow field of an ideal fluid along a streamline in the S-direction, as shown in Fig.4.3.1. We shall consider the forces acting in the direction of the streamline on a small element of the fluid in the stream tube. The cross-sectional area of the element at right angle to the streamline may have the area dA.

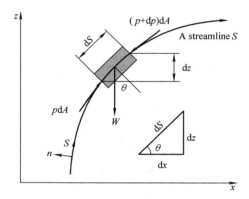

Fig.4.3.1   The forces acting on a fluid particle along a streamline.

The mass of the fluid element is $m = \rho dSdA$. The forces tending to accelerate or decelerate this mass along S are (a) the pressure forces, $pdA - (p + dp)dA = -dpdA$, and (b) the

weight component in the direction of motion, which is $-W \sin \theta = -\rho gdAdS \dfrac{dz}{dS} = -\rho gdAdz$,

where θ is the angle between the streamline and the horizontal x-axis at that point, and $\sin \theta = dz / dS$.

Applying Newton's second law, $\Sigma F = ma$, along the streamline, we get

$$-dpdA - \rho gdAdz = \rho dSdAa_S$$

Dividing by the volume $dAdS$,

$$-\frac{dp}{dS} - \rho g \frac{dz}{dS} = \rho a_S \qquad\qquad (4.3.1)$$

Recalling from Eq.3.1.14 that, $a_S = \dfrac{\partial u}{\partial t} + u \dfrac{\partial u}{\partial S}$, for a steady flow $\partial u/\partial t = 0$, and since the acceleration change along streamline S only, the partial differential symbol is replaced by the full differential symbol, that is $a_S = u \dfrac{du}{dS}$, and dividing by ρ, we get

4.3.1   沿流线的流体运动微分方程

在沿流线 S 方向做稳定无黏流动的理想流体中选取一段流体微元，如图 4.3.1 所示。我们沿流线方向来分析这个圆柱形流体微元的受力情况。假设流体微元与流线垂直的截面面积为 dA。

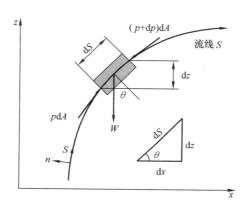

图 4.3.1   沿流线方向流体微元所受的力

流体微元的质量 $m = \rho \mathrm{d}S\mathrm{d}A$。沿流线 S 方向使流体微元运动加速或减速的力包括（1）压力，$p\mathrm{d}A - (p + \mathrm{d}p)\mathrm{d}A = -\mathrm{d}p\mathrm{d}A$，和（2）重力在流线方向的分量，等于 $-W\sin\theta = -\rho g\mathrm{d}A\mathrm{d}S\dfrac{\mathrm{d}z}{\mathrm{d}S} = -\rho g\mathrm{d}A\mathrm{d}z$，式中 θ 是在这一点流线方向与 x 轴的夹角，且 $\sin\theta = \mathrm{d}z/\mathrm{d}S$。

应用牛顿第二定律，$\Sigma F = ma$，沿流线可得

$$-\mathrm{d}p\mathrm{d}A - \rho g\mathrm{d}A\mathrm{d}z = \rho \mathrm{d}S\mathrm{d}A a_s$$

等式两边除以流体微元的体积 dAdS，得

$$-\frac{\mathrm{d}p}{\mathrm{d}S} - \rho g\frac{\mathrm{d}z}{\mathrm{d}S} = \rho a_s \tag{4.3.1}$$

回顾式 3.1.14，$a_s = \dfrac{\partial u}{\partial t} + u\dfrac{\partial u}{\partial S}$，对于稳定流动 $\partial u/\partial t = 0$，且流体质点的加速度仅沿流线 S 变化，所以把偏微分符号改为全微分符号，则有 $a_s = u\dfrac{\mathrm{d}u}{\mathrm{d}S}$，等式两边都除以 ρ，得

$$\frac{1}{\rho}\frac{\mathrm{d}p}{\mathrm{d}S} + g\frac{\mathrm{d}z}{\mathrm{d}S} + u\frac{\mathrm{d}u}{\mathrm{d}S} = 0 \tag{4.3.2}$$

This is the kinetic differential equation of incompressible ideal fluid in steady flow along a streamline, also called one-dimensional Euler equation, because it was firstly derived in about 1750 by Euler.

4.3.2 The Bernoulli Equation

Divide both side of Eq.4.3.2 by g multiply $\mathrm{d}S$ and then integrate, we have, along the streamline of a steady flow

$$\int\frac{\mathrm{d}p}{\rho g} + \mathrm{d}z + \frac{u}{g}\mathrm{d}u = \text{Constant} \tag{4.3.3}$$

and integration gives

$$z + \frac{p}{\rho g} + \frac{u^2}{2g} = \text{Constant} \tag{4.3.4}$$

This is the famous *Bernoulli equation*, in honor of Daniel Bernoulli (1700—1782), the Swiss physicist who presented this equation in 1738, also called *energy equation*. There are so many basic assumptions involved in Bernoulli equation, it is important when applying it. They are: steady, incompressible flow along a streamline in regions where viscous effects are negligible, and no energy is added to or removed from the fluid along the streamline.

Note that the velocity represented by u is the true value of the velocities of different fluid particles crossing a section, which are not the same, therefore, the velocity distribution must be accounted for. If the assumption of uniform velocity profiles is acceptable for a problem of interest, it is convenient to express the true value in terms of the mean velocity V. We can account for nonuniform velocity distributions by modifying Eq.4.3.4 to

$$z + \frac{p}{\rho g} + \alpha\frac{V^2}{2g} = \text{Constant} \tag{4.3.5}$$

where α is *kinetic energy correction factor*, defined by

$$\alpha = \frac{\int u^3 \mathrm{d}A}{V^3 A} \tag{4.3.6}$$

We simply let $\alpha = 1$ since it is so close to unity, this will always be done unless otherwise stated. More details about kinetic energy correction factor will be discussed later in this section. And Eq.4.3.5 always gives the form

$$z + \frac{p}{\rho g} + \frac{V^2}{2g} = \text{Constant} \tag{4.3.7}$$

4.3.3 Mechanical Energies of a Flowing Fluid

The Bernoulli equation can be viewed as the law of conservation of mechanical ener-

$$\frac{1}{\rho}\frac{\mathrm{d}p}{\mathrm{d}S} + g\frac{\mathrm{d}z}{\mathrm{d}S} + u\frac{\mathrm{d}u}{\mathrm{d}S} = 0 \qquad (4.3.2)$$

这是理想不可压缩流体沿流线做稳定流动时的运动微分方程，也称为欧拉一元运动方程，由瑞士数学家 Leonhard Euler（1707—1783）于 1750 年首次提出。

4.3.2　伯努利方程

把式 4.3.2 两边都乘以 $\mathrm{d}S$ 且除以 g 并积分，我们可以得出沿流线稳定流动时

$$\int \frac{\mathrm{d}p}{\rho g} + \mathrm{d}z + \frac{u}{g}\mathrm{d}u = \mathrm{C} \qquad (4.3.3)$$

积分后可得

$$z + \frac{p}{\rho g} + \frac{u^2}{2g} = \mathrm{C} \qquad (4.3.4)$$

这就是著名的伯努利方程，也称为能量方程，为纪念瑞士物理学家 Daniel Bernoulli（1700—1782）于 1738 年提出这一方程而命名。伯努利方程是基于多个基本假设而成立的，这一点在应用时必须注意。这些假设包括：不可压缩流体、沿流线运动、黏性力的影响可以忽略，且流体沿流线运动过程中没有增加或失去能量。

另外还需注意，我们习惯用 u 表示流体质点的实际运动速度，各质点在截面上的速度往往不一致，所以必须考虑速度分布的不均匀性。如果假设流速均匀分布，并且对式 4.3.4 中速度的不均匀性加以修正，我们就可以方便地用平均速度 V 来代替实际速度

$$z + \frac{p}{\rho g} + \alpha\frac{V^2}{2g} = \mathrm{C} \qquad (4.3.5)$$

式中，α 是**动能修正系数**，定义为

$$\alpha = \frac{\int u^3 \mathrm{d}A}{V^3 A} \qquad (4.3.6)$$

因为 α 值接近 1，所以通常没有特别说明的情况下都取 $\alpha = 1$。在本节稍后的内容中将对动能修正系数做更多的介绍。这样，式 4.3.5 通常写为

$$z + \frac{p}{\rho g} + \frac{V^2}{2g} = \mathrm{C} \qquad (4.3.7)$$

4.3.3　流动流体的机械能

伯努利方程表达的是机械能守恒定律的观点，如果系统的机械能和内能之间不

gy, for systems that do not involve any conversion of mechanical energy and internal energy to each other, and thus the mechanical energy is conserved.

We recognize z as potential energy, $p/\rho g$ as flow pressure energy, and $V^2/2g$ as kinetic energy, all per unit weight, as shown in Fig.4.3.2. Therefore, the Bernoulli equation can be viewed as an expression of mechanical energy balance and can be stated as: The sum of the kinetic, potential, and pressure energies of a fluid particle is constant along a streamline during steady flow when compressibility and viscous effects are negligible.

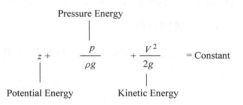

Fig.4.3.2　The mechanical energy balance stated by Bernoulli Equation

The potential energy of a fluid particle depends on its elevation above an arbitrary datum plane. We are usually interested only in differences of elevation, and therefore the location of the datum plane used is determined solely by convenience. A fluid particle of weight W situated a distance z above datum possesses a potential energy of Wz. Thus its potential energy per unit weight is z. The particle potential energy per unit mass is gz, its potential energy per unit volume is ρgz.

Pressure energy containing in fluid of volume Ω is $p\Omega$, the pressure energy per unit weight is $p\Omega/\rho g\Omega = p/\rho g$, measured in units of N·m/N = m.

A fluid flowing with all particles moving at the same velocity V, its kinetic energy would be

$$E_k = \frac{mV^2}{2} = \frac{\rho V^3 A}{2} \tag{4.3.8}$$

where A is the area of cross section, for unit weight of the fluid we can write this as

$$\frac{E_k}{W} = \frac{mV^2/2}{mg} = \frac{V^2}{2g} \tag{4.3.9}$$

similarly, the kinetic energy per unit mass is $V^2/2$, and kinetic energy per unit volume is $\rho V^2/2$.

Now we come back to the definition of kinetic energy correction factor α, the true flow of kinetic energy per unit of time for a fluid parcel across area dA is

$$E_{k\,True} = \frac{mu^2}{2} = \frac{(\rho u dA)u^2}{2} = \frac{\rho u^3 dA}{2} \tag{4.3.10}$$

here the mass rate of flow through dA is $\rho u dA$.

For the entire section A

$$E_{k\,True} = \frac{1}{2}\int_A (\rho u^3)\,dA \tag{4.3.11}$$

Comparing Eq.4.3.8 with Eq.4.3.11, we get the definition of kinetic energy correction factor

$$\alpha = \frac{E_{k\,True}}{E_k} = \frac{\int_A (\rho u^3/2)\,dA}{\rho V^3 A/2} = \frac{\int_A u^3\,dA}{V^3 A} \tag{4.3.12}$$

存在互相转换，那么机械能就是守恒的。

如图 4.3.2 所示，伯努利方程中各项：z 代表位势能，$p/\rho g$ 代表压力能，$V^2/2g$ 则是动能，各项能量均为单位重量的流体所包含。所以，伯努利方程的机械能守恒可以描述为：沿流线稳定流动的理想、不可压缩流体，其动能、位势能和压力能的总和恒定不变。

图 4.3.2　伯努利方程表达的机械能守恒

流体质点所具有的位势能取决于其相对某一基准平面的高度，我们通常关注二者间的高度差，所以基准面的位置可依方便选取。位于基准面之上 z 高度位置重量为 W 的流体微元，其位势能为 Wz，因此单位重量的流体含有的位势能大小为 z，单位质量流体的势能为 gz，单位体积流体的势能为 ρgz。

体积为 Ω 的流体内含压力能 $p\Omega$，单位重量流体内的压力能为 $p\Omega/\rho g\Omega = p/\rho g$，单位为 N·m/N = m。

流体微元内的全部质点都以速度 V 运动时，其动能为

$$E_k = \frac{mV^2}{2} = \frac{\rho V^3 A}{2} \tag{4.3.8}$$

式中，A 是有效截面的面积，单位重量流体的动能为

$$\frac{E_k}{W} = \frac{mV^2/2}{mg} = \frac{V^2}{2g} \tag{4.3.9}$$

同样，单位质量流体内的动能为 $V^2/2$，单位体积流体内的动能为 $\rho V^2/2$。

再来讨论动能修正系数 α，流体质点单位时间通过面积为 $\mathrm{d}A$ 截面时的实际动能

$$E_{k\text{True}} = \frac{mu^2}{2} = \frac{(\rho u \mathrm{d}A)u^2}{2} = \frac{\rho u^3 \mathrm{d}A}{2} \tag{4.3.10}$$

此时通过截面 $\mathrm{d}A$ 的质量流量是 $\rho u \mathrm{d}A$。

对整个截面 A

$$E_{k\text{True}} = \frac{1}{2}\int_A (\rho u^3) \mathrm{d}A \tag{4.3.11}$$

比较式 4.3.8 和式 4.3.11，动能修正系数可以定义为

$$\alpha = \frac{E_{k\text{True}}}{E_k} = \frac{\int_A (\rho u^3/2)\mathrm{d}A}{\rho V^3 A/2} = \frac{\int_A u^3 \mathrm{d}A}{V^3 A} \tag{4.3.12}$$

The value of α will always be more than 1, the greater the variation in velocity across the section, the larger will be the value of α. For laminar flow in a circular pipe $\alpha = 2$ (see section 6.3), but the velocity is usually so small that the kinetic energy per unit weight of fluid is negligible. For turbulent flow in pipes, α ranges usually between 1.03 and 1.06, it is customary to assume that $\alpha = 1$.

4.3.4 Geometric Interpretation of Bernoulli Equation

Each term in Bernoulli equation has the dimension of length and represents some kind of "head", which make it convenient to represent the level of mechanical energy graphically.

The first term, z, called the "*elevation head*" or "*potential head*", represents the elevation the fluid particle above the datum plane.

The second term, $p/\rho g$, called the "*pressure head*", represents the height a unit weight fluid rises by static pressure p inside a vacuum tube, or the height of a fluid column that produces the static pressure p.

Recall from Chapter 2 that $z + p/\rho g$ is called the static head.

And $V^2/2g$, called the "*velocity head*", it represents the elevation needed for unit weight of fluid to reach the velocity V during frictionless free fall, or the elevation of unit weight fluid may rise against gravity at initial velocity V.

We call the sum of these three terms the *total head*, usually denoted by H, so that

$$z + \frac{p}{\rho g} + \frac{V^2}{2g} = H = \text{Constant} \tag{4.3.13}$$

Therefore, the essence of Bernoulli equation is about conservation and conversion of energy, it is expressed in terms of heads as: The sum of the pressure, velocity, and elevation heads along a streamline is constant during steady flow when compressibility and frictional effects are negligible.

The value of the total head in Eq.4.3.13 can be evaluated at any point on the streamline where the pressure, density, velocity, and elevation are known. The Bernoulli equation can also be written between any two sections where gradually varied flow maintaining

$$z_1 + \frac{p_1}{\rho g} + \frac{V_1^2}{2g} = z_2 + \frac{p_2}{\rho g} + \frac{V_2^2}{2g} \tag{4.3.14}$$

That means, in an idealized Bernoulli-type flow, the various forms of mechanical energy are convertible to each other, but their sum remains constant. The total head line of the stream flow shown in Fig.4.3.3 is horizontal, the elevation head at center of section A_2 raises by $(z_2 - z_1)$, while the velocity head $V_2^2/2g$ drops as the velocity at section A_2 decreases by continuum equation $V_2 = (A_1/A_2) V_1$, obviously, $V_2 < V_1$.

The Bernoulli equation can also be viewed in the light of Newton's second law of motion as: The work done by the pressure and gravity forces on the fluid particle is equal to the increasing in the kinetic energy of the particle.

The energy equation, in the form of Eq.4.3.14, applied between two sections of a control volume where the flow crosses, is useful in many applications and is, perhaps, the most often used form of the energy equation.

The control volume is usually selected such that there is one entrance and one exit, as

α 的值总是大于 1，截面上速度分布越不均匀，α 值越大。圆管内层流的 $\alpha = 2$（见 6.3 节），但流速通常很小，以致单位重量流体内的动能可以忽略。而圆管内紊流，α 取值范围为 1.03 ~ 1.06，也通常假设为 $\alpha = 1$。

4.3.4　伯努利方程的几何意义

伯努利方程中各项的量纲都是长度，常称为某种"水头"，以便于用图表中的"水头"高度表达对应机械能的大小。

第一项 z，称为"位置水头"或"势能水头"，代表流体质点高于基准面的高度。

第二项 $p/\rho g$，称为"压力水头"，代表单位重量的流体在真空管内由于静压强 p 的作用能够上升的高度，或者静压强 p 能够产生的液柱高度。

在第 2 章中，我们把 $z + p/\rho g$ 称为静压头。

$V^2/2g$ 项称为"速度水头"，表示单位重量流体做无摩擦的自由落体运动时，欲达到速度 V 所需的高度，或者，单位重量的流体以初速度 V 克服重力，能够上升的最大高度。

我们把三项水头之和称为"总水头"，常用 H 表示，所以有

$$z + \frac{p}{\rho g} + \frac{V^2}{2g} = H = C \tag{4.3.13}$$

因此，也可以用"水头"来说明伯努利方程关于能量的守恒和转换：在不可压缩流体沿流线稳定流动的过程中，如果忽略黏性摩擦力的影响，则压力水头、速度水头和位置水头的总和保持不变。

流线上任意一点，只要其压强、密度、流速和位置高度已知，就可以用来表示式 4.3.13 中的总水头。所以，伯努利方程也可以用缓变流上两截面表示

$$z_1 + \frac{p_1}{\rho g} + \frac{V_1^2}{2g} = z_2 + \frac{p_2}{\rho g} + \frac{V_2^2}{2g} \tag{4.3.14}$$

上式说明，伯努利方程条件下的流动，不同形式的机械能可以相互转化，但其总和保持不变。图 4.3.3 所示的总水头线是一条水平线，截面 A_2 中心处的高度上升了（$z_2 - z_1$），而速度水头 $V_2^2/2g$ 则有所下降，这是由连续性方程 $V_2 = (A_1/A_2)V_1$ 决定的，显然，$V_2 < V_1$。

以牛顿第二运动定律的思想，同样可以解读伯努利方程：压力和重力在流体质点上所做的功，等于其带来的流体质点动能的增量。

用流体流经控制体上的两个截面表示的伯努利方程，如式 4.3.14 的形式，是伯努利方程应用最广泛的形式，甚至是能量方程最常用的形式。

为处理问题方便，选取控制体时，尽量选择只有一个入口和一个出口的部分，且进口和出口的总水头一致。因为自由液面的表压强为零，因此如果能够将一个控

while as the entrance and exit sections have a uniform total head. However, a convenient choice would be the points at the free surface, where the pressures have been set to zero.

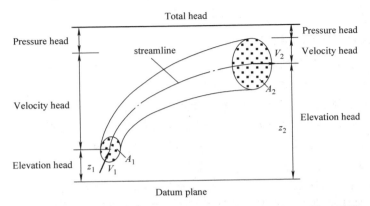

Fig.4.3.3   Conservation and conversion of heads of a Bernoulli type flow

Example 4.3.1   A large tank open to the atmosphere is filled with water to a height of 1.5m from the outlet (Fig.X4.3.1). An outlet of 20mm diameter is now opened, and water flows out from the smooth and rounded outlet. Determine the water velocity at the outlet. Neglect friction.

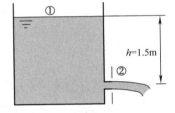

Fig.X4.3.1

Solution   We select the control volume to extend from section ① to section ② on the tank and outlet surfaces, where we know or we can obtain the velocities, pressures, and elevations. We consider the water surface of the tank to be the entrance and the water surface of the outlet to be the exit.

Now, consider the Bernoulli equation. We will use gage pressures so that $p_1 = p_2 = 0$. The datum is placed through the lower section ② so that $z_2 = 0$.

By continuum equation on control volume between section ① to section ②

$$V_1 A_1 = V_2 A_2 \quad \because A_1 >> A_2 \quad \therefore V_1 << V_2$$

The average velocity V_1 on the free surface is negligibly small compare to the outlet. The Bernoulli equation then becomes

$$z_1 + \frac{\cancel{p_1}}{\rho g} + \frac{\cancel{V_1^2}}{2g} = \cancel{z_2} + \frac{\cancel{p_2}}{\rho g} + \frac{V_2^2}{2g} \rightarrow z_1 = \frac{V_2^2}{2g}$$

Solving for V_2 and substituting

$$V_2 = \sqrt{2gz_1} = \sqrt{2gh} = 5.42 \text{m/s}$$

Example 4.3.2   Water is spraying out from a fountain nozzle of 5mm diameter, and the flow rate going out is 6L/min, the pressure just inside the nozzle is measured 200kPa (Fig.X4.3.2), what is the maximum height that the jet could achieve?

制面选在自由液面上则会更加方便。

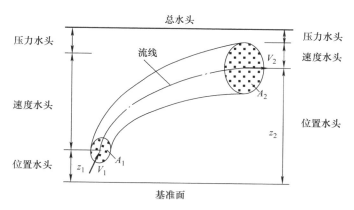

图 4.3.3　伯努利流动的 "水头" 守恒和转化

例 4.3.1　一个开口通大气的水池，内部水位至底部出口中心的高度保持为 1.5m（见图 X.4.3.1），底部光滑的圆形出口直径为 20mm。水自出口流出，忽略摩擦，计算出口处水的流速。

解　选取池内截面①和出口处截面②间的部分为控制体。这两个截面处的流速、压强和高度都是已知或待求的。池内液面为控制体的进口而出水口截面为控制体的出口。在伯努利方程中使用表压强，那么 $p_1 = p_2 = 0$。取较低的截面②为基准面，所以 $z_2 = 0$。

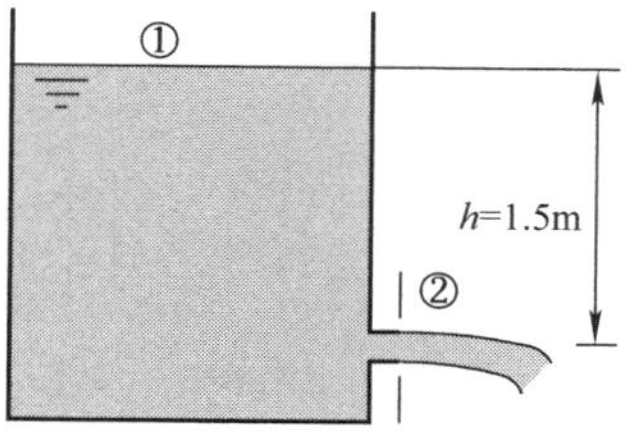

在控制体中对截面①和截面②应用连续性方程

$$V_1 A_1 = V_2 A_2 \qquad \text{因为 } A_1 \gg A_2 \qquad \text{所以 } V_1 \ll V_2$$

自由液面上的流速 V_1 与出口处流速相比非常小，可以忽略。

伯努利方程可写为

$$z_1 + \frac{\cancel{p_1}}{\rho g} + \frac{\cancel{V_1^2}}{2g} = \cancel{z_2} + \frac{\cancel{p_2}}{\rho g} + \frac{V_2^2}{2g} \ \to \ z_1 = \frac{V_2^2}{2g}$$

代入已知量并求解 V_2

$$V_2 = \sqrt{2g z_1} = \sqrt{2gh} = 5.42 \text{m/s}$$

例 4.3.2　水从直径 5mm 的喷泉喷嘴中喷出，流量达 6L/min，靠近喷嘴内部位置测得的水压为 200kPa（见图 X4.3.2），试计算水柱可以达到的最大高度。

图 X4.3.1

Solution   At the top of the water trajectory $V_2 = 0$, and atmospheric pressure pertains. Then the Bernoulli equation along a streamline from ① to ② simplifies to

$$\cancel{z_1} + \frac{p_1}{\rho g} + \frac{V_1^2}{2g} = z_2 + \cancel{\frac{p_2}{\rho g}} + \cancel{\frac{V_2^2}{2g}} \quad \rightarrow \quad \frac{p_1}{\rho g} + \frac{V_1^2}{2g} = z_2$$

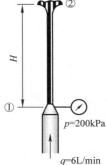

The average velocity at section ① determined by the flow rate and the area of the section,

$$V_1 = \frac{4q}{\pi d^2} = \frac{4 \times 6 \times 10^{-3}}{60\pi (0.005)^2} \text{m/s} = 5.1\text{m/s}$$

Solving for z_2 and substituting

$$z_2 = \frac{p_1}{\rho g} + \frac{V_1^2}{2g} = \frac{200 \times 10^3}{9810} \text{m} + \frac{5.1^2}{2 \times 9.81} \text{m} = 21.71\text{m}$$

Fig.X4.3.2

Example 4.3.3   A fire hydrant shown in Fig.X4.3.3, where $d_1 = 150\text{mm}$, $d_2 = 50\text{mm}$, gage pressure measured near the nozzle is 300kPa, find (a) the velocity of discharge, (b) the maximum height that the jet could achieve and (c) the diameter of the jet at the highest point. Neglect friction.

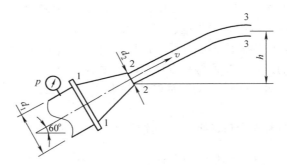

Fig.X4.3.3

Solution
(a) We take first control volume from section 1 to section 2, the height difference between this two sections is neglected. By continuum equation and simplifying the Bernoulli equation, we have

$$\frac{V_1}{V_2} = \frac{d_2^2}{d_1^2} \quad \rightarrow \quad V_2 = 9V_1$$

$$\frac{p_1}{\rho g} + \frac{V_1^2}{2g} = \frac{V_2^2}{2g} \quad \rightarrow \quad V_1 = \sqrt{\frac{p_1}{40\rho}} = 2.74\text{m/s} \quad \rightarrow \quad V_2 = 9V_1 = 24.65\text{m/s}$$

解　水柱顶点 $V_2 = 0$，且为大气压。截面①和②之间沿流线的伯努利方程可以化简为

$$\cancel{z_1} + \frac{p_1}{\rho g} + \frac{V_1^2}{2g} = z_2 + \frac{\cancel{p_2}}{\rho g} + \frac{\cancel{V_2^2}}{2g} \quad \rightarrow \quad \frac{p_1}{\rho g} + \frac{V_1^2}{2g} = z_2$$

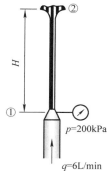

截面①处的平均流速可由流量和截面积计算

$$V_1 = \frac{4q}{\pi d^2} = \frac{4 \times 6 \times 10^{-3}}{60\pi (0.005)^2} \, \text{m/s} = 5.1 \text{m/s}$$

代入流速并解得 z_2

图 X4.3.2

$$z_2 = \frac{p_1}{\rho g} + \frac{V_1^2}{2g} = \frac{200 \times 10^3}{9810} \, \text{m} + \frac{5.1^2}{2 \times 9.81} \, \text{m} = 21.71 \text{m}$$

例 4.3.3　如图 X4.3.3 所示的消防喷头，直径 $d_1 = 150 \text{mm}$，$d_2 = 50 \text{mm}$，靠近喷嘴处测得表压强 300kPa，计算（1）出口流速，（2）水柱可达的最大高度，（3）水柱在最高点时的直径。忽略摩擦损失。

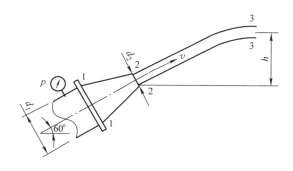

图 X4.3.3

解　（1）取截面 1 和截面 2 中间部分为控制体，忽略两截面间高度差。应用连续性方程对伯努利方程进行化简，可得

$$\frac{V_1}{V_2} = \frac{d_2^2}{d_1^2} \quad \rightarrow \quad V_2 = 9V_1$$

$$\frac{p_1}{\rho g} + \frac{V_1^2}{2g} = \frac{V_2^2}{2g} \quad \rightarrow \quad V_1 = \sqrt{\frac{p_1}{40\rho}} = 2.74 \text{m/s} \quad \rightarrow \quad V_2 = 9V_1 = 24.65 \text{m/s}$$

(b) The jet angle is 60°, the outlet velocity elements in x and z direction are

$$u_2 = V_2 \cos 60° = 12.33\text{m/s}, \; w_2 = V_2 \sin 60° = 21.35\text{m/s}$$

At the highest point 3, the vertical velocity element reduced to zero, $w_3 = 0$, the horizontal element dose not slow down, $V_3 = u_3 = u_2 = 12.33\text{m/s}$.

Using Bernoulli equation on the control volume from section 2 to section 3, therefore

$$\frac{V_2^2}{2g} = h + \frac{V_3^2}{2g} \quad \text{that results } h = 23.22\text{m}$$

(c) Using continuum equation on the control volume from section 2 to section 3, that is

$$V_2 A_2 = V_3 A_3 \quad \rightarrow \quad A_3 = \frac{V_2 A_2}{V_3} = \frac{V_2}{V_3} \frac{\pi d_2^2}{4} = \frac{\pi d_3^2}{4} \quad \rightarrow \quad d_3 = 70.7\text{mm}$$

The water jet diffused.

Example 4.3.4 Pumping water from a water pool as shown in Fig.X4.3.4, the diameter of inlet pipe is 100mm and 60mm for outlet pipe. The measured actual output flow rate is 108L/min, the volumetric efficiency of the pump is given as 0.9. The inlet end of the pipe is located 3m below the inlet port of the pump and 2.5m below the free surface of the water pool. Find the vacuum pressure at inlet port of the pump. What if the water level in the pool dropped by 2m.

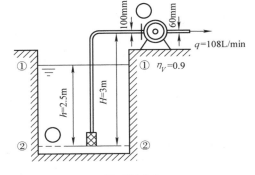

Fig.X4.3.4

Solution The actual output flow rate is given as 108L/min, and the volumetric efficiency of the pump is 0.9, the inlet flow rate of the pump can be calculated

$$q_{\text{pumpin}} = \frac{q_{\text{out}}}{\eta_V} = \frac{108}{0.9}\text{L/min} = 120\text{L/min} = 2 \times 10^{-3}\text{m}^3/\text{s}$$

Average velocity on section ① is

$$V_1 = \frac{4q}{\pi D^2} = 0.2548\text{m/s}$$

The velocity V_2 on the pool section 2.5m below the free surface is negligibly small compare to the inlet port, that is $V_2 = 0$.

Static pressure at section ② where 2.5m below the free surface is

$$p_2 = \rho g h = 24.525\text{kPa}$$

Using Bernoulli equation on the control volume from section ① to section ②, then

（2）射流角度为 60°，出流速度在 x 和 z 方向的分量为

$$u_2 = V_2 \cos 60° = 12.33 \text{m/s}, \quad w_2 = V_2 \sin 60° = 21.35 \text{m/s}$$

在最高点 3，竖直方向速度分量降为零，$w_3 = 0$，水平方向速度分量保持不变，即

$$V_3 = u_3 = u_2 = 12.33 \text{m/s}$$

对截面②—③间的控制体应用伯努利方程，有

$$\frac{V_2^2}{2g} = h + \frac{V_3^2}{2g} \quad 解得 \quad h = 23.22 \text{m}$$

（3）对截面②—③间的控制体应用连续性方程，可得

$$V_2 A_2 = V_3 A_3 \quad \rightarrow \quad A_3 = \frac{V_2 A_2}{V_3} = \frac{V_2}{V_3} \frac{\pi d_2^2}{4} = \frac{\pi d_3^2}{4} \quad \rightarrow \quad d_3 = 70.7 \text{mm}$$

水流发生了扩散。

例 4.3.4　如图 X4.3.4 所示用水泵从水池中抽水，吸水管直径为 100mm，出水管直径为 60mm。出口实际测得的体积流量为 108L/min，已知泵的容积效率为 0.9。吸水管的进口端位于泵进水口下 3m 处且低于池内水面 2.5m。计算此时泵入口处的真空压强，如果池内水位继续下降 2m 后泵入口的真空压强又是多少？

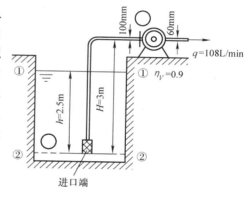

图 X4.3.4

解　泵出口处的体积流量为 108L/min，而泵的容积效率为 0.9，可知泵入口的体积流量为

$$q_{\text{pumpin}} = \frac{q_{\text{out}}}{\eta_V} = \frac{108}{0.9} \text{L/min} = 120 \text{L/min} = 2 \times 10^{-3} \text{m}^3/\text{s}$$

截面①的平均流速为

$$V_1 = \frac{4q}{\pi D^2} = 0.2548 \text{m/s}$$

位于水池自由液面下 2.5m 处的截面②，其流速与泵入口处相比很小，可忽略不计，所以 $V_2 = 0$。

低于自由液面 2.5m 的截面②处流体静压强为

$$p_2 = \rho g h = 24.525 \text{kPa}$$

对截面①—②间的控制体应用伯努利方程，有

$$z_1 + \frac{p_1}{\rho g} + \frac{V_1^2}{2g} = z_2 + \frac{p_2}{\rho g} + \frac{V_2^2}{2g} \quad \rightarrow \quad p_1 = p_2 - z_1\rho g - \frac{\rho V_1^2}{2} = -4.94\text{kPa}$$

That means pressure at section ① is 4.94kPa below the atmospheric pressure, therefore, the vacuum pressure at section 1 is 4.94kPa.

After pumping for some time, the water level dropped by 2m, it is $h = 0.5$m, while $p_2 = 4.905$kPa.

The pressure at inlet port of the pump is

$$p_{v1} = -\left(p_2 - z_1\rho g - \frac{\rho V_1^2}{2} \right) = 24.56\text{kPa}$$

Example 4.3.5 Hydraulic oil ($\rho = 860\text{kg/m}^3$) with a vapor pressure of 26.2kPa being pumped through a pump from an oil tank at a flow rate of 60L/min, as shown in FigX4.3.5, diameter of inflow pipe is 10mm. Find the maximum installation height H of the pump. The atmospheric pressure is 100kPa. Neglect frictional effective.

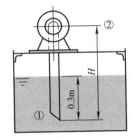

Solution Take the opening of the pipe as the datum plane, the area of the cross section of oil tank is very large relative to the inlet of the pump, thus the velocity at section ① is $V_1 = 0$.

Absolute pressure at inlet of the pump should be higher than the vapor pressure, that is $p_{2a} \geqslant 26.2$kPa. Take the minimum value of $p_{2a} = 26.2$kPa, where the gage pressure is

Fig.X4.3.5

$$p_2 = p_a - p_{2a} = -73.8\text{kPa}$$

Average velocity of hydraulic oil at inlet of pump is

$$V_2 = \frac{4q}{\pi d^2} = \frac{4 \times 60 \times 10^{-3}}{60 \times 3.142 \times 100 \times 10^{-6}} 12.74\text{m/s} = 12.74\text{m/s}$$

Using Bernoulli equation from section 1 to 2, it is

$$z_1 + \frac{p_1}{\rho g} + \frac{V_1^2}{2g} = z_2 + \frac{p_2}{\rho g} + \frac{V_2^2}{2g} \quad \rightarrow \quad 0.3\text{m} = z_2 - \frac{73.8 \times 10^3}{\rho g} + \frac{V_2^2}{2g}$$

Substituting and calculating, gives the result

$$H_{\min} = z_2 = 0.78\text{m}$$

Noting that the cavitation will happen when the installation height of the pump is higher than 0.78m, this may be harmful to the performance of the pump and hydraulic system. The cavitation can be avoid effectively if we decrease the velocity head and then enlarge the elevation head by increasing the diameter of inlet pipe.

$$z_1 + \frac{p_1}{\rho g} + \frac{V_1^2}{2g} = \cancel{z_2} + \frac{p_2}{\rho g} + \frac{\cancel{V_2^2}}{2g} \quad \rightarrow \quad p_1 = p_2 - z_1\rho g - \frac{\rho V_1^2}{2} = -4.94\text{kPa}$$

负号说明泵入口处的压强低于大气压 4.94kPa，所以截面 1 的真空压强为 4.94kPa。

抽水一段时间后，水位下降 2m，则 $h = 0.5\text{m}$，此时 $p_2 = 4.905\text{kPa}$，泵入口处的真空压强为

$$p_{v1} = -\left(p_2 - z_1\rho g - \frac{\rho V_1^2}{2} \right) = 24.56\text{kPa}$$

例 4.3.5　液压泵从油箱中以 60L/min 的流量吸取液压油（$\rho = 860\text{kg/m}^3$），如图 X4.3.5 所示，此时油液饱和蒸气压为 26.2kPa，吸油管直径 10mm。计算泵的最大安装高度 H。当地大气压为 100kPa，忽略摩擦损失。

解　取吸油管开口端为基准面，油箱截面积与泵入口面积相比非常大，所以忽略截面①处的流速，有 $V_1 = 0$。

泵入口处的绝对压强应该高于饱和蒸气压，所以 $p_{2a} \geqslant 26.2\text{kPa}$。取最小值 $p_{2a} = 26.2\text{kPa}$，换成相对压强

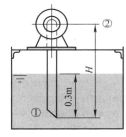

图 X4.3.5

$$p_2 = p_a - p_{2a} = -73.8\text{kPa}$$

液压泵入口处的平均流速

$$V_2 = \frac{4q}{\pi d^2} = \frac{4 \times 60 \times 10^{-3}}{60 \times 3.142 \times 100 \times 10^{-6}}\text{m/s} = 12.74\text{m/s}$$

在①—②截面间应用伯努利方程，有

$$\cancel{z_1} + \frac{p_1}{\rho g} + \frac{\cancel{V_1^2}}{2g} = z_2 + \frac{p_2}{\rho g} + \frac{V_2^2}{2g} \quad \rightarrow \quad 0.3\text{m} = z_2 - \frac{73.8 \times 10^3}{\rho g} + \frac{V_2^2}{2g}$$

代入流速及密度，计算得到

$$H_{\min} = z_2 = 0.78\text{m}$$

当液压泵的安装高度大于 0.78m 时，就可能在泵内发生气穴现象，这对液压泵的性能和液压系统都是有害的。加大吸油管直径降低吸油管内的速度水头，可以提高位置水头，有效避免气穴现象。

The Bernoulli equation of a whole-stream as Eq.4.3.14 can be applied to any control volume. For example, consider steady, uniform incompressible flow through a T-section in a pipe (Fig. 4.3.4) in which there is one entrance and two exits. The energy equation can be applied along a streamline going from ① to ② or a streamline going from ① to ③, as shown in Eq.4.3.15, each equation represents the energy balance on unit weight of fluid.

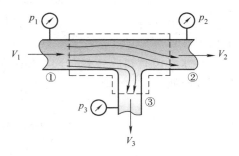

$$z_1 + \frac{p_1}{\rho g} + \frac{V_1^2}{2g} = z_2 + \frac{p_2}{\rho g} + \frac{V_2^2}{2g}$$

$$z_1 + \frac{p_1}{\rho g} + \frac{V_1^2}{2g} = z_3 + \frac{p_3}{\rho g} + \frac{V_3^2}{2g}$$

(4.3.15)

Fig.4.3.4   Application of the Bernoulli equation to a T-section.

Example 4.3.6   A 500mm diameter horizontal water pipe measured 8kPa of pressure at a flow rate of 0.35m³/s, separated into two branches in the same horizontal plane somewhere shown in Fig.X4.3.6, the diameter of exit ① is 400mm and 300mm at exit ②, the outlet flow rate of 0.2m³/s at exit ① and 0.15m³/s at exit ②. Find the pressure at branch ① and ②.

Solution   The energy equation can be applied to each of two control volumes, one for the mass flux that exits section ① and the other for the mass flux that exits section ②:

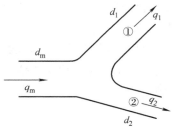

$$z_m + \frac{p_m}{\rho g} + \frac{V_m^2}{2g} = z_1 + \frac{p_1}{\rho g} + \frac{V_1^2}{2g}$$

$$z_m + \frac{p_m}{\rho g} + \frac{V_m^2}{2g} = z_2 + \frac{p_2}{\rho g} + \frac{V_2^2}{2g}$$

Fig.X4.3.6

Where

$$V_m = \frac{4q_m}{\pi d_m^2} = 1.78\text{m/s}, \qquad V_1 = \frac{4q_1}{\pi d_1^2} = 1.25\text{m/s}, \qquad V_2 = \frac{4q_2}{\pi d_2^2} = 2.12\text{m/s}$$

Therefor

$$p_2 = 8.322\text{kPa}, \qquad p_3 = 7.337\text{kPa}$$

4.3.5   Distribution of pressure and velocity normal to the streamline

Let us now consider how pressure varies over a cross section of flow in a uniform conduit.

Fluid element show in Fig.4.3.5 experience a corresponding centripetal force and centripetal acceleration due to a pressure gradient, forces balance in the direction normal to the streamline yields the following relation applicable for steady, incompressible flow:

$$p\mathrm{d}A - (p + \mathrm{d}p)\mathrm{d}A - \rho g\mathrm{d}r\mathrm{d}A\cos\theta = -\rho\mathrm{d}r\mathrm{d}A\frac{V^2}{R} \qquad (4.3.16)$$

式 4.3.14 的总流伯努利方程可用于各种形式的控制体。如图 4.3.4 所示，经
T 形管的不可压缩流体稳定流动，包含
一个入口和两个出口。因为伯努利方程
中的每一项都是单位重量流体的能量平
衡关系，所以在截面①—②间或截面
①—③间均可建立能量方程。

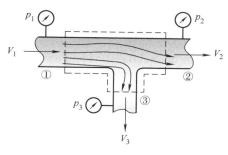

$$z_1 + \frac{p_1}{\rho g} + \frac{V_1^2}{2g} = z_2 + \frac{p_2}{\rho g} + \frac{V_2^2}{2g}$$

$$z_1 + \frac{p_1}{\rho g} + \frac{V_1^2}{2g} = z_3 + \frac{p_3}{\rho g} + \frac{V_3^2}{2g}$$

（4.3.15）

图 4.3.4　伯努利方程应用于 T 形管

例 4.3.6　如图 X4.3.6 所示直径为 500mm 的水管，测得管内压强为 8kPa，流
量为 0.35m³/s。水管在某处分岔为直径 400mm 的出口①和直径 300mm 的出口②，
两出口处的流量分别为 0.2m³/s 和 0.15m³/s。水管各进、出口在同一水平面内，计
算两分支管内的压强。

解　在入口和两个分支出口截面分别建立伯
努利方程：

$$z_m + \frac{p_m}{\rho g} + \frac{V_m^2}{2g} = z_1 + \frac{p_1}{\rho g} + \frac{V_1^2}{2g}$$

$$z_m + \frac{p_m}{\rho g} + \frac{V_m^2}{2g} = z_2 + \frac{p_2}{\rho g} + \frac{V_2^2}{2g}$$

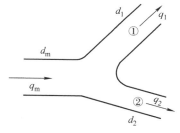

图 X4.3.6

式中

$$V_m = \frac{4q_m}{\pi d_m^2} = 1.78\text{m/s}, \qquad V_1 = \frac{4q_1}{\pi d_1^2} = 1.25\text{m/s}, \qquad V_2 = \frac{4q_2}{\pi d_2^2} = 2.12\text{m/s}$$

解得

$$p_2 = 8.322\text{kPa}, \qquad p_3 = 7.337\text{kPa}$$

4.3.5　沿流线法向的压强和流速分布

接下来我们讨论在均匀管道流动的截面上，压强是如何分布的。

图 4.3.5 所示的不可压缩流体微元沿流线做稳定流动，受到由压力梯度而产生
的向心力和向心加速度作用，在流线法向的受力平衡关系为

$$p\text{d}A - (p + \text{d}p)\text{d}A - \rho g\text{d}r\text{d}A\cos\theta = -\rho\text{d}r\text{d}A\frac{V^2}{R}$$

（4.3.16）

where R is the local radius of curvature of the streamline, the weight component in the normal direction is $-W\cos\theta = -\rho gdAdr\cos\theta =$

$-\rho gdAdr\dfrac{dz}{dr} = -\rho gdAdz$, here θ is the angle be-

tween the normal direction and the vertical z-axis at the central point, and $\cos\theta = dz/dr$. Eq.4.3.16 is reduced to

$$\frac{dz}{dr} + \frac{1}{\rho g}\frac{dp}{dr} = \frac{V^2}{Rg} \qquad (4.3.17)$$

Eq.4.3.17 can be rearranged as

$$\frac{d}{dr}\left(\frac{p}{\rho g} + z\right) = \frac{V^2}{Rg} \qquad (4.3.18)$$

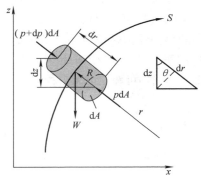

Fig.4.3.5 Force balance of a fluid element normal to the streamline

For flow along a straight line, $R\to\infty$ and

Eq.4.3.18 reduces to $\dfrac{d}{dr}\left(\dfrac{p}{\rho g} + z\right) = 0$, and then $\dfrac{p}{\rho g} + z = $ constant, which is an expression

for the variation of hydrostatic pressure with vertical distance for a stationary fluid body.

Therefore, the variation of pressure with elevation in steady, incompressible flow along a straight line in an inviscid region of flow is the same as that in the stationary fluid. In other words, in any plane perpendicular to the direction of a parallel and steady flow the pressure varies according to the hydrostatic law. For large conduits, it may be necessary to consider hydrostatic variation in the pressure normal to the streamlines. The pressure is lowest near the top of the conduit, and cavitation, if it were to occur, would appear there first, the average pressure is then the pressure at the centroid of such an area. For pipes that diameter is not so large, however, the pressure at cross sections can be considered uniform.

4.4 Applications of the Bernoulli Equation

So far, we have discussed the fundamental aspects of the Bernoulli equation. Now, we introduce some of its applications.

4.4.1 Velocity measurement by a Pitot tube

The Bernoulli equation states that: the sum of the potential, pressure, and kinetic energies of a fluid particle along a streamline is constant. Therefore the kinetic and potential energies of the fluid can be converted to pressure energy during flow, causing the pressure to change. This can be made more visible by multiplying the Bernoulli equation by the density and acceleration of gravity.

$$\rho gz + p + \frac{\rho V^2}{2} = \text{Constant} \qquad (4.4.1)$$

Each term in this equation has pressure unit, therefore, the Bernoulli equation also states that the total pressure along a streamline is constant. Each term in Eq.4.4.1 represents some kind of pressure: p is the static pressure, represents the actual pressure in

式中，R 是流线在此处的半径。重力在此处沿流线法向的分量为 $-W\cos\theta = -\rho g\mathrm{d}A\mathrm{d}r\cos\theta =$ $-\rho g\mathrm{d}A\mathrm{d}r\dfrac{\mathrm{d}z}{\mathrm{d}r} = -\rho g\mathrm{d}A\mathrm{d}z$，其中 θ 是 z 坐标

轴和流线法向的夹角，且 $\cos\theta = \mathrm{d}z/\mathrm{d}r$。

式 4.3.16 可化简为

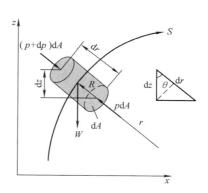

$$\frac{\mathrm{d}z}{\mathrm{d}r} + \frac{1}{\rho g}\frac{\mathrm{d}p}{\mathrm{d}r} = \frac{V^2}{Rg}\qquad (4.3.17)$$

将式 4.3.17 改写为

$$\frac{d}{\mathrm{d}r}\left(\frac{p}{\rho g} + z\right) = \frac{V^2}{Rg}\qquad (4.3.18)$$

对于沿直线流动，$R\to\infty$，式 4.3.18 化为

图 4.3.5　流线法向流体微元的力平衡

$\dfrac{d}{\mathrm{d}r}\left(\dfrac{p}{\rho g} + z\right) = 0$，则可以得出 $\dfrac{p}{\rho g} + z = C$，这正是绝对静止流体的静压强在竖直方向分布规律的表达式。

所以，对于非黏性、不可压缩流体沿直线运动，其压强在竖直方向的分布规律与静止流体内的分布规律相同。或者说，在任何与稳定流动方向垂直的平面上，流体内部压强都按照静压强规律分布。对于大型管道，可能需要考虑流线法向的压强分布。管道截面上的平均压强等于管道中心处的压强，靠近管道顶部压强最小，如果有气穴现象，会首先在此处出现。对于直径不太大的管路，截面压强可以视为均匀分布。

4.4　伯努利方程的应用

之前，我们讨论了伯努利方程的基本内容，现在我们来介绍该方程的应用。

4.4.1　毕托管测流速

伯努利方程表明流体沿流线运动时，其势能、压力能和动能的总和是恒定不变的。所以在流动过程中，流体的势能和动能可以转化为压力能，导致压力的改变。将伯努利方程等式两边都乘以密度和重力加速度可以看得更加清楚。

$$\rho gz + p + \frac{\rho V^2}{2} = C\qquad (4.4.1)$$

上式中各项的单位都是压强，所以伯努利方程也可以解释为沿流线的总压强保持恒定。用压强来解读式 4.4.1 中的各项：p 是静压强，代表流体内不会由于测量仪器而改变的实际压强。流体静压强项 ρgz，是受高度位置影响的部分，随液体深度增加而减小（淹深 h 增大意味着 z 坐标值减小）。$\rho V^2/2$ 是动压强，表示当流体的运

the fluid unchanged by the measuring instrument. The hydrostatic pressure term $\rho g z$, accounts for the elevation effects and decreases with fluid depth (the submergence depth h increases with the z-ordinate value decreases). $\rho V^2/2$ is the dynamic pressure, it represents the pressure rise when the fluid in motion is brought to a stop isentropically.

A combination of a Piezometer tube and a Pitot tube is used in the measurement of flow velocity, as illustrated in Fig.4.4.1a. A Piezometer is simply a small tube opening at a hole drilled into the wall of a conduit such that the plane of the hole is parallel to the flow direction and flush with the wall surface. The Piezometer tube, or simply Piezometer, measures the static pressure. A Pitot tube, named after its inventor Henri Pitot (1695—1771), is a small tube with its open end aligned into the flow and facing upstream, so as to sense the full impact pressure of the flowing fluid. It measures the total pressure.

A point at nose of the Pitot tube, such as point 0, as show in Fig.4.4.1a where the fluid is brought to a complete stop with unchanging entropy, i.e. all the kinetic energy is converted into pressure energy, called the stagnation point. Pressure at stagnation point is the total of the static and dynamic pressures.

Consider point 1 and at some distance away from stagnation point 0, in Fig.4.4.1a, along the same streamline and of the same height. We can derive the formula for the Pitot tube for incompressible flow by writing the Bernoulli equation between points 1 and 0

$$\frac{p_1}{\rho g} + \frac{V_1^2}{2g} = \frac{p_0}{\rho g} \tag{4.4.2}$$

when static and stagnation pressures are measured at these two points, or, the pressure difference is measured, the fluid velocity at that location is calculated from

$$V = \sqrt{\frac{2(p_0 - p_1)}{\rho}} \tag{4.4.3}$$

In situations which the static and stagnation pressure of a flowing liquid are greater than atmospheric pressure, the liquid rises in the vertical tube to a head that is proportional to the pressure being measured. If the pressures to be measured are below atmospheric, or if measuring pressures in gases, Piezometer tubes do not work. However, the static pressure tap and Pitot tube must be connected to some other kind of pressure measurement device such as a U-tube.

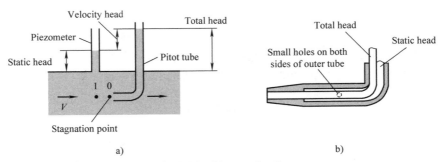

Fig.4.4.1   Pitot-static tube
a) The principle of Pitot tube   b) The structure of a Pitot tube

Sometimes it is convenient to integrate static pressure holes on a Pitot tube. The result

动忽然停止时会带来的压强陡增。

图 4.4.1a 表示的是把静压管和毕托管组合而成的流速测量装置。静压管是在被测管道壁上钻一个小孔而后连接的小管，孔口与该处管道壁平齐且与流体流动方向平行。静压管顾名思义，测量的是静压。毕托管，以其发明人 Henri Pitot（1695—1771）命名，小管的开口端置于流体内并径直迎向流体，以感受流动流体的全部压力冲击。毕托管测量的是流体的总压。

毕托管的前端，如图 4.4.1a 中的 0 点，运动的流体在此发生等熵停驻，也就是全部的动能都转化为压力能，因此 0 点也称为驻点。驻点处的压强是静压和动压的总和。

我们来分析图 4.4.1a 中点 1 和靠近驻点 0 的两个位置，它们处于同一流线且高度相同。我们通过在这两点间建立伯努利方程，来说明毕托管测量不可压缩流体流速的原理

$$\frac{p_1}{\rho g} + \frac{V_1^2}{2g} = \frac{p_0}{\rho g} \tag{4.4.2}$$

测量这两点的静压和总压，或者测量两点压强差，就可以计算此处流体的流速

$$V = \sqrt{\frac{2(p_0 - p_1)}{\rho}} \tag{4.4.3}$$

对于液体，静压和总压都高于大气压时，竖管内的液柱会升高，高度与被测压强成正比。如果测量的是气体流速，或被测液体的压强低于大气压，静压管不适用。这时，毕托管需要与一些可以测静压的装置，如 U 形管，连接使用。

通常毕托管和静压管是集成在一起的，其结构如图 4.4.1b 所示。流体通过外管壁上的两个或多个孔进入环形间隙，从而测量静压。

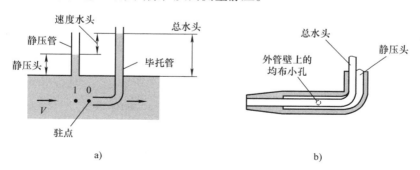

图 4.4.1　毕托管

a）毕托管原理　b）毕托管结构

例 4.4.1　用毕托管测量水平管道内流体平均流速，静压管及总压管液柱高度如图 X4.4.1 所标注。

is a Pitot-static tube, as shown in Fig.4.4.1b. We measure the static pressure through two or more holes drilled through an outer tube into an annular space.

Example 4.4.1 A Piezometer and a Pitot tube are tapped into a horizontal water pipe, as shown in Fig.X4.4.1, to measure static and total (static + dynamic) pressures. For the indicated water column heights, determine the average velocity at the center of the pipe.

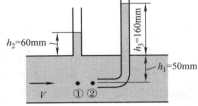

Solution This is a steady flow with straight and parallel streamlines, and the gage pressures at points ① and ② can be expressed as

$$p_1 = \rho g\left(h_1 + h_2\right), \ p_2 = \rho g\left(h_1 + h_3\right)$$

Fig.X4.4.1

Noting that $z_1 = z_2$, point ② is a stagnation point and thus $V_2 = 0$, the application of the Bernoulli equation between points ① and ② gives

$$\cancel{z_1} + \frac{p_1}{\rho g} + \frac{V_1^2}{2g} = \cancel{z_2} + \frac{p_2}{\rho g} + \frac{\cancel{V_2^2}}{2g} \ \rightarrow \ \frac{p_1}{\rho g} + \frac{V_1^2}{2g} = \frac{p_2}{\rho g} \ \rightarrow \ V_1 = \sqrt{\frac{2\left(p_2 - p_1\right)}{\rho}}$$

Substituting the p_1 and p_2 expressions gives

$$V_1 = \sqrt{2g\left(h_3 - h_2\right)} = 1.4\text{m/s}$$

Note that to determine the average flow velocity, all we need is to measure the height of the fluid column in the Pitot tube compared to that in the Piezometer tube.

Example 4.4.2 Air of density 1.2kg/m³ flows in a pipeline, a manometer which measuring fluid is water is used to measure the velocity of the air. Height differential of the two water column is 60mm (Fig.X4.4.2). Find the velocity of air in the pipeline.

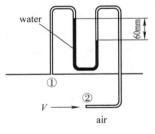

Solution The height distance between point ① and ② is negligible because air is very light.

The air velocity in the pipe is expressed by Eq.4.4.2

$$V = \sqrt{\frac{2\left(p_2 - p_1\right)}{\rho_{\text{air}}}}$$

Fig.X4.4.2

where the pressure difference measured in U-tube manometer is

$$p_2 - p_1 = \rho_{\text{water}} g \Delta h \quad \text{(Also, the weight of } \Delta h \text{ air column is neglected)}$$

therefore, the velocity of air flowing in the pipe is calculated by

$$V = \sqrt{\frac{2\rho_{\text{water}} g \Delta h}{\rho_{\text{air}}}} = 31.3\text{m/s}$$

Note that, fluid flowing in the pipe is air, the static pressure maybe below the atmosphere, and the U-tube manometer is used to measure the static pressure. The density of water is applied when pressures are calculated.

解　管内流体沿平直流线做稳定流动，点①和点②的表压强可以表示为

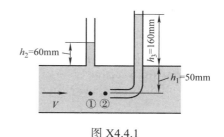

图 X4.4.1

$$p_1 = \rho g\left(h_1 + h_2\right),\ p_2 = \rho g\left(h_1 + h_3\right)$$

由于①、②两点高度相同 $z_1 = z_2$，且点②为驻点，所以有 $V_2 = 0$，在点①和②间建立伯努利方程，得

$$\cancel{z_1} + \frac{p_1}{\rho g} + \frac{V_1^2}{2g} = \cancel{z_2} + \frac{p_2}{\rho g} + \frac{\cancel{V_2^2}}{2g} \ \rightarrow\ \frac{p_1}{\rho g} + \frac{V_1^2}{2g} = \frac{p_2}{\rho g} \ \rightarrow\ V_1 = \sqrt{\frac{2\left(p_2 - p_1\right)}{\rho}}$$

代入 p_1 和 p_2 的表达式，得

$$V_1 = \sqrt{2g\left(h_3 - h_2\right)} = 1.4\,\text{m/s}$$

由例题可知，测量流体平均流速时仅需测得毕托管和静压管中的液柱高就可以完成。

例 4.4.2　管道内流动的空气密度为 $1.2\,\text{kg/m}^3$，测压管中工作液体为水，现测量管道内气体流速。测压管内两段水柱高度差为 60mm（见图 X4.4.2）。计算管道内气体的流速。

解　因为空气质量很轻，所以忽略点①和点②间的高度差。

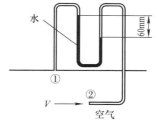

图 X4.4.2

用式 4.4.2 表示管道内气体流速

$$V = \sqrt{\frac{2\left(p_2 - p_1\right)}{\rho_{\text{气}}}}$$

式中两点压差由 U 形测压管测得

$$p_2 - p_1 = \rho_{\text{水}} g \Delta h \quad (\text{同样忽略高度为 } \Delta h \text{ 的空气柱重量})$$

所以，计算管道内气体流速可得

$$V = \sqrt{\frac{2\rho_{\text{water}} g \Delta h}{\rho_{\text{air}}}} = 31.3\,\text{m/s}$$

注意，管道内的流体为气体，其静压强可能低于大气压，所以使用 U 形管来测量静压强。计算时需考虑水的密度。

例 4.4.3　如图 X4.4.3 所示，直径 600mm 的管道内气体密度为 $1.2\,\text{kg/m}^3$，水

Example 4.4.3 Air of density 1.2kg/m³ flowing in a conduit of diameter 600mm as shown in Fig.X4.4.3, the height of water column h is 45mm. What is the mass flow rate of air in the conduit?

Solution Using the Bernoulli equation on the control volume from section ① to ②, where section ② at a little distance away from the opening of the pipe.

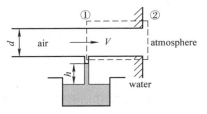

Note that, at opening of the pipe, gage pressure is $p_2 = 0$, and air spreads when flow out from the opening, that is $V_2 = 0$.

$$\frac{p_1}{\rho_{air}g} + \frac{V_1^2}{2g} = 0$$

Fig.X.4.4.3

The water column is above the free surface in the tank, it implies that the pressure inside the conduit is below the atmosphere.

$$p_1 = -\rho_{water}gh$$

The velocity of air flowing in the conduit is given by

$$V_1 = \sqrt{2g\frac{\rho_{water}}{\rho_{air}}h}$$

The mass flow rate is

$$q_m = \frac{\pi D^2}{4}\rho_{air}V_1 = \frac{\pi D^2}{4}\sqrt{2g\rho_{water}\rho_{air}h} = 9.20\text{kg/s}$$

Example 4.4.4 A Piezometer and a Pitot tubes are attached to a horizontal water nozzle, as shown in FigX.4.4.4. Velocity of jet outside the nozzle is measured of 20m/s. For the indicated mercury column height differential $h = 50$mm, determine the average velocity and static pressure in the pipe.

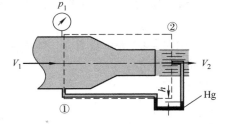

Solution The tube outside the nozzle is the Pitot tube, measuring the total head; the tube at wall of the pipe is the Piezometer tube which measuring the static pressure head.

The pressure at jet section ② is atmospheric, where all the static pressure energy conversed

Fig.X4.4.4

into the kinetic energy, the total head is velocity head. By using Bernoulli equation on control volume between section ① and ②, the total head keeps constant

$$H_{total} = \frac{V_2^2}{2g} = \frac{p_1}{\rho g} + \frac{V_1^2}{2g}$$

The static head is given by

$$H_1 = \frac{p_1}{\rho_{water}g}$$

Therefore the average velocity in the pipe can be calculated from

柱高度 h 为 45mm。计算管道内气体的质量流量。

解　对截面①—②间的控制体应用伯努利方程，截面②位于管道出口处稍向外一点的位置。

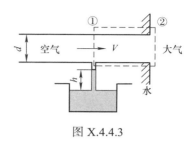

图 X.4.4.3

由于管道出口开放，所以此处表压强 $p_2 = 0$，且空气从出口流出后立即分散开，所以有 $V_2 = 0$。

$$\frac{p_1}{\rho_{\text{气}}g} + \frac{V_1^2}{2g} = 0$$

测压管内水柱高于水箱液水面，说明管道内的气压低于大气压。

$$p_1 = -\rho_{\text{水}}gh$$

管道内空气的流速可以表示为

$$V_1 = \sqrt{2g\frac{\rho_{\text{水}}}{\rho_{\text{气}}}h}$$

质量流量为

$$q_{\text{m}} = \frac{\pi D^2}{4}\rho_{\text{气}}V_1 = \frac{\pi D^2}{4}\sqrt{2g\rho_{\text{水}}\rho_{\text{气}}h} = 9.20\text{kg/s}$$

例 4.4.4　如图 X.4.4.4 所示，将毕托管安装于喷水管的喷嘴外，喷嘴外水流喷射速度为 20m/s。测压管两侧水银柱高度差 $h = 50$mm，计算管道内水的平均流速和静压强。

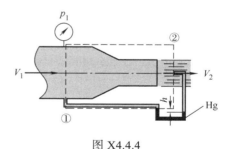

图 X4.4.4

解　管外喷嘴处为总压管，测得的是总压；管壁处的是静压管，测量静压头。

截面 2 处为大气压，管内静压在此处全部转化为流体的动压，所以总压头就是速度水头。在截面①—②间的控制体上应用伯努利方程，总压头保持恒定

$$H_{\text{total}} = \frac{V_2^2}{2g} = \frac{p_1}{\rho g} + \frac{V_1^2}{2g}$$

静压头表示为

$$H_1 = \frac{p_1}{\rho_{\text{水}}g}$$

管道内气体平均流速可通过下式计算

$$H_{\text{total}} - H_1 = \frac{\left(\rho_{\text{Hg}} - \rho_{\text{water}}\right)gh}{\rho_{\text{water}}g} = \frac{V_1^2}{2g} \quad \rightarrow \quad V_1 = 3.51\text{m/s}$$

Substituting the V_2 and V_1 gives

$$p_1 = 193.8\text{kPa}$$

Example 4.4.5 A manometer and Pitot-piezometer tube are used to measure the velocity of oil flow shown in Fig.X4.4.5. Density of oil is known as 800kg/m^3. The mercury column height differential of U-tube is 60mm, find the velocity of oil flow.

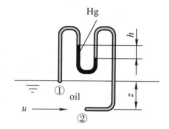

Solution The mercury column heights represent the total head and static pressure respectively. Note that point ② at tip of the Pitot probe is the stagnation point, that is $V_2 = 0$.

Fig.X4.4.5

This is a steady flow with straight and parallel streamlines, the application of the Bernoulli equation between points ① and ② gives

$$z + \frac{p_1}{\rho g} + \frac{V_1^2}{2g} = \frac{p_2}{\rho g} \quad \rightarrow \quad \frac{V^2}{2g} = \frac{p_2 - p_1}{\rho g} - z$$

That $p_2 - p_1$ measured by the U-tube manometer gives

$$p_2 - p_1 = \rho gz + \left(\rho' - \rho\right)gh$$

where ρ' and ρ are density of mercury and oil respectively. Substituting the pressure differential expression gives

$$V = \sqrt{2gh\frac{\rho' - \rho}{\rho}} = 4.34\text{m/s}$$

Example 4.4.6 Water flows in a wide open channel as shown in Fig.X4.4.6. Two pitot tubes are connected to a differential manometer containing a liquid ($\rho = 680\text{kg/m}^3$). Find u_A and u_B.

Solution u_A is measured by the upper pitot tube

$$\frac{u_A^2}{2g} = 50 \times 10^{-3}\text{m/s} \quad \rightarrow \quad u_A = 0.99\text{m/s}$$

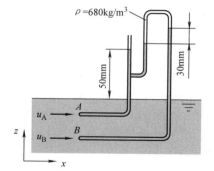

Fig.X4.4.6

The pressure differential between point A and point B is

$$p_B - p_A = \rho gz_{A-B} + \left(\rho - \rho'\right)g\Delta h \quad \rightarrow \quad \frac{p_B - p_A}{\rho g} = z_{A-B} + \frac{\left(\rho - \rho'\right)}{\rho}\Delta h$$

From Bernoulli equation between point A and B

$$H_{\text{total}} - H_1 = \frac{(\rho_{水银} - \rho_{水})gh}{\rho_{水}g} = \frac{V_1^2}{2g} \quad \rightarrow \quad V_1 = 3.5\text{lm/s}$$

代入 V_2 和 V_1, 得

$$p_1 = 193.8\text{kPa}$$

例 4.4.5　用毕托管测量如图 X4.4.5 所示的油液流速。油液密度已知为 800kg/m^3。U 形管内水银柱高度差为 60mm, 计算油液流速。

解　测压管内两段水银柱的高度分别代表总压和静压。毕托管前端处的点②为驻点, 所以可知 $V_2 = 0$。

流场内油液沿平直流线流动, 在点①和点②应用伯努利方程, 得

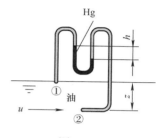

图 X4.4.5

$$z + \frac{p_1}{\rho g} + \frac{V_1^2}{2g} = \frac{p_2}{\rho g} \quad \rightarrow \quad \frac{V^2}{2g} = \frac{p_2 - p_1}{\rho g} - z$$

压强差 $p_2 - p_1$ 由 U 形测压管测得

$$p_2 - p_1 = \rho g z + (\rho' - \rho)gh$$

式中, ρ' 和 ρ 分别代表水银和油液的密度, 代入压强差, 得

$$V = \sqrt{2gh\frac{\rho' - \rho}{\rho}} = 4.34\text{m/s}$$

例 4.4.6　开放沟渠内的水流如图 X4.4.6 所示, 现使用两段毕托管与测压管连接, 测压管内工作液体密度 $\rho = 680\text{kg/m}^3$。试确定渠内不同深度处的流速 u_A 和 u_B。

解　u_A 由上方的毕托管测量

$$\frac{u_A^2}{2g} = 50 \times 10^{-3}\text{m/s} \quad \rightarrow \quad u_A = 0.99\text{m/s}$$

A、B 两点间的压差为

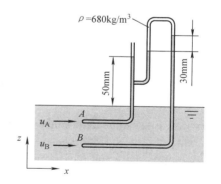

图 X4.4.6

$$p_B - p_A = \rho g z_{A-B} + (\rho - \rho')g\Delta h \quad \rightarrow \quad \frac{p_B - p_A}{\rho g} = z_{A-B} + \frac{(\rho - \rho')}{\rho}\Delta h$$

在 A、B 两点间应用伯努利方程

$$z_{A-B} + \frac{p_A}{\rho g} + \frac{u_A^2}{2g} = \frac{p_B}{\rho g} + \frac{u_B^2}{2g}$$

Here substituting for $(p_B - p_A)/\rho g$

$$\frac{u_A^2}{2g} - \frac{u_B^2}{2g} = \frac{(\rho - \rho')}{\rho}\Delta h \rightarrow \frac{u_B^2}{2g} = \frac{u_A^2}{2g} - \frac{(\rho - \rho')}{\rho}\Delta h = 50 \times 10^{-3} \text{m/s} - \frac{(1000 - 680)}{1000} 30 \times 10^{-3} \text{m/s}$$

$$u_B = 0.89 \text{m/s}$$

4.4.2 Rate of flow measurement by a Venturi meter

We know from the preceding analysis, that kinetic and pressure energies can be inter-converted without loss. There is a definite relation between the pressure differential and the velocity of flow. A decrease in velocity head will equals the increase in pressure head for an incompressible steady flow of a non-viscous fluid in a horizontal pipe.

Consider incompressible steady flow of a fluid in a horizontal pipe of diameter D that is constricted to a flow area of diameter d, as shown in Fig.4.4.2a. The Bernoulli equation between point1 and point 2, where the locations before the constriction and constriction occurs, is written as

$$\frac{p_1}{\rho g} + \frac{V_1^2}{2g} = \frac{p_2}{\rho g} + \frac{V_2^2}{2g} \qquad (4.4.4)$$

Continuity allows us to relate V_2 to V_1 by

$$V_1 = \left(\frac{d}{D}\right)^2 V_2 = \beta^2 V_2 \qquad (4.4.5)$$

where $\beta = d/D$, is the diameter ratio.

Combining Eq.4.4.4 and 4.4.5 and solving for velocity V_2 gives

$$V_2 = \sqrt{\frac{2(p_1 - p_2)}{\rho(1 - \beta^4)}} \qquad (4.4.6)$$

the flow rate can be determined from $q_V = V_2 A_2 = V_2 \pi d^2 / 4$.

This simple analysis shows that the flow rate through a pipe can be determined by constricting the flow and measuring the decrease in pressure due to the increase in velocity at the obstruction site. Noting that the pressure drop between two points along the flow is measured easily by a pressure manometer, it appears that a simple flow rate measurement device can be built by constricting the flow. The venturi meter, named after Italian physicist Giovanni B. Venturi (1746—1822), based on this principle is used for measuring the flow rates of gases and liquids.

As shown in Fig.4.4.2b, the venturi meter consists of a tube with a constricted throat, which produces an increased velocity accompanied by a reduction in pressure, followed by a gradually diverging portion in which the velocity is transformed back into pressure with slight friction loss.

$$z_{A-B} + \frac{p_A}{\rho g} + \frac{u_A{}^2}{2g} = \frac{p_B}{\rho g} + \frac{u_B{}^2}{2g}$$

上式代入 $(p_B - p_A)/\rho g$，得

$$\frac{u_A{}^2}{2g} - \frac{u_B{}^2}{2g} = \frac{(\rho - \rho')}{\rho}\Delta h \rightarrow \frac{u_B{}^2}{2g} = \frac{u_A{}^2}{2g} - \frac{(\rho - \rho')}{\rho}\Delta h = 50 \times 10^{-3}\,\text{m/s} - \frac{(1000 - 680)}{1000}30 \times 10^{-3}\,\text{m/s}$$

$$u_B = 0.89\,\text{m/s}$$

4.4.2　文丘里流量计测流量

从前述内容可知，动能和压力能可以无损失的互相转换。流体的流速与压强间存在明确的关联。水平管道内不可压缩、无黏流体速度水头的降低等于压强水头的升高。

考虑图 4.4.2a 所示的不可压缩流体在直径为 D 的水平管道内稳定流动，与另一直径 d 的约束面相通。在约束前、后的点 1 和点 2 应用伯努利方程，可以写为

$$\frac{p_1}{\rho g} + \frac{V_1^2}{2g} = \frac{p_2}{\rho g} + \frac{V_2^2}{2g} \tag{4.4.4}$$

应用连续性方程由 V_2 表示 V_1

$$V_1 = \left(\frac{d}{D}\right)^2 V_2 = \beta^2 V_2 \tag{4.4.5}$$

式中，$\beta = d/D$，为两截面直径比

用式 4.4.4 和式 4.4.5 解 V_2 得到

$$V_2 = \sqrt{\frac{2(p_1 - p_2)}{\rho(1 - \beta^4)}} \tag{4.4.6}$$

可由流速计算体积流量 $q_V = V_2 A_2 = V_2 \pi d^2/4$。

由以上分析可知，在流场内制造一处约束，可以在限流处形成压降从而产生流速的提高，我们能够应用这种现象测量流体的流量。由于沿流动路径上两点间的压差是很容易测得的，所以通过人为制造限流的简单装置就可以测量流量。文丘里流量计，以意大利物理学家 Giovanni B.Venturi（1746—1822）命名，就是根据这一原理测量气体和液体的流量。

如图 4.4.2b 所示，文丘里流量管有一处收缩的喉部，这里会导致压强下降，同时引起该处流速增加，喉部之后是渐扩段，流体在这一段区域逐渐恢复到原来的流速并且没有明显的摩擦损失。

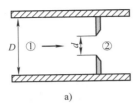 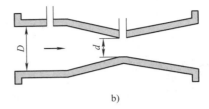

a)   b)

Fig.4.4.2   Principle and structure of venture meter
a) Flow through a constriction in a pipe   b) a venturi meter

The Venturi tube provides an accurate means for measuring flow in pipelines. With a suitable recording device, we can integrate the flow rate so as to give the total quantity of flow. The only disadvantage of the venturi meter is that it introduces a permanent frictional resistance in the pipeline. Practically all this loss occurs in the diverging part, and is ordinarily from $0.1h$ to $0.2h$, where h is the static-head differential between the upstream section and the throat.

Example 4.4.7   The venturi meter shown reduces the pipe diameter from 100mm to a minimum of 50mm (Fig. X4.4.7). Heights differential of fluid column is 60mmHg. Calculate the flow rate assuming ideal conditions.

Solution   The control volume is selected as shown in Fig.X4.4.7, the Bernoulli Equation assuming ideal conditions on section ① and ② takes the form

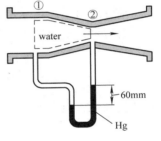

Fig.X4.4.7

$$\frac{p_1}{\rho g}+\frac{V_1^2}{2g}=\frac{p_2}{\rho g}+\frac{V_2^2}{2g}$$

Continuity allows us to relate V_2 to V_1 by

$$V_1A_1=V_2A_2 \ \rightarrow \ \frac{V_2}{V_1}=\left(\frac{d_1}{d_2}\right)^2=4 \ \rightarrow \ V_2=4V_1$$

The manometer's reading is interpreted as follows

$$p_1-p_2=(\rho'-\rho)g\Delta h$$

where ρ' is the density of mercury, and Δh is the column heights differential. Solving for velocity V_1 gives

$$V_1=\sqrt{\frac{2(\rho'-\rho)\Delta h}{15\rho}}=0.1\text{m/s}$$

The flow rate is

$$q=\frac{\pi d_1^2}{4}V_1=0.785\times10^{-3}\,\text{m}^3\text{/s}$$

Example 4.4.8   The flow rate of oil at 20℃ ($\rho = 850\text{kg/m}^3$) through a 260mm diameter pipe to be measured with a 180mm diameter throat venturi meter equipped with a

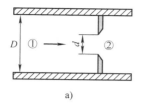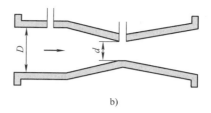

图 4.4.2　文丘里流量计原理和结构

a）流体经过管内限流处　b）文丘里流量管

文丘里流量计可以精确测量管道内流体的流速。使用适当的记录装置，通过计算就可以测得系统的流量。文丘里流量计唯一的不足在于其在管路上产生了摩擦阻力。实验证实，全部的损失都发生在渐缩段，损失水头在（0.1～0.2）h 之间，这里 h 是喉部前后压强水头之差。

例 4.4.7　管路直径 100mm 文丘里流量计喉部直径 50mm（见图 X4.4.7）。测压管水银柱高度差 60mmHg。计算理想条件下管内流量。

解　选取图 X4.4.7 所示控制体，流动符合伯努利方程条件，在截面①—②间应用伯努利方程

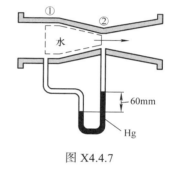

图 X4.4.7

$$\frac{p_1}{\rho g}+\frac{V_1^2}{2g}=\frac{p_2}{\rho g}+\frac{V_2^2}{2g}$$

依据连续性方程用 V_1 表示 V_2

$$V_1 A_1 = V_2 A_2 \quad \rightarrow \quad \frac{V_2}{V_1}=\left(\frac{d_1}{d_2}\right)^2=4 \quad \rightarrow \quad V_2=4V_1$$

通过测压管读数计算压强差

$$p_1-p_2=\left(\rho'-\rho\right)g\Delta h$$

式中，ρ' 是水银的密度；Δh 是两段水银柱高度差。
解 V_1 得

$$V_1=\sqrt{\frac{2\left(\rho'-\rho\right)\Delta h}{15\rho}}=0.1\text{m/s}$$

流量为

$$q=\frac{\pi d_1^2}{4}V_1=0.785\times10^{-3}\,\text{m}^3/\text{s}$$

例 4.4.8　用喉部直径 180mm 的文丘里流量计测量直径 260mm 管道内油液的

piston cylinder to balance the pressure drop, as shown in Fig.X4.4.8. The diameter of
cylinder is 200mm and the area of piston
rod is negligible, force needed to keep
the piston stationary is 80N. Assuming
all ideal conditions. Determine the flow
rate of oil through the pipe.

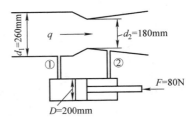

Fig.X4.4.8

Solution   The Bernoulli Equation assuming ideal conditions on section 1 and
2 takes the form

$$\frac{p_1}{\rho g} + \frac{V_1^2}{2g} = \frac{p_2}{\rho g} + \frac{V_2^2}{2g}$$

The continuum equation is applied to relate V_1 and V_2

$$V_2 = \left(\frac{260}{180}\right)^2 V_1$$

Balance between pressure differential and piston force gives

$$p_1 \frac{\pi D^2}{4} = p_2 \frac{\pi D^2}{4} + F \quad \rightarrow \quad p_1 - p_2 = \frac{4F}{\pi D^2}$$

From this the velocity of oil flow is found

$$V_1 = 1.79 \text{m/s}$$

The flow rate of oil through the pipe is

$$q_V = V_1 \frac{\pi d_1^2}{4} = 0.095 \text{m}^3/\text{s}$$

4.4.3   Method of Solution of Liquid Flow Problems

For the solutions of problems of incompressible liquid steady flow, there are two
fundamental equations: the equation of continuity Eq.4.2.11 and the energy equation
Eq.4.3.14. We may employ the following procedure:

1. Choose a datum plane through any convenient point, usually the lower one, especially the one of atmospheric pressure.

2. Note at what sections we know or must assume the velocity. If at any point the section area is great compared with its value elsewhere, the velocity head is so small that we
may disregard it.

3. Note at what points we know or must assume the pressure. In a volume of liquid at
rest with a free surface we know the pressure at every point within the volume. The gage
pressure at free surface is atmospheric pressure. The pressure in a jet is the same as that
of the medium (atmosphere) surrounding the jet.

4. Note whether or not there is any point where we know all three terms, pressure, elevation, and velocity.

5. Note whether or not there is any point where there is only one unknown quantity.

6. Write an energy equation that will fulfill conditions 4 and 5.

7. If there are two unknowns in the equation then we must also use the continuity
equation.

流量，如图 X4.4.8 所示，油液温度 20℃，密度 $\rho = 850\text{kg/m}^3$，流量计通过活塞缸加压，活塞直径 200mm，活塞杆面积可忽略，保持活塞固定所需力为 80N。计算理想条件下管内油液流量。

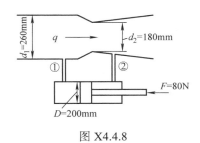

图 X4.4.8

解　符合伯努利方程假设条件下，在截面①—②上应用伯努利方程

$$\frac{p_1}{\rho g} + \frac{V_1^{\,2}}{2g} = \frac{p_2}{\rho g} + \frac{V_2^{\,2}}{2g}$$

根据连续性方程用 V_1 表示 V_2

$$V_2 = \left(\frac{260}{180}\right)^2 V_1$$

压强差与活塞上外力的平衡关系为

$$p_1 \frac{\pi D^2}{4} = p_2 \frac{\pi D^2}{4} + F \;\;\rightarrow\;\; p_1 - p_2 = \frac{4F}{\pi D^2}$$

由此解得油液流速为

$$V_1 = 1.79\text{m/s}$$

管道内油液流量为

$$q_V = V_1 \frac{\pi d_1^{\,2}}{4} = 0.095\text{m}^3/\text{s}$$

4.4.3　求解液体流动问题的方法

求解关于不可压缩液体、稳定流动的相关问题，主要有两种方法：式 4.2.11 连续性方程和式 4.3.14 能量方程。可以按照以下步骤求解：

1. 依方便选取基准面，通常选择位置更低的那一个截面，最好是大气压面。

2. 找出流速已知或可以假设的截面，如果某截面的面积与其他截面相比非常大，那么可以假设此处流速为零。

3. 找出压强已知或可以假设的点，有自由液面的静止流体内部各处压强是可知的，射流内压强与其周围介质压强（如大气压）相同。

4. 查找是否有某位置，其压强、高度及流速均为已知。

5. 查找是否有某位置，上述三项中仅有一项未知。

6. 在满足条件 4 和 5 的两个位置间列能量方程。

7. 如果方程中有两个未知量，则使用连续性方程替换其中一个。

4.5 The Momentum Equation

In this section, we use Newton's second law, i.e. the momentum conservation principle to determine the forces and torques associated with fluid flow. The momentum equation is primarily used to calculate the forces induced by the flow.

4.5.1 Forces Acting on a Control Volume

The momentum principle deals with the forces that act on the fluid mass in a designated control volume. The forces acting on a control volume consist of mass forces that act throughout the entire control volume and to be proportional to the mass of fluid (such as gravity, electric, and magnetic forces) and surface forces that act on the control surface (such as pressure, viscous forces and reaction forces at points of contact).

Mass forces act on each volumetric portion of the control volume. The mass force acting on a differential element of fluid of volume dV within the control volume, and we must perform a volume integral to account for the net mass force on the entire control volume. The most common mass force is that of gravity, the only mass force we considered here, which exerts a downward force on every differential element of the control volume.

The differential gravity acting on a small fluid element is simply its weight. In Cartesian coordinates we adopt the convention that g acts in the negative z-direction

$$dF_z = -\rho g d\Omega \qquad (4.5.1)$$

where $d\Omega$ is the volume of the small fluid element, ρ is average density of this element.

Since gravity is the only mass force being considered, integration of Eq4.5.1 yields

$$F_z = -\rho \mathcal{V} g \qquad (4.5.2)$$

here $\mathcal{V}$ is volume of the control volume.

Surface forces act on each portion of the control surface and to be proportional to the force area. For ideal fluid in this section, the only surface force we study is pressure. The pressure normal to a differential surface element of area dA is $dF_p = pdA$, for a control surface where the pressure can be considered uniform, mathematically we write $F_p = pA$.

Pressure always acts inwardly normal to a surface, we conveniently define pressure along one axis direction positive where velocity vector inwardly normal to the surface and negative for pressure where the velocity vector is outwardly normal to the surface of fluid element. Components of the pressure in Cartesian coordinates on the left and right faces of an element of fluid shown in Fig.4.5.1, the net pressure force at x-direction is

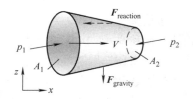

Fig.4.5.1 Forces on a fluid element

$$F_{p_x} = p_1 A_1 - p_2 A_2 \qquad (4.5.3)$$

A common simplify in the application of momentum equation is to work with gage pressures. This is because atmospheric pressure acts in all directions, and its effect cancels out in every direction. This means we can also ignore the pressure forces at outlet sections where the fluid is discharged at relative low velocities to the atmosphere and the discharge pressures in such cases are very near atmospheric pressure.

4.5 动量方程

本节我们应用牛顿第二定律，即动量守恒定律来确定与流体运动相关的力和力矩。动量定律主要应用于计算使流体流动的各项力。

4.5.1 控制体上的力

动量定律针对的对象为控制体内部流体所受的力。控制体上的力包括：控制体内全部流体受到的质量力，其大小与流体质量成正比，如重力、电磁力等；作用于控制面上的表面力，如压力、黏性力和接触点的反作用力等。

质量力作用于控制体内部各处，体积为 $\mathrm{d}\mathcal{V}$ 的微体积内流体所受的质量力对控制体积分，得到整个控制体内的质量力合力。最常遇到的质量力是重力，也是本书中唯一考虑的质量力，对控制体内每一个微元的作用方向都是竖直向下。

单位质量流体所受质量力等于其重量，用 g 表示，在直角坐标系中指向 z 轴负方向

$$\mathrm{d}F_z = -\rho g \mathrm{d}\Omega \qquad (4.5.1)$$

式中，$\mathrm{d}\Omega$ 是流体微元的体积；ρ 是微元的平均密度。

需考虑的唯一质量力，即式 4.5.1 表示的微元上重力，对控制体积分得到

$$F_z = -\rho \mathcal{V} g \qquad (4.5.2)$$

这里 $\mathcal{V}$ 是控制体的体积。

表面力作用于控制面的各处，其大小与力的作用面积成正比。对于本节所分析的理想流体，表面力仅是压力。流体微元表面的压力与微元面 $\mathrm{d}A$ 垂直，大小为 $\mathrm{d}F_p = p\mathrm{d}A$，如果假设压强均匀分布，那么压力值 $F_p = pA$。

压力总是沿表面的内法线方向，通常我们取流体与控制面垂直的流入方向为坐标轴正方向，沿此方向的压力为正；控制体流出面上的压力垂直指向流出面，为负。图 4.5.1 所示直角坐标系内的流体微元，左侧面和右侧面上沿 x 轴方向的净压力为

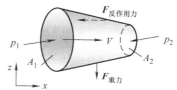

图 4.5.1 流体微元受力

$$F_{p_x} = p_1 A_1 - p_2 A_2 \qquad (4.5.3)$$

应用动量方程时，通常使用表压强，这是因为流体各方向都受到大气压的作用，各方向的大气压力可以互相抵消。那么，对于某些出流截面，流体以较低的流速排到大气中，这时流体内部的压强接近大气压，我也可以忽略其压力。

Other two terms combine with the pressure force to form the surface force are the viscous force tangential to the surface it acts which is discussed in Chp.6 and reaction forces which are equal in magnitude and opposite in direction with the forces the fluid exert on external objects.

In the light of these, the total external force on the element in Fig.4.5.1 consists of weight, pressure forces, and reaction forces, we write the total force on control volume for analysis of nonviscous fluid

$$\sum F = F_{\text{gravity}} + F_{\text{pressure}} + F_{\text{reaction}} \qquad (4.5.4)$$

All these surface forces arise as the control volume is isolated from its surroundings for analysis, and the effect of any detached object is accounted for by a force at that location.

4.5.2   The Linear Momentum Equation

The Newton's second law expressed as Eq.4.1.3 states that: The sum of all external forces acting on a system is equal to the time rate of change of momentum of the system, and this must be measured in an inertial reference frame. Noting that both the density and velocity may change from point to point within the system, Eq.4.1.3 can be expressed more generally as

$$\sum F = \frac{\mathrm{d}(mV)_{\text{sys}}}{\mathrm{d}t} \qquad (4.5.5)$$

The use of Eq. 4.5.5 is limited since most flow systems are analyzed using control volumes. The Reynolds transport theorem developed in Section 3.3 provides the necessary tools to shift from the system formulation to the control volume formulation. The Reynolds transport theorem (Eq.3.3.13) is expressed for linear momentum as

$$\sum F = \frac{\mathrm{d}(mV)_{\text{sys}}}{\mathrm{d}t} = \frac{\mathrm{d}(mV)_{\text{CV}}}{\mathrm{d}t} + \frac{\mathrm{d}(mV)_{\text{CS}}^{\text{out}}}{\mathrm{d}t} - \frac{\mathrm{d}(mV)_{\text{CS}}^{\text{in}}}{\mathrm{d}t} \qquad (4.5.6)$$

here the momentum flux is positive at outlet and negative at inlet control surface, since the velocity vector outward normal to the control surface at exit and inward normal at entrance.

Eq.4.5.6 is interpreted as: The sum of all external forces acting on a system (fluid) equals the sum of the time rate of change of linear momentum of the contents of control volume and the net flow rate of linear momentum out of the control surface by mass flow.

In the case of steady flow, conditions within the control volume do not change with time, so $\mathrm{d}(mV)_{\text{CV}}/\mathrm{d}t = 0$, and the equation becomes

$$\sum F = \frac{\mathrm{d}(mV)_{\text{sys}}}{\mathrm{d}t} = \frac{\mathrm{d}(mV)_{\text{CS}}^{\text{out}}}{\mathrm{d}t} - \frac{\mathrm{d}(mV)_{\text{CS}}^{\text{in}}}{\mathrm{d}t} \qquad (4.5.7)$$

thus, for steady flow the net force on the fluid mass is equal to the net rate of outflow of momentum across the control surface.

If we select a control volume so that the control surface is normal to the velocity where it cuts the flow, and we further specify the velocity is uniform where it across the control surface, then, for a device has finite number of entrances and exits, across which the steady flow

$$\sum F = \sum_{i=1}^{N} \dot{m}_i V_i \qquad (4.5.8)$$

　　构成表面力的另外两种力分别是黏性力和反作用力。黏性剪切力与流体表面相切，我们将在第 6 章讨论；流体所受的反作用力与流体作用于外界物体的力大小相同、方向相反。

　　由此，图 4.5.1 中的流体微元所受的外力包括重力、压力和反作用力。对非黏性流体，控制体上的合外力可以写为

$$\sum \boldsymbol{F} = \boldsymbol{F}_{重力} + \boldsymbol{F}_{压力} + \boldsymbol{F}_{反作用力} \tag{4.5.4}$$

　　分析表面力时要把控制体从其周围分离出来，其他可剥离物体的影响都视为外力。

4.5.2　线性动量方程

　　牛顿第二定律式 4.1.3 说明：惯性参照系内，系统所受合外力等于系统动量随时间的变化率。因为流体内部各处的密度和流速可能不同，式 4.1.3 的广义形式为

$$\sum \boldsymbol{F} = \frac{\mathrm{d}(m\boldsymbol{V})_{\text{sys}}}{\mathrm{d}t} \tag{4.5.5}$$

　　因为我们对大多数系统的分析是在控制体上进行的，所以式 4.5.5 的应用具有局限性。而 3.3 节介绍的雷诺输运定理可以帮助我们完成从系统到控制体的转换。用式 3.3.13 的雷诺输运定理表示动量的转换，可以有

$$\sum \boldsymbol{F} = \frac{\mathrm{d}(m\boldsymbol{V})_{\text{sys}}}{\mathrm{d}t} = \frac{\mathrm{d}(m\boldsymbol{V})_{\text{CV}}}{\mathrm{d}t} + \frac{\mathrm{d}(m\boldsymbol{V})_{\text{CS}}^{\text{out}}}{\mathrm{d}t} - \frac{\mathrm{d}(m\boldsymbol{V})_{\text{CS}}^{\text{in}}}{\mathrm{d}t} \tag{4.5.6}$$

因为出口处速度向量为外法线方向，而进口处速度向量为内法线方向，所以规定出口处动量通量为正，进口为负。

　　式 4.5.6 可以解释为：系统（流体）所受合外力等于控制体内动量随时间的变化率与控制体表面随质量流出的动量净通量。

　　对于稳定流动，控制体内部条件不随时间变化，所以 $\mathrm{d}(m\boldsymbol{V})_{\text{CV}}/\mathrm{d}t = 0$，上式可改写为

$$\sum \boldsymbol{F} = \frac{\mathrm{d}(m\boldsymbol{V})_{\text{sys}}}{\mathrm{d}t} = \frac{\mathrm{d}(m\boldsymbol{V})_{\text{CS}}^{\text{out}}}{\mathrm{d}t} - \frac{\mathrm{d}(m\boldsymbol{V})_{\text{CS}}^{\text{in}}}{\mathrm{d}t} \tag{4.5.7}$$

因此稳定流动的流体所受合外力等于经过控制体表面的动量净通量。

　　如果我们选取的控制体，其控制面与经过此处的流速垂直，且假设流速在其所流经的控制面上均匀分布，则对于一个稳定流动、仅包含有限个进出口的装置，可以列出

$$\sum F = \sum_{i=1}^{N} \dot{m}_i V_i \tag{4.5.8}$$

where N is the number of inlets and outlets, $\dot{m}$ denotes mass flow rate per unit time.

Unfortunately, the velocity across most inlets and outlets of practical engineering interest is not uniform, and a dimensionless correction factor β, called the *momentum correction factor*, is required for the accuracy of equation 4.5.8

$$\sum F = \sum_{i=1}^{N} \beta_i \dot{m}_i V_i \tag{4.5.9}$$

For the case in which V is in the same direction as average velocity $\bar{V}$ over the inlet or outlet, and differential area dA is control surface slices normal to the inlet or outlet flow, we solve β,

$$\beta = \frac{\int d(\dot{m}V)}{\dot{m}\bar{V}} = \frac{\int_A (\rho V) V dA}{(\rho \bar{V} A) \bar{V}} = \frac{1}{A} \int_A \left(\frac{V}{\bar{V}}\right)^2 dA \tag{4.5.10}$$

where densities cancel and since $\bar{V}$ is constant, it can be brought inside the integral.

It may be shown that $\beta = 1$ for the case of uniform flow over an inlet or outlet, and β is always greater than or equal to unity. For fully developed turbulent pipe flow, β ranges from about 1.01 to 1.04, is so close to unity, that many practicing engineers disregard the momentum correction factor. While β is not very close to unity for fully developed laminar pipe flow, and ignoring β could potentially lead to significant error. The momentum correction factor for laminar flow is discussed in section 6.3.

Substituting $\dot{m}$ by $\dot{m} = \rho VA = \rho q_V$, thus the Eq.4.5.9 takes the form

$$\sum F = \sum_{i=1}^{N} V_i \rho_i q_{V_i} \tag{4.5.11}$$

conveniently the overbar is omitted and V represents the average velocity.

Note that the unit vector points to the volume inwardly at the inlet, a negative symbol is taken for accounted. If there is only one entrance and one exit, the momentum equation becomes

$$\sum F = \rho_2 q_{V_2} V_2 - \rho_1 q_{V_1} V_1 \tag{4.5.12}$$

Using continuum equation for incompressible fluid $\rho_1 q_{V_1} = \rho_2 q_{V_2} = \rho q_V$, the momentum equation takes the simplified form

$$\sum F = \rho q_V (V_2 - V_1) \tag{4.5.13}$$

The momentum equation is a vector equation, each term should be treated as a vector. Also, the components of this equation can be resolved along orthogonal coordinates for convenience.

$$\begin{aligned} \sum F_x &= \rho q_V (u_2 - u_1) \\ \sum F_y &= \rho q_V (v_2 - v_1) \\ \sum F_z &= \rho q_V (w_2 - w_1) \end{aligned} \tag{4.5.14}$$

where u, v and w are velocity components in three directions.

Example 4.5.1 A free-surface flow in a rectangular channel of 5m wide is shown in FigX.4.5.1. When the upstream level is $h_1 = 6$m and downstream level is $h_2 = 1.2$m, the force of the gate on the flow is desired. The flow is steady, and the frictional effects are negligible.

式中，N 是进、出口的个数；$\dot{m}$ 是单位时间内流出（流入）的质量。

　　但在工程实际中，大多数进、出口截面上流速分布并不是均匀的，那么可以应用无量纲的修正系数 β，称为**动量修正系数**，来确保式 4.5.8 不失其精度

$$\sum F = \sum_{i=1}^{N} \beta_i \dot{m}_i V_i \qquad (4.5.9)$$

　　取一处进口或出口，速度向量 V 与平均流速 $\bar{V}$ 方向相同，控制面上的面积微元 $\mathrm{d}A$ 与进、出口流动方向垂直，解修正系数 β

$$\beta = \frac{\int \mathrm{d}(\dot{m}V)}{\dot{m}\bar{V}} = \frac{\int_A (\rho V) V \mathrm{d}A}{(\rho \bar{V} A) \bar{V}} = \frac{1}{A} \int_A \left(\frac{V}{\bar{V}}\right)^2 \mathrm{d}A \qquad (4.5.10)$$

式中约去了密度，且由于平均流速 $\bar{V}$ 保持不变，所以可移入积分符号内。

　　可见，对于均匀流动的进、出口 $\beta = 1$，且 β 取值总是大于等于 1。对于充分发展的管道内紊流，β 取值为 $1.01 \sim 1.04$，非常接近 1。因此，工程实践中往往忽略动量修正系数。但是对于充分发展的管道内层流，β 数值与 1 相差很多，所以被忽略可能导致较大误差。关于层流的动量修正系数的数值，会在 6.3 节详细讨论。

　　把式 4.5.9 中的质量 $\dot{m}$ 项替换为 $\dot{m} = \rho V A = \rho q_V$，则可以改写成

$$\sum F = \sum_{i=1}^{N} V_i \rho_i q_{V_i} \qquad (4.5.11)$$

为书写方便，此后都省略平均流速符号上方的横线，用 V 表示平均流速。

　　对仅有一个入口和一个出口的流动，因为进口处流速为控制体内法线方向，所以加负号，则动量方程为

$$\sum \boldsymbol{F} = \rho_2 q_{V_2} \boldsymbol{V}_2 - \rho_1 q_{V_1} \boldsymbol{V}_1 \qquad (4.5.12)$$

根据连续性方程，对于不可压缩流体 $\rho_1 q_{V_1} = \rho_2 q_{V_2} = \rho q_V$，动量方程化简为

$$\sum \boldsymbol{F} = \rho q_V (\boldsymbol{V}_2 - \boldsymbol{V}_1) \qquad (4.5.13)$$

　　动量方程为向量方程，式中各项均为向量，也可以写成直角坐标系的分量形式。

$$\begin{aligned}
\sum F_x &= \rho q_V (u_2 - u_1) \\
\sum F_y &= \rho q_V (v_2 - v_1) \\
\sum F_z &= \rho q_V (w_2 - w_1)
\end{aligned} \qquad (4.5.14)$$

式中，u、v 和 w 分别为三个坐标方向的速度分量。

　　例 4.5.1　水面为自由液面的矩形水渠宽 5m，如图 X.4.5.1 所示。上游水位高 $h_1 = 6\text{m}$，下游水位高 $h_2 = 1.2\text{m}$。求闸门所受水的冲击力。假设稳定流动，且忽略摩擦影响。

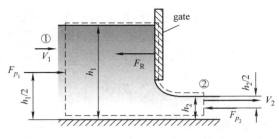

Fig.X4.5.1

Solution   We take the control volume in the dotted box as shown in Fig.X4.5.1, and designate the inlet by ① and the outlet by ② .

Forces acting on the control volume are upstream pressure Force F_{p_1}, downstream pressure Force F_{p_2}, the resistance from the gate F_R.

Taking the centers of inlet and outlet cross section as the acting points of static pressure, we have

$$P_1 = \rho g \frac{h_1}{2} \;\rightarrow\; F_{p_1} = \rho g \frac{h_1}{2} h_1 b = \rho g \frac{h_1^2 b}{2}$$

$$P_2 = \rho g \frac{h_2}{2} \;\rightarrow\; F_{p_2} = \rho g \frac{h_2}{2} h_2 b = \rho g \frac{h_2^2 b}{2}$$

The total external force on the control volume gives

$$\sum F = F_{p_1} - F_{p_2} - F_R$$

The momentum equation for steady flow is

$$\sum F = F_{p_1} - F_{p_2} - F_R = \rho q (V_2 - V_1)$$

We use the Bernoulli equation as the next approximation to calculate the velocities.

$$\frac{h_1}{2} + \frac{P_1}{\rho g} + \frac{V_1^2}{2g} = \frac{h_2}{2} + \frac{P_2}{\rho g} + \frac{V_2^2}{2g}$$

where $\dfrac{P_1}{\rho g} = \dfrac{h_1}{2}, \quad \dfrac{P_2}{\rho g} = \dfrac{h_2}{2}.$

and by continuity

$$V_1 h_1 b = V_2 h_2 b \;\rightarrow\; V_1 = 0.032 \text{m/s}, \qquad V_2 = 0.160 \text{m/s}, \qquad q = 0.958 \text{m}^3/\text{s}$$

The resistance acting on the control volume by the gate is given by

$$F_R = F_{p_1} - F_{p_2} - \rho q (V_2 - V_1) = 847.5 \text{kN} \leftarrow$$

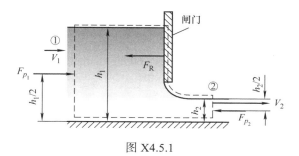

图 X4.5.1

解　取控制体如图 X4.5.1 中虚线框所示，指定截面①为流入面、截面②为流出面。

控制体受力为上游来水的压力 F_{p_1}，下游水形成的压力 F_{p_2} 和闸门对水产生的阻力 F_R。

取流入面和流出面中心处的静压强，得

$$p_1 = \rho g \frac{h_1}{2} \quad \rightarrow \quad F_{p_1} = \rho g \frac{h_1}{2} h_1 b = \rho g \frac{h_1^2 b}{2}$$

$$p_2 = \rho g \frac{h_2}{2} \quad \rightarrow \quad F_{p_2} = \rho g \frac{h_2}{2} h_2 b = \rho g \frac{h_2^2 b}{2}$$

控制体上的全部外力为

$$\sum F = F_{p_1} - F_{p_2} - F_R$$

稳定流动的动量方程为

$$\sum F = F_{p_1} - F_{p_2} - F_R = \rho q (V_2 - V_1)$$

应用伯努利方程计算流速

$$\frac{h_1}{2} + \frac{p_1}{\rho g} + \frac{V_1^2}{2g} = \frac{h_2}{2} + \frac{p_2}{\rho g} + \frac{V_2^2}{2g}$$

式中，$\dfrac{p_1}{\rho g} = \dfrac{h_1}{2}$，$\quad \dfrac{p_2}{\rho g} = \dfrac{h_2}{2}$。

继续应用连续性方程

$$V_1 h_1 b = V_2 h_2 b \quad \rightarrow \quad V_1 = 0.032\text{m/s}, \qquad V_2 = 0.160\text{m/s}, \qquad q = 0.958\text{m}^3/\text{s}$$

计算可得闸门对控制体内的水形成的阻力

$$F_R = F_{p_1} - F_{p_2} - \rho q (V_2 - V_1) = 847.5\text{kN} \leftarrow$$

The force on the gate by water flow is the same magnitude but opposite direction with F_R.

$$F_{gate} = 847.5kN$$

Example 4.5.2   A nozzle that discharges a 60mm-diameter water jet into the air is on the right end of a horizontal 120mm-diameter pipe (Fig.X4.5.2). Water in the pipe has a velocity of 4m/s and a gage pressure of 400kPa. Find the resultant axial force the water exerts on the nozzle. Neglect the friction.

Solution   The nozzle is selected to be the control volume, the pressure p_2 is zero because the flow discharges into the atmosphere.

The velocity at outlet gives

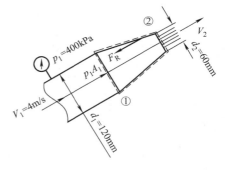

$$\frac{V_2}{V_1} = \left(\frac{d_1}{d_2}\right)^2 \quad \rightarrow \quad V_2 = 16m/s$$

Fig.X4.5.2

Applying Eq.4.5.13 on the control volume from section ① to ② , velocities are parallel to the axis, the external forces acting on the water in control volume are the pressure force at inlet and the resistant force by the reducer, then,

$$\sum F = p_1 A_1 - F_R = \rho q (V_2 - V_1)$$

where $A_1 = \dfrac{\pi d_1^2}{4}$ and $q = V_1 A_1$

Solving for F_R, we have the result

$$F_R = 3.98kN$$

The force the water exert on the nozzle is of the same magnitude and opposite direction.

Example 4.5.3   A reducing elbow is used to deflect water flow as shown in Exp.4.5.2 such that the fluid makes an $180°$ U-turn before it is discharged into the atmosphere, as shown in Fig.X4.5.3. The axes of the pipe and elbow lie in a horizontal plane. The weight of the elbow and the water in it is considered to be negligible. Determine the anchoring force needed to hold the elbow in place. Neglect the effects of friction.

Solution   The elbow discharges water to the atmosphere, that is $p_2 = 0$.

The direction of water at outlet is opposite to it at the inlet, the inlet and outlet velocities of water are $V_1 = 4m/s$, and $V_2 = -16m/s$.

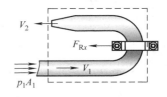

Fig.X4.5.3

The momentum equation for steady flow is

$$p_1 A_1 - F_R = \rho q (V_2 - V_1) \quad \rightarrow \quad F_R = 5.68kN$$

The reaction force is larger than that of Example 4.5.2 since the pipe turns the water over a much great angle.

Example 4.5.4   The horizontal flow to the right through the reducer of Fig.X4.5.4a, find the net force of the reducer on the fluid. Neglect the friction.

水对闸门的冲击力与闸门对水的阻力 F_R 是一对大小相等方向相反的力。

$$F_{闸门} = 847.5\text{kN}$$

例 4.5.2　直径 120mm 的水管通过前端直径 60mm 的喷头出水，如图 X4.5.2 所示。已知水管内水的流速 4m/s，表压强 400kPa。计算水流对喷头的冲击力。

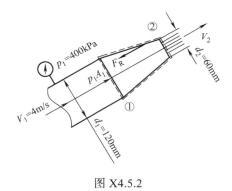

图 X4.5.2

解　选取喷头部分为控制体，因为出口为大气，所以出口处压强 $p_2 = 0$。
出口流速

$$\frac{V_2}{V_1} = \left(\frac{d_1}{d_2}\right)^2 \quad \rightarrow \quad V_2 = 16\text{m/s}$$

对截面①—②间的控制体应用式 4.5.13，流速与管道轴线平行，控制体内水所受外力为进水方向的压力和喷头的阻力，可得

$$\sum F = p_1 A_1 - F_R = \rho q(V_2 - V_1)$$

式中，$A_1 = \dfrac{\pi d_1^2}{4}$，且 $q = V_1 A_1$。

求解 F_R，得

$$F_R = 3.98\text{kN}$$

水作用于喷头的力与此阻力大小相同方向相反。

例 4.5.3　用如图 X4.5.3 所示的弯头，使例 4.5.2 水管内的水在喷出前做 180° 的 U 形转向，水管和弯头的中心线均在同一水平面内。忽略弯管和管内水的重量，计算固定弯管所需的力。

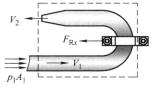

图 X4.5.3

解　弯管放出的水流入大气，所以 $p_2 = 0$。

出口水流动方向与进口相反，进、出口水的流速分别为 $V_1 = 4\text{m/s}$，和 $V_2 = -16\text{m/s}$。

对此稳定流动的动量方程为

$$p_1 A_1 - F_R = \rho q(V_2 - V_1) \quad \rightarrow \quad F_R = 5.68\text{kN}$$

显然，管路大角度转弯形成的冲击力明显大于例 4.5.2 中的情况。

例 4.5.4　水平管路内的水流径直流过渐缩接头，忽略摩擦，各参数如图 X4.5.4a 所示，试计算变径接头上的力。

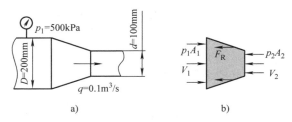

Fig.X4.5.4

Solution Applying Eq.4.5.13 and assuming the fluid is ideal as shown in Fig.X4.5.4b, since the entry and exit velocities are parallel to the axis, we get

$$\sum F = p_1A_1 - p_2A_2 - F_R = \rho q(V_2 - V_1)$$

Rewriting the equivalent, it is

$$F_R = p_1A_1 - p_2A_2 - \rho q(V_2 - V_1)$$

The average velocities are found to be

$$V_1 = \frac{4q}{\pi D^2} = 3.18\text{m/s}, \qquad V_2 = V_1\left(\frac{D}{d}\right)^2 = 12.74\text{m/s}$$

Neglecting losses between the entrance and the exit of the reducer, the energy equation gives

$$\frac{p_1}{\rho g} + \frac{V_1^2}{2g} = \frac{p_2}{\rho g} + \frac{V_2^2}{2g} \quad \rightarrow \quad p_2 = 424\text{kPa}$$

Solving for F_R, the result is

$$F_R = 11.42\text{kN}$$

Example 4.5.5 Water flows through a horizontal pipe bend and exits into the atmosphere (Fig.X4.5.5a). The flow rate is 0.01m³/s. Calculate the force in each of the supports holding the pipe bend in position. Neglect mass forces and viscous effects and shear force in the supports.

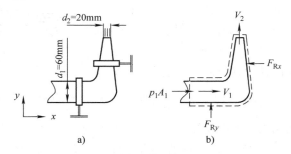

Fig.X4.5.5

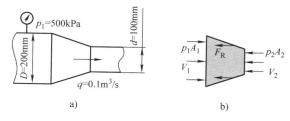

图 X4.5.4

解　假设水为理想流体，如图 X4.5.4b 所示变径段进出口流速方向平行，对控制体应用动量方程，有

$$\sum F = p_1 A_1 - p_2 A_2 - F_R = \rho q(V_2 - V_1)$$

改写上式，得

$$F_R = p_1 A_1 - p_2 A_2 - \rho q(V_2 - V_1)$$

解得平均流速

$$V_1 = \frac{4q}{\pi D^2} = 3.18\text{m/s}, \qquad V_2 = V_1\left(\frac{D}{d}\right)^2 = 12.74\text{m/s}$$

忽略进、出口间由于管路截面缩小而导致的损失，能量方程可以写为

$$\frac{p_1}{\rho g} + \frac{V_1^2}{2g} = \frac{p_2}{\rho g} + \frac{V_2^2}{2g} \quad \rightarrow \quad p_2 = 424\text{kPa}$$

求解 F_R，得

$$F_R = 11.42\text{kN}$$

例 4.5.5　水经水平管道及弯头喷出（见图 X.4.5.5a）至空气中，流量为 0.01m³/s。忽略质量力、黏性力及支座处摩擦，计算每个弯管固定支座的受力。

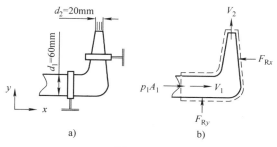

图 X4.5.5

Solution   We have selected a control volume that surrounds the bend, as shown in Fig.X4.5.5b. Since the supports are detached, the forces that the supports exert on the control volume are included.

The average velocities are calculated

$$V_1 = \frac{4q}{\pi d_1^{\,2}} = 3.54\text{m/s}, \qquad V_2 = \frac{4q}{\pi d_2^{\,2}} = 31.84\text{m/s}$$

The pressure p_2 is zero because the flow exits into the atmosphere. The pressure p_1 can be determined using the Bernoulli equation. Neglecting losses between sections ① and ②, it gives

$$\frac{p_1}{\rho g} + \frac{V_1^{\,2}}{2g} = \frac{V_2^{\,2}}{2g} \quad \rightarrow \quad p_1 = 501\text{kPa}$$

Applying the momentum equation in the x-direction to find F_{Rx} and in the y-direction to find F_{Ry}.

x-direction:

$$p_1 A_1 - F_{Rx} = \rho q (V_{2x} - V_{1x}) \quad \rightarrow \quad V_{2x} = 0 \quad \rightarrow \quad F_{Rx} = 1450\text{N}$$

y-direction:

$$F_{Ry} = \rho q \left(V_{2y} - V_{1y} \right) \quad \rightarrow \quad V_{1y} = 0 \quad \rightarrow \quad F_{Ry} = 318\text{N}$$

The forces exerted on the supports are of the same magnitude and opposite direction with the resistant forces to the control volume.

Example 4.5.6   The bend in Example 4.5.5 is replaced by a bend of 45° as shown in Fig.X4.5.6a, calculate the force in each of the supports holding the pipe bend in position.

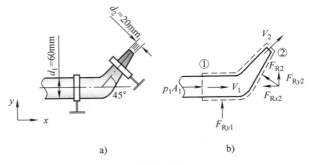

Fig.X4.5.6

Solution   The average velocities are calculated

$$V_1 = \frac{4q}{\pi d_1^{\,2}} = 3.54\text{m/s}, \qquad V_2 = \frac{4q}{\pi d_2^{\,2}} = 31.84\text{m/s} \quad \rightarrow \quad V_{2x} = V_{2y} = 22.51\text{m/s}$$

$$p_1 = 501\text{kPa} \quad p_2 = 0$$

解 选取弯头周围部分为控制体，如图 X4.5.5b 所示。支座为可分离物体，所以将支座作用于控制体的力视为外力。

计算平均流速

$$V_1 = \frac{4q}{\pi d_1^2} = 3.54\mathrm{m/s}, \qquad V_2 = \frac{4q}{\pi d_2^2} = 31.84\mathrm{m/s}$$

出口为大气压，所以 $p_2 = 0$。应用伯努利方程计算压强 p_1。忽略截面①—②间损失，有方程

$$\frac{p_1}{\rho g} + \frac{V_1^2}{2g} = \frac{V_2^2}{2g} \quad \rightarrow \quad p_1 = 501\mathrm{kPa}$$

应用动量方程计算 x 方向阻力 F_{Rx} 及 y 方向阻力 F_{Ry}。

x 方向：

$$p_1 A_1 - F_{Rx} = \rho q (V_{2x} - V_{1x}) \quad \rightarrow \quad V_{2x} = 0 \quad \rightarrow \quad F_{Rx} = 1450\mathrm{N}$$

y 方向：

$$F_{Ry} = \rho q \left(V_{2y} - V_{1y} \right) \quad \rightarrow \quad V_{1y} = 0 \quad \rightarrow \quad F_{Ry} = 318\mathrm{N}$$

支座所受水流冲击力与其施加在控制体上的阻力大小相等、方向相反。

例 4.5.6 例 4.5.5 中的弯头改为如图 X4.5.6a 所示的 45°，计算图中弯头支座受力。

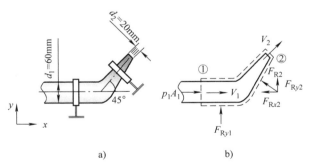

图 X4.5.6

解 计算平均流速

$$V_1 = \frac{4q}{\pi d_1^2} = 3.54\mathrm{m/s}, \qquad V_2 = \frac{4q}{\pi d_2^2} = 31.84\mathrm{m/s} \quad \rightarrow \quad V_{2x} = V_{2y} = 22.51\mathrm{m/s}$$

$$p_1 = 501\mathrm{kPa} \quad p_2 = 0$$

The resisting force F_{R2} is placed at $-45°$ to the y-direction, and is decomposed into F_{Rx2} and F_{Ry2}.

The momentum equation is applied in the x-direction to find F_{Rx} and then in the y-direction for F_{Ry}.

x-direction:

$$p_1 A_1 - F_{Rx2} = \rho q(V_{2x} - V_{1x}) \quad \rightarrow \quad F_{Rx2} = 1225\text{N} \quad \rightarrow \quad F_{R2} = \frac{F_{Rx2}}{\sin 45°} = 1733\text{N}$$

y-direction:

$$F_{Ry1} + F_{Ry2} = \rho q(V_{2y} - V_{1y}), \quad F_{Ry2} = F_{R2}\cos 45° = 1225\text{N}$$

$$F_{Ry1} = -1000\text{N}$$

The negative result for F_{Ry1} indicates that the assumed direction is wrong, and it should be reversed. Therefore, the force by the support ① is 1000N acting in the negative y-direction. And the supporting force on support ① is 1000N acting in the y-direction.

Example 4.5.7   A nozzle discharges water at a mass flow rate of 15kg/s and strikes a stationary vertical plate with a normal velocity of 20m/s (Fig.X4.5.7a). After the strike, the water stream splatters off in all directions in the plane of the plate. Determine the force needed to prevent the plate from moving horizontally due to the water stream.

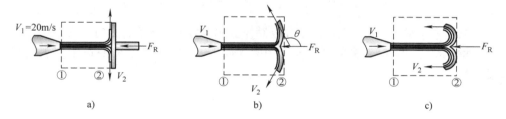

a)   b)   c)

Fig.X4.5.7

Solution   The water jet is exposed to the atmosphere on the entire system, this gives $p_1 = p_2 = 0$.

The vertical forces and momentum fluxes are not considered since they have no effect on the horizontal reaction force, then, $V_{2x} = 0$.

The momentum correction factor is negligible, this is $\beta = 1$.

The momentum equation for steady one-dimensional flow is given as

$$F_R = \rho q(V_{2x} - V_1) \quad \rightarrow \quad F_R = 300\text{N}$$

Furtherly consider the similar conditions as shown in Fig.4.5.7b and Fig.4.5.7c, the water jets strike the 2-dimenssional curve surfaces.

For the control volume in Fig.4.5.7b, the water stream splatters off into two parts on the surface of the plate, the velocity component in x-direction of V_2 is

$$V_{2x} = V_2 \cos\theta$$

The momentum equation in x-direction gives

阻力 F_{R2} 位于 y 方向 $-45°$，可分解为 F_{Rx2} 和 F_{Ry2}。

应用动量方程在 x 方向求解 F_{Rx}，在 y 方向求解 F_{Ry}。

x 方向：

$$p_1 A_1 - F_{Rx2} = \rho q (V_{2x} - V_{1x}) \quad \rightarrow \quad F_{Rx2} = 1225\text{N} \quad \rightarrow \quad F_{R2} = \frac{F_{Rx2}}{\sin 45°} = 1733\text{N}$$

y 方向：

$$F_{Ry1} + F_{Ry2} = \rho q (V_{2y} - V_{1y}), \quad F_{Ry2} = F_{R2} \cos 45° = 1225\text{N}$$

$$F_{Ry1} = -1000\text{N}$$

F_{Ry1} 的负号说明其假设方向是错误的，应该反向。所以，支座①对水的阻力为沿 y 轴负方向 1000N。即，弯头对支座①的作用力为沿 y 轴正方向 1000N。

例 4.5.7　如图 X4.5.7a 所示的喷嘴以 15kg/s 的流量向竖直放置的固定平板喷射水流，水流垂直于平板且流速 20m/s。水流冲击平板后沿平板表面向各方向发散。计算在水流作用下保持平板水平方向固定所需的力。

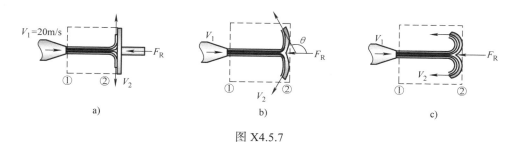

图 X4.5.7

解　整个系统水流都处于空气中，所以 $p_1 = p_2 = 0$。

因为只需计算水平方向的反作用力，因此 $V_{2x} = 0$，不考虑竖直方向的动量。动量修正系数取 $\beta = 1$。

对此一元稳定流动列动量方程，得

$$F_R = \rho q (V_{2x} - V_1) \quad \rightarrow \quad F_R = 300\text{N}$$

进一步分析类似情况，如图 4.5.7b、c 所示，水流冲击 2 元曲面。

对于图 4.5.7b 中的控制体，水流在曲面处发散成两部分，速度 V_2 在 x 方向的分量为

$$V_{2x} = V_2 \cos \theta$$

x 方向的动量方程为

$$F_R = \rho q(V_{2x} - V_1) = \rho q(V_2 \cos\theta - V_1)$$

For the control volume in Fig.4.5.7c, the water stream splatters back totally, the velocity component in x-direction of V_2 is

$$V_{2x} = V_2 = -V_1$$

The momentum equation in x-direction gives

$$F_R = \rho q(V_{2x} - V_1) = -2\rho q V_1$$

Example 4.5.8   Water flows through the double nozzle as shown in Fig.X4.5.8a. Determine the magnitude and direction of the resultant force the water exerts on the nozzle. The velocity of both nozzle jets is 15m/s. The axes of the pipe and both nozzles lie in a horizontal plane. Neglect friction.

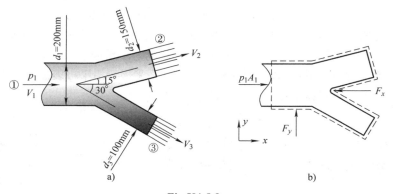

Fig.X4.5.8

Solution   Jets ② and ③ are in the atmosphere, then $p_2 = p_3 = 0$

$$V_{2x} = V_2 \cos 15° = 14.49\text{m/s}, \qquad V_{2y} = V_2 \sin 15° = 3.88\text{m/s}$$

$$V_{3x} = V_3 \cos(-30°) = 12.99\text{m/s}, \qquad V_{3y} = V_3 \sin(-30°) = -7.5\text{m/s}$$

Continuity

$$A_1 V_1 = A_2 V_2 + A_3 V_3 \quad \rightarrow \quad V_1 = 12.1875\text{m/s} \quad \rightarrow \quad V_{1x} = V_1, \qquad V_{1y} = 0$$

$$q_1 = A_1 V_1 = 0.3827\text{m}^3/\text{s}, \quad q_2 = A_2 V_2 = 0.2649\text{m}^3/\text{s}, \quad q_3 = A_3 V_3 = 0.1178\text{m}^3/\text{s}$$

Writing the Bernoulli Equation along a streamline, at A_1 and A_2, or, at A_1 and A_3 for p_1 as shown in Fig.4.5.8b

$$\frac{p_1}{\rho g} + \frac{V_1^2}{2g} = \frac{V_2^2}{2g} \quad \rightarrow \quad p_1 = 38.23\text{kPa}$$

By using Eq.4.5.11, the momentum equation can be manipulated as follows in x- and y-direction respectively

$$\sum F_x = p_1 A_1 - F_x = \rho q_2 V_{2x} + \rho q_3 V_{3x} - \rho q_1 V_1 \quad \rightarrow \quad F_x = 496\text{N} \leftarrow$$

$$F_{\mathrm{R}} = \rho q (V_{2x} - V_1) = \rho q (V_2 \cos\theta - V_1)$$

对于图 4.5.7c 中的控制体，水流完全溅射返回，速度 V_2 在 x 方向的分量为

$$V_{2x} = V_2 = -V_1$$

x 方向的动量方程为

$$F_{\mathrm{R}} = \rho q (V_{2x} - V_1) = -2\rho q V_1$$

例 4.5.8 如图 X4.5.8a 所示，水流通过双喷嘴时流速均为 15m/s，管道及两个喷嘴中心线在同一水平面内。忽略摩擦，计算喷嘴上水流的合力大小及方向。

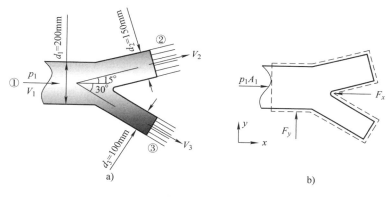

图 X4.5.8

解 水流②和水流③在空气中，所以 $p_2 = p_3 = 0$

$$V_{2x} = V_2 \cos 15° = 14.49\mathrm{m/s}, \qquad V_{2y} = V_2 \sin 15° = 3.88\mathrm{m/s}$$

$$V_{3x} = V_3 \cos(-30°) = 12.99\mathrm{m/s}, \qquad V_{3y} = V_3 \sin(-30°) = -7.5\mathrm{m/s}$$

连续性方程

$$A_1 V_1 = A_2 V_2 + A_3 V_3 \;\rightarrow\; V_1 = 12.1875\mathrm{m/s} \;\rightarrow\; V_{1x} = V_1, \qquad V_{1y} = 0$$

$$q_1 = A_1 V_1 = 0.3827\mathrm{m^3/s}, \quad q_2 = A_2 V_2 = 0.2649\mathrm{m^3/s}, \quad q_3 = A_3 V_3 = 0.1178\mathrm{m^3/s}$$

在图 4.5.8b 所示的截面①—②或截面①—③间列出沿流线的伯努利方程

$$\frac{p_1}{\rho g} + \frac{V_1^2}{2g} = \frac{V_2^2}{2g} \;\rightarrow\; p_1 = 38.23\mathrm{kPa}$$

应用式 4.5.11，分别列出 x 方向和 y 方向的动量方程

$$\sum F_x = p_1 A_1 - F_x = \rho q_2 V_{2x} + \rho q_3 V_{3x} - \rho q_1 V_1 \;\rightarrow\; F_x = 496\mathrm{N} \leftarrow$$

$$\sum F_y = F_y = \rho q_2 V_{2y} + \rho q_3 V_{3y} - \rho q_1 V_{1y} \quad \rightarrow \quad F_y = 144.3\text{N}\uparrow$$

$$F = \sqrt{F_x^2 + F_y^2} = 490.7\text{N}$$

$$\theta = \arctan\frac{144.3}{-496} = 180° - 16.22°$$

The resultant force on the double nozzle is 490.7N at a 16.22° to the negative x-direction.

As we have observed in the examples and problems of this section, and we know that it is extremely important that the boundaries of the control volume are well defined during an analysis. A control volume and its bounding control surface can be selected as any arbitrary region in space through which fluid flows. A well-chosen control volume exposes only the forces that are to be determined (such as reaction forces) and a minimum number of other forces.

When choosing a control volume, you are not limited to the fluid alone. The control surfaces are selected at locations either where information is known or where the unknowns appear. Often it is more convenient to slice the control surface through solid objects such as walls, struts, or bolts. A control volume may even surround an entire object.

We always draw our control surface such that it slices normal to the velocity at each such inlet or outlet, where the velocity is constant. It is also very helpful to locate a control surface at the free surface.

Fluid forces acting on a structure are equal and opposite of boundary pressure forces acting on the fluid. To avoid confusion over signs, we may first solve for the magnitude and direction of the reaction force of the structure on the fluid, and only in the last step to find the equal and opposite force of the fluid on the structure.

4.5.3 Applications Steady state flow force acting on the hydraulic valves.

Let us first consider the spool valve, as shown in Fig.A4.5.1a, hydraulic oil enters the valve through an opening of width x_T between the valve core and the seat at a speed of V_1, and flows out at a speed of V_2. The angle θ between inflow velocity V_1 and the center line of the valve is called injection angel.

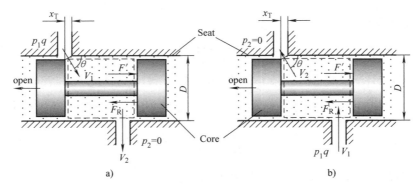

Fig.A4.5.1　The steady state flow force acting on a spool valve

a) The hydraulic oil flow in through the opening　b) The hydraulic oil flow out from the opening

$$\sum F_y = F_y = \rho q_2 V_{2y} + \rho q_3 V_{3y} - \rho q_1 V_{1y} \quad \rightarrow \quad F_y = 144.3\text{N}\uparrow$$

$$F = \sqrt{F_x^2 + F_y^2} = 490.7\text{N}$$

$$\theta = \arctan \frac{144.3}{-496} = 180° - 16.22°$$

双喷头上的合力大小为 490.7N，与 x 轴负方向夹角 16.22°。

从以上内容和例题中我们可以看到，合理选择控制体及其边界面对分析问题是非常重要的。控制体及其控制面可以选在流动区域内任意位置，但合理的选择应该尽可能减少模型中力的数量，仅考虑必须计入的力。

选择控制体时不仅限于流体本身，甚至可以包括整个结构。控制面应该尽量选在变量已知或包含待解变量的位置。通常把控制面选在固体，如壁面、支撑物或螺栓等表面，更便于分析计算。

控制面应尽量取在与流速垂直的流入、流出面，且流速保持不变的位置。控制面选在自由液面处也会更便于计算。

流体作用于固体结构上的力与流 - 固边界面上的压力大小相等、方向相反。为避免在正负号上造成混乱，我们可以先考虑固体作用于流体的反作用力大小和方向，求解的最后步骤再于固体上找出数值相同而方向相反、来自流体的作用力。

4.5.3   工程应用   液压阀芯上的稳态液动力

首先我们分析图 A4.5.1a 所示的滑阀。液压油从阀芯和阀体间宽度为 x_T 的开口，以流速 V_1 流入阀内，以速度 V_2 流出。进口流速 V_1 和阀芯中心线间的夹角 θ 称为入射角。

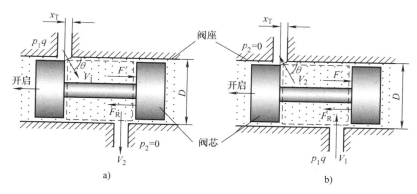

图 A4.5.1   滑阀上的稳态液动力

a）油液自阀口流入阀体   b）油液自阀体流出阀口

233

Assuming steady flow and neglect the effects of viscous, the momentum equation along the direction of center line of the valve takes the form

$$p_1 \pi D x_T \cos\theta - F_R = \rho q \left(V_{2x} - V_1 \cos\theta \right)$$

here F_R is the resistant force of core to the oil. Assume that the inflow pressure is almost atmospheric pressure, the equation is reduced to

$$F_R = \rho q V_1 \cos\theta$$

the value of F_R is positive, that is the direction of F_R is the same as we assumed, i.e., pointing to the left.

$$F' = -F_R = -\rho q V_1 \cos\theta$$

The force exerted on the core by hydraulic oil is the same magnitude but opposite in direction, that means, the steady state flow force tends to close the valve.

Now we consider the condition of Fig.A4.5.1b, the direction of hydraulic oil flow changes, exiting at the speed of V_2 through the x_T wide opening, and neglect the inflow pressure force, the momentum equation is written as

$$F_R = \rho q \left(V_2 \cos\theta - V_{1x} \right) \quad \rightarrow \quad F_R = \rho q V_2 \cos\theta$$

$$F' = -F_R = -\rho q V_2 \cos\theta$$

Also, the directions of the resistance are the same as illustrated in Fig.A4.5.1b and opposite to the flow force. This shows, the steady state flow force of the spool valve always try to keep the valve closing, which has a significant impact on performance of the valve. Design must be considered to balance the flow force for eliminating the difficult manipulate.

Let us consider the poppet valve, as shown in Fig.A4.5.2a, hydraulic oil enters the valve through the gap between core and seat at speed of V_1 and flows out at a speed of V_2. Tapper angle of the core is 2ϕ. The control volume is selected as shown.

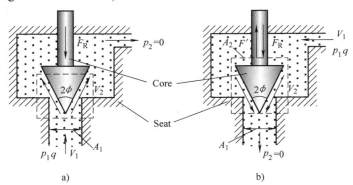

Fig.A4.5.2   The steady state flow force acting on a poppet valve

Assuming steady flow and neglect the effects of viscous, the momentum equation on the control volume in vertical direction takes the form

$$p_1 A_1 - F_R = \rho q \left(V_2 \cos\phi - V_{1y} \right)$$

忽略黏性且仅考虑稳定流动，沿液压阀中心线方向列出动量方程，得

$$p_1 \pi D x_T \cos\theta - F_R = \rho q (V_{2x} - V_1 \cos\theta)$$

式中，F_R 是阀芯对液压油的阻力。假设液流内部压强近似为大气压，该方程可以化简为

$$F_R = \rho q V_1 \cos\theta$$

F_R 值为正，说明阻力的方向与我们假设的一致，即，向左。

$$F' = -F_R = -\rho q V_1 \cos\theta$$

液压油作用于阀芯的力与阀芯给液压油的阻力大小相同、方向相反，所以，阀芯上的稳态液动力总是倾向于使液压阀关闭。

再来分析图 A4.5.1b 的情况，改变液压油流动方向，油液从开口 x_T 以速度 V_2 流出，忽略流入端压力，则动量方程可写为

$$F_R = \rho q (V_2 \cos\theta - V_{1x}) \quad \rightarrow \quad F_R = \rho q V_2 \cos\theta$$

$$F' = -F_R = -\rho q V_2 \cos\theta$$

同样，油液所受阻力方向与图示相同，阀芯所受液动力方向与 F_R 相反。这表明，滑阀上的稳态液动力总是具有使阀门关闭的倾向，这对液压阀的工作性能有很大的影响。设计液压阀时必须考虑如何平衡阀芯上的液动力以便于操纵。

接下来分析图 A4.5.2a 所示的锥阀。液压油以流速 V_1 从阀芯和阀体间的间隙流入并以流速 V_2 流出。锥阀阀芯的锥角为 2ϕ，控制体选取如图 A4.5.2 中虚线所示。

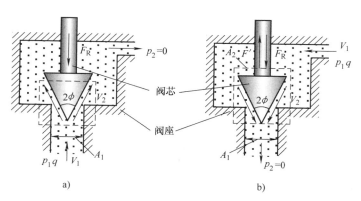

图 A4.5.2　锥阀上的稳态液动力

同样忽略黏性考虑稳定流动，在竖直方向对控制体列出动量方程

$$p_1 A_1 - F_R = \rho q (V_2 \cos\phi - V_{1y})$$

The inflow speed V_1 can be considered zero since the area A_1 is so large relative to the area of the outlet orifice. The inflow pressure p_1 is assumed be almost atmospheric. The equation is reduce to

$$F_R = -\rho q V_2 \cos\phi$$

The tapper angle $2\phi < 90°$, it is $\cos\phi > 0$, the negative sign indicated that the assumed direction for F_R is wrong and should be reversed.

So that, the stream flow force on the core which is the same magnitude with F_R, acts downward, tends to close the valve.

Consider the condition shown in Fig.A4.5.2b, the oil enters the poppet valve through the opening and exits through the annular orifice. The momentum equation on the control volume indicated by the dashed line, is written as

$$p_1 A_2 + F_R = \rho q \left(V_2 \cos\phi - V_{1y} \right)$$

The equation is reduced by neglecting the very small V_1 and p_1

$$F_R = \rho q V_2 \cos\phi$$

This shows the direction of F_R is the same with that we assumed. Obviously, the reaction force, i.e., the stream state flow force F', which is in the opposite direction with F_R, tries to open the valve.

4.5.4　Moment of Momentum Equation

The linear momentum equation discussed in the preceding is useful for determining the relationship between the linear momentum of flow streams and resultant forces. Many engineering problems involve the moment of momentum of streams, and the rotational effects caused by them. Such problems are best analyzed by the moment of momentum equation. An important class of fluid devices, which include centrifugal pumps, turbines, and fans, are analyzed by the moment of momentum equation, also called the angular momentum equation.

The moment of momentum equation for a steady and uniform flow is expressed as

$$\sum M = \rho q \left(r_2 \times V_2 \right) - \rho q \left(r_1 \times V_1 \right) \tag{4.5.15}$$

where $\sum M = \sum (r \times F)$ is the net torque or moment applied on the system, which is the vector sum of the moments of all forces acting on the system.

The vector product $\rho q \left(r \times V \right)$ can be written as $r \times \left(\rho q V \right) = r \times mV$, gives the moment of momentum, also called the angular momentum.

Note that only the normal component of the velocity is take into account for the flow rate, it is $q = \int_A V \cdot n \, \mathrm{d}A = \int_A u_n \, \mathrm{d}A$, for uniform flow, $q = V_n A$, only for the area of control surface that normal to the direction of flow, we have $q = VA$. Then the normal velocity components contribute to mass flux, the tangential velocity components contribute to torque.

Eq.4.5.15 is stated as the net torque acting on the control volume during steady flow is equals to the difference between the outgoing and incoming angular momentum flow rates. Note that velocity V in Eq.4.5.15 is the velocity relative to an inertial coordinate system, this statement can be expressed for any specified direction.

The comparison of momentum equation and moment of momentum equation is shown in Fig.4.5.2.

因为入口截面面积 A_1 与环形间隙出口面积相比非常大，所以假设进口流速约为 0、进口压力近似为大气压，该方程可化简为

$$F_{\mathrm{R}} = -\rho q V_2 \cos\phi$$

阀芯锥角 $2\phi < 90°$，所以有 $\cos\phi > 0$，则阻力 F_{R} 为负，表明其方向与假设方向相反。

可见，阀芯上的液动力大小与 F_{R} 相等，方向指向下，倾向关闭液压阀。

改变液压油流动方向，如图 A4.5.2b 所示，液压油从开口流入从环形缝隙流出。对虚线内的控制体列出动量方程，写为

$$p_1 A_2 + F_{\mathrm{R}} = \rho q \left(V_2 \cos\phi - V_{1y}\right)$$

忽略方程中很小的 V_1 和 p_1 值，该方程化简为

$$F_{\mathrm{R}} = \rho q V_2 \cos\phi$$

这说明 F_{R} 方向与假设相同，显然液动力 F' 与 F_{R} 反向，倾向于使锥阀开启。

4.5.4　动量矩方程

前面介绍的动量方程用于确定沿流线的动量和合力，但许多工程应用不仅涉及流动的动量，还有动量引起的转动，这些问题更适合用动量矩方程分析。像离心泵、涡轮机、扇叶等常用的流体机械装置，都需要用动量矩方程，也称为角动量方程，来进行分析。

对于均匀的稳定流动，动量矩方程的形式为

$$\sum M = \rho q \left(r_2 \times V_2\right) - \rho q \left(r_1 \times V_1\right) \tag{4.5.15}$$

式中，$\sum M = \sum \left(r \times F\right)$，是作用于系统的全部动量矩或力矩，是系统所受全部外力的力矩之和。

向量的乘积 $\rho q \left(r \times V\right)$ 可记为 $r \times \left(\rho q V\right) = r \times m V$，是动量矩，也称为角动量。

需要注意，计算流量时只取速度的法向分量，即 $q = \int_A V \cdot n \mathrm{d}A = \int_A u_n \mathrm{d}A$，对于均匀流动，$q = V_n A$，只有当控制面与流动方向垂直时，才有 $q = VA$。也就是说，速度法向分量决定流量，速度切向分量产生力矩。

式 4.5.15 说明，稳定流动中作用于系统的合动量矩等于系统流入和流出的角动量之差。式 4.5.15 中的速度 V 是相对于惯性系的，可在任意方向上。

动量方程和动量矩方程的关系比较如图 4.5.2 所示。

控制体所受外力包括作用于整个体积的质量力，如重力，以及作用于控制面的表面力，包括压力和接触点（面）上的反作用力。合力矩就是这些外力作用于控制体的力矩总和。通过转动中心轴的力，由于力臂长度为零，所以不产生力矩。

All the forces acting on the system | The momentum outflow from the control volume | The momentum inflow into the control volume

$$\sum F = \rho q V_2 - \rho q V_1$$

Angular momentum per unit mass

$$\sum (r \times F) = \rho q (r_2 \times V_2) - \rho q (r_1 \times V_1)$$

The moments of all forces acting on the system | The angular momentum outflow from the control volume | The angular momentum inflow into the control volume

Fig.4.5.2   Comparison of momentum equation and the moment of momentum equation

The forces acting on the control volume consist of mass forces that act throughout the entire control volume such as gravity, and surface forces that act on the control surface such as the pressure and reaction forces at points of contact. The net torque consists of the moments of these forces as well as the torques applied on the control volume. Forces that pass through the shaft center do not contribute to torque about the origin, since the momentum arm is zero.

In many problems, all the significant forces and momentum flows are in the same plane, and thus all give rise to moments in the same plane and about the same axis. For such cases, Eq4.5.15 can be expressed in scalar form as

$$\sum M = \rho q r_2 V_2 - \rho q r_1 V_1, \quad \text{or,} \quad \sum M = r_2 m V_2 - r_1 m V_1 \tag{4.5.16}$$

When there are no external moments applied, the difference of angular momentum between the incoming and outgoing fluxes rotates the control volume.

$$0 = M_1 + \rho q (r_2 \times V_2) - \rho q (r_1 \times V_1) \tag{4.5.17}$$

When the moment of inertia of the control volume remains constant, the control volume can be treated as a solid body, and Eq.4.5.15 takes the form

$$M_1 = \rho q (r_1 \times V_1) - \rho q (r_2 \times V_2) \tag{4.5.18}$$

where M_1 is the moment of inertia of the control volume.

Example 4.5.9   Underground water is pumped through a 100mm-diameter pipe that consists of a 1.5-m-long vertical and 0.8-m-long horizontal section, as shown in Fig.X4.5.9. Water discharges to atmospheric air at an average velocity of 5 m/s, and the mass of the horizontal pipe section when filled with water is 18 kg per meter length. The pipe is anchored on the ground by flange and bolts. The diameter of the centerline of bolt holes is 200mm.Determine the bending moment acting at the bolt A and B of the flange and the required length of the horizontal section that would make the moment at center of the flange (point C) zero.

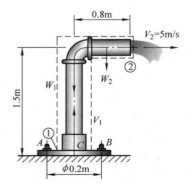

Fig.X4.5.9

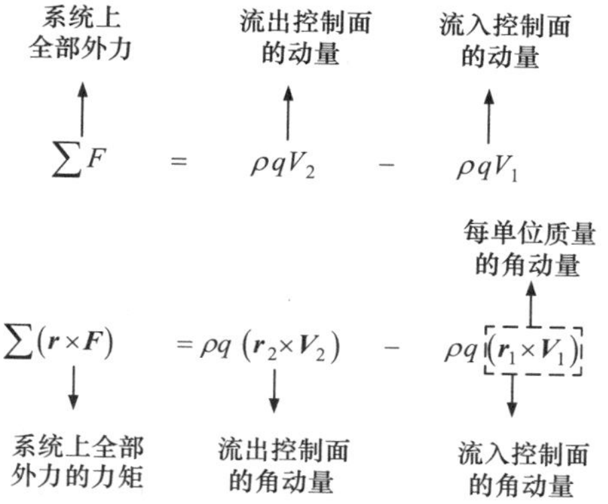

系统上
全部外力
流出控制面
的动量
流入控制面
的动量
$\sum F = \rho q V_2 - \rho q V_1$
每单位质量
的角动量
$\sum(\boldsymbol{r}\times\boldsymbol{F}) = \rho q(\boldsymbol{r}_2\times\boldsymbol{V}_2) - \rho q(\boldsymbol{r}_1\times\boldsymbol{V}_1)$
系统上全部
外力的力矩
流出控制面
的角动量
流入控制面
的角动量

图 4.5.2　动量方程和动量矩方程比较

很多应用中需要考虑的主要力和动量在同一平面内，所以力矩也都处于同一平面且几乎同轴。这种情况下，式 4.5.15 可以写为标量形式

$$\sum M = \rho q r_2 V_2 - \rho q r_1 V_1,\quad 或，\quad \sum M = r_2 m V_2 - r_1 m V_1 \tag{4.5.16}$$

当系统不受外力矩作用而转动时，控制体流入及流出的角动量之差使其维持流动。

$$0 = M_{\mathrm{I}} + \rho q(\boldsymbol{r}_2\times\boldsymbol{V}_2) - \rho q(\boldsymbol{r}_1\times\boldsymbol{V}_1) \tag{4.5.17}$$

控制体的惯性矩保持不变时，可将控制体看作刚性物体，式 4.5.15 可改写为

$$\boldsymbol{M}_{\mathrm{I}} = \rho q(\boldsymbol{r}_1\times\boldsymbol{V}_1) - \rho q(\boldsymbol{r}_2\times\boldsymbol{V}_2) \tag{4.5.18}$$

式中，$\boldsymbol{M}_{\mathrm{I}}$是控制体的转动惯量。

例 4.5.9　如图 X4.5.9 所示，通过一条直径为 100mm，竖直部分长度为 1.5m，水平部分长度为 0.8m 的水管把水从地下引至地面。水以 5m/s 的流速排至空中，且水平段内每米长度管道及水的质量为 18kg。管道用法兰和螺栓固定在地面。法兰上螺栓孔中心线圆的直径为 200mm。假设为稳定流动，并忽略摩擦，求

（1）法兰上固定螺栓 A 所受的弯曲力矩。

（2）法兰上固定螺栓 B 所受的弯曲力矩。

（3）水平段水管长度为多少时法兰中心点 C 处的力矩可以为零。

0.8m
V_2=5m/s
②
W_2
W_1
1.5m
V_1
①
A
C
B
ϕ0.2m

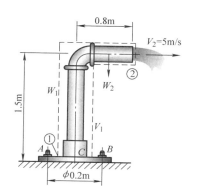

图 X4.5.9

解　规定逆时针方向为力矩的正方向，顺时针方向为负方向。

与力臂长度相比，管道直径很小，所以取管道截面的平均流速。

The flow is steady. Neglect friction.

Solution   All moments in the counterclockwise direction are positive, and all moments in the clockwise direction are negative.

a. The pipe diameter is small compared to the moment arm, and thus we use average velocity.

The forces that yield a moment about point A is the weight W_1 of vertical pipe section and W_2 of the horizontal pipe section, and the momentum flow that yield moments are the inlet and outlet stream.

Then the angular momentum equation about point A becomes

$$M_A - W_1 r_0 - W_2 (r_2 + r_0) = \rho q (-hV_2 - r_0 V_1)$$

where $r_0 = 0.2/2\text{m} = 0.1\text{m}$, $r_2 = 0.8/2\text{m} = 0.4\text{m}$, $h = 1.5\text{m}$, $V_1 = V_2 = 5\text{m/s}$

$$M_A = W_1 r_0 + W_2 (r_2 + r_0) - \rho q V (h + r_0) = -216.88\text{N·m}$$

The negative sign indicates that M_A on the pipe structure is in clockwise direction, trends to push the bolt A in.

b. The angular momentum equation about point B becomes

$$M_B + W_1 r_0 - W_2 (r_2 - r_0) = \rho q (-hV_2 + r_0 V_1)$$

$$M_B = -W_1 r_0 + W_2 (r_2 - r_0) - \rho q V (h - r_0) = -258.9\text{N·m}$$

Bending moment acting by the bolt B of the flange is 258.9N·m in clockwise direction, trends to pull the bolt B out.

c. The moments of all forces and momentum flows passing through point C are zero, the only force that yields a moment about point C is the weight of the horizontal pipe section, and the only momentum flow that yields a moment is the outlet stream. The angular momentum equation about point C gives as

$$M_C - W_2 r_2 = -\rho q h V_2$$

Setting $M_C = 0$ and substituting, the length $L = 2r_2$ of the horizontal pipe that would cause the moment at the center of flange to vanish is determined to be

$$W_2 \frac{L}{2} = \rho q h V_2 \quad \rightarrow \quad L = 1.827\text{m}$$

Note that the pipe weight and the momentum of the exit stream cause opposing moments at the flange. This example shows the importance of accounting for the moments of momentums of flow streams when performing a dynamic analysis and evaluating the stresses in piping system at critical cross sections.

Example 4.5.10   A sprinkler has four 0.5m-long arms with nozzles at right angles with the arms and 45° with the ground (Fig.X4.5.10). If the total flow rate is 0.01m³/s and a nozzle exit diameter is 10mm, find the rotational speed of the sprinkler. Neglect friction.

Solution   There are no external moments applied on the nozzles, the angular momentum of outgoing fluxes rotates the sprinkler at an angular velocity ω.

The incoming flow that pass through the shaft center do not contribute to the rotation. Then, Eq.4.5.18 reduces to

$$M_1 = -\rho q (r_2 \times V_2)$$

here M_1 is the moment of momentum of the four arms due to the mass of water in them. It has the magnitude of

（1）可对 A 点形成转矩的力包括：竖直段管路的重量 W_1、水平段管路的重量 W_2 以及水流入和流出时产生的动量矩。

A 点的角动量方程为

$$M_A - W_1 r_0 - W_2 (r_2 + r_0) = \rho q (-hV_2 - r_0 V_1)$$

式中，$r_0 = 0.2/2\text{m} = 0.1\text{m}$，$r_2 = 0.8/2\text{m} = 0.4\text{m}$，$h = 1.5\text{m}$，$V_1 = V_2 = 5\text{m/s}$

$$M_A = W_1 r_0 + W_2 (r_2 + r_0) - \rho q V (h + r_0) = -216.88\text{N·m}$$

力矩 M_A 的负号表明其作用于管道的方向为顺时针，是使螺栓 A 被压入的趋势。

（2）B 点的角动量方程为

$$M_B + W_1 r_0 - W_2 (r_2 - r_0) = \rho q (-hV_2 + r_0 V_1)$$

$$M_B = -W_1 r_0 + W_2 (r_2 - r_0) - \rho q V (h - r_0) = -258.9\text{N·m}$$

来自于法兰上螺栓 B 的弯曲力矩为 258.9N·m，顺时针方向，是螺栓被拔出的趋势。

（3）对 C 点产生力矩的外力只有水平段管道的重量和水流出时的角动量。当作用于 C 点的全部力矩为零时，C 点的角动量方程为

$$M_C - W_2 r_2 = -\rho q h V_2$$

令 $M_C = 0$，水平管道长度 $L = 2r_2$，法兰中心力矩为零时，解得

$$W_2 \frac{L}{2} = \rho q h V_2 \quad \rightarrow \quad L = 1.827\text{m}$$

我们看到，管道的重量和管道出口水流对法兰引起的力矩方向是相反的。这一例题证明了对管流系统及系统中重要位置进行动力学分析和应力校核及验算动量矩的重要性。

例 4.5.10 如图 X4.5.10 所示，一喷淋器有四个 0.5m 的长臂，臂末端喷头与长臂垂直且与地面夹角为 45°。如果喷淋器的总出流量为 0.01m³/s，且喷头出水口直径为 10mm，试计算喷淋器的转速。忽略摩擦损失。

解 无外力矩作用于喷淋系统，所以是出流的角动量导致喷淋器以角速度 ω 转动。

流入的水流经过中心轴，不产生转动力矩，所以式 4.5.17 化简为

$$M_1 = -\rho q (\boldsymbol{r}_2 \times \boldsymbol{V}_2)$$

式中，M_1 是四个喷淋臂由于其自身质量而产生的转动惯量，其大小为

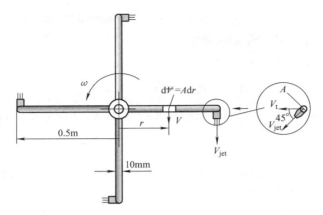

Fig.X4.5.10

$$M_1 = 4\int rVdm = 4\int_0^{0.5} rV\rho A dr$$

where V is the tangential velocity at this location, and we have $V = \omega r$, and A represents the area of cross section of arms

Therefore

$$M_1 = 4\int_0^{0.5} \rho A\omega r^2 dr = \frac{1}{6}\rho A\omega$$

The vector product of rotational radius r and vector of velocity V means that we must calculate only that water which has its momentum changed, hence the velocity used is the tangential component of jet velocity

$$V_t = V_{jet}\cos 45° = \frac{q}{4A}\cos 45° = 22.52\text{m/s}$$

The velocity V of the fluid relative to the fixed centerline of the shaft is

$$V = V_t - \omega R = 22.52\text{m/s} - 0.5\omega$$

Adding these equations together, we have

$$\frac{1}{6}\rho A\omega = -0.5\rho q(22.52\text{m/s} - 0.5\omega)$$

The result is

$$\omega = 45.28\text{rad/s}$$

Example 4.5.11　The sprinkler similar to which in Example 4.5.10 is to be converted into a turbine to generate electric power by attaching a generator to its rotating base, as shown in Fig.X4.5.11. Water enters the sprinkler from the top along the axis of rotation at a flow rate of 30L/s and leaves the nozzles in the tangential direction. The sprinkler rotates at a rate of 300r/min in a horizontal plane about the shaft. Estimate the electric power produced. Neglect friction.

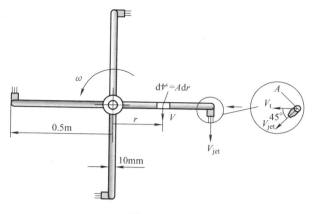

图 X4.5.10

$$M_I = 4 \int rV \mathrm{d}m = 4 \int_0^{0.5} rV \rho A \mathrm{d}r$$

式中，V 是半径 r 处的切向线速度，$V = \omega r$；A 是喷淋壁通流截面积。
所以

$$M_I = 4 \int_0^{0.5} \rho A \omega r^2 \mathrm{d}r = \frac{1}{6} \rho A \omega$$

式 4.5.18 中，转动半径 r 与速度 V 的向量积，意味着仅需考虑水流动量发生改变的那部分速度，所以仅需要取出流速度的切向分量 V_t

$$V_t = V_{jet} \cos 45° = \frac{q}{4A} \cos 45° = 22.52 \mathrm{m/s}$$

水流与固定转动轴中线的相对速度为

$$V_2 = V_t - \omega R = 22.52 \mathrm{m/s} - 0.5\omega$$

把各等式代入动量矩平衡方程，有

$$\frac{1}{6} \rho A \omega = -0.5 \rho q (22.52 \mathrm{m/s} - 0.5\omega)$$

解得

$$\omega = 45.28 \mathrm{rad/s}$$

例 4.5.11 如图 X4.5.11 所示，将涡轮安装于与例 4.5.10 类似的同尺寸结构上，其转动带动基座上的发电机产生电能。水从中心轴上方流入，从切向的喷嘴流出，流量为 30L/s。喷射装置以 300r/min 的转速在水平面内绕中心轴旋转，忽略摩擦损失，估算装置的发电量。

Solution   We take the disk that encloses the sprinkler arms as the stationary control volume.

The average jet exit velocity relative to the rotating nozzle is

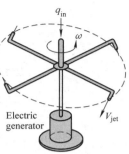

$$V_{jet} = \frac{q}{4A} = \frac{30 \times 10^{-3}}{\pi 0.01^2} \text{m/s} = 95.5 \text{m/s}$$

The angular velocity of the nozzles is $\omega = \frac{300}{60} 2\pi \text{rad/s} = 10\pi \text{rad/s}$

The tangential velocity of nozzles is $V_t = \omega R = 5\pi \text{m/s}$

The average absolute velocity of the water jet relative to the fixed center on the basement is

Fig.X4.5.11

$$V = V_{jet} + V_t$$

The magnitude is

$$V = V_{jet} - V_t = 79.8 \text{m/s}$$

The water jets leaving the nozzles yield a moment, and the angular momentum equation about the axis of rotation becomes

$$T_{shaft} = \rho q (r_2 \times V_2) = \rho q R V = 1197 \text{N} \cdot \text{m}$$

Then the power generated is calculated from

$$W = \omega T = 10\pi \times 1197 \text{kW} = 37.59 \text{kW}$$

This sprinkler-type turbine has the potential to produce 37.59 kW of power.

Exercises 4

4.1   A vacuum cleaner is capable of creating a vacuum of 2kPa just inside the hose. The maximum average velocity in the hose would be nearest _____ .
  A. 2m/s   B. 14m/s   C. 45m/s   D. 57m/s
4.2   Water flows through a 60mm-diameter pipe with an average velocity of 20m/s. The flow rate is about _____ .
  A. 0.012m³/s   B. 0.040m³/s   C. 0.057m³/s   D. 0.072m³/s
4.3   Water jet of 20mm diameter impact a huge plate at 10m/s velocity, and the impacting force exerted on the plate is _____ .
  A. 10N   B. 20N   C. 31.4N   D. 62.8N
4.4   A submarine travelling 200m below the sea surface at 5m/s, take the density of sea water as $\rho = 1025 \text{kg/m}^3$, the pressure at stagnation point at the nose of this submarine is about _____ .
  A. 12.8kPa   B. 1962kPa   C. 2000kPa   D. 2024kPa
4.5   A pitot tube is used to measure _____ of flow.
  A. pressure   B. velocity   C. flow rate   D. power
4.6   The velocity of water at the 200mm-diameter section of a converging duct is

解　选取包括喷射臂在内的圆形固定区域为控制体。
射流与喷嘴的相对速度为

$$V_{\text{jet}} = \frac{q}{4A} = \frac{30 \times 10^{-3}}{\pi 0.01^2}\,\text{m/s} = 95.5\,\text{m/s}$$

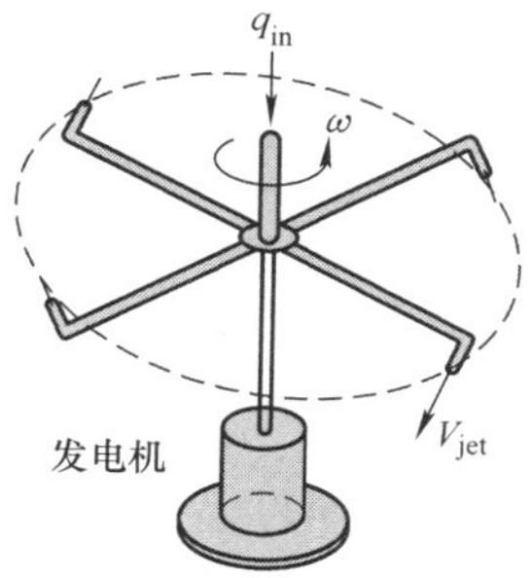

喷嘴转动的角速度　　　$\omega = \dfrac{300}{60} 2\pi\,\text{rad/s} = 10\pi\,\text{rad/s}$

喷嘴的切向速度为　　　$V_{\text{t}} = \omega R = 5\pi\,\text{m/s}$
水流与固定转动中心基座的相对速度为

$$V = V_{\text{jet}} + V_{\text{t}}$$

速度值

$$V = V_{\text{jet}} - V_{\text{t}} = 79.8\,\text{m/s}$$

图 X4.5.11

水喷出时产生了转矩，其关于转动轴的角动量方程为

$$T_{\text{shaft}} = \rho q\left(r_2 \times V_2\right) = \rho q R V = 1197\,\text{N·m}$$

根据转矩计算功率

$$W = \omega T = 10\pi \times 1197\,\text{kW} = 37.59\,\text{kW}$$

这一喷射涡轮装置的最大可发电量为 37.59kW。

习题四

4.1　一台吸尘器可产生 2kPa 的吸力，则吸尘器管内空气的平均流速约为_____。

A. 2m/s　　　　　B. 14m/s　　　　　C. 45m/s　　　　　D. 57m/s

4.2　直径为 60mm 的管内水的平均流速为 20m/s，则水的体积流量为_____。

A. 0.012m³/s　　B. 0.040m³/s　　C. 0.057m³/s　　D. 0.072m³/s

4.3　直径 20mm 的水流以 10m/s 的速度垂直冲击一巨大平板，则平板受到水流冲击力为_____。

A. 10N　　　　　B. 20N　　　　　C. 31.4N　　　　　D. 62.8N

4.4　取海水密度 $\rho = 1025\,\text{kg/m}^3$，以 5m/s 速度潜行于海面下 200 米深度的潜艇前部的驻点压强约为_____。

A. 12.8kPa　　　B. 1962kPa　　　C. 2000kPa　　　D. 2024kPa

4.5　毕托管用于测量流体的_____。

A. 压强　　　　　B. 流速　　　　　C. 流量　　　　　D. 功率

4.6　渐缩管直径 200mm 截面处测量流速为 2m/s，则在直径 100mm 截面处的

measured of 2m/s, then the velocity at the 100mm-diameter section is _____ .
    A. 0.5m/s   B. 1m/s   C. 4m/s   D. 8m/s

4.7   Among the value of the momentum correction factor for turbulent flow is about

_____ .
    A. 0   B. 1   C. 4   D. 2320

4.8   The following forces, _____ and _____ are mass force; _____ and _____ are surface force.
    A. pressure force   B. weight   C. viscous friction   D. inertial force

4.9   State the major assumptions for Bernoulli equation.

4.10 Define mass and volume flow rates. How are they related to each other?

4.11 When is the flow through a control volume steady?

4.12 Whether a fluid particle can accelerate in steady flow?

4.13 Whether the velocity of fluid particle can vary with time in a uniform flow?

4.14 Express the Bernoulli equation in three different ways using (a) heads, (b) pressures, and (c) energies.

4.15 Express the pressure head, velocity head, and elevation head for a fluid stream whose pressure is p, velocity is V, and elevation is z.

4.16 What is the kinetic energy correction factor? Whether is it significant?

4.17 What is the momentum correction factor? Whether is it significant?

4.18 State Newton's first, second, and third laws.

4.19 Express the conservation of momentum principle.

4.20 Water is flowing in a 40mm-diameter pipe at 15m/s. If the pipe enlarges to a diameter of 100mm, calculate the reduced velocity. Also, calculate the flow rate.

4.21 Water is being pumped into a pool whose cross section is 4m×5m while water is discharged through a 40mm-diameter orifice at a constant average velocity of 5m/s. If the water level in the pool rises at a rate of 10mm/min, determine the flow rate at which water is supplied to the pool.

4.22 A pipe transports 270kg/s of liquid ($S_g = 0.9$). The pipe branches into a 100mm-diameter pipe and a 60mm-diameter pipe (Fig.E4.22). If the average velocity in the larger-diameter pipe is 20m/s, calculate the flow rate in the smaller pipe.

4.23 The incompressible flow of water through the short contraction of Fig.E4.23 is assumed to be inviscid. If a pressure drop of 20kPa is measured, estimate the velocity just downstream of the contraction.

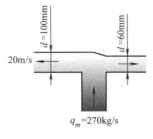

Fig.E4.22

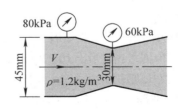

Fig.E4.23

4.24 The water tank on the ground is open to the atmosphere and a hose is connected to the bottom of the tank. A nozzle at the end of the hose is pointed straight up. The wa-

流速为_____。

　　A. 0.5m/s　　　　B. 1m/s　　　　C. 4m/s　　　　D. 8m/s

4.7　紊流的动量修正系数值约为_____。

　　A. 0　　　　　B. 1　　　　　C. 4　　　　　D. 2320

4.8　以下力中,_____和_____是质量力;_____和_____是表面力。

　　A. 压力　　　B. 重力　　　C. 黏性力　　　D. 惯性力

4.9　简述伯努利方程的基本假设条件。

4.10　什么是质量流量和体积流量?二者之间的关系如何?

4.11　什么条件下经过控制体的流动为稳定流动?

4.12　稳定流动中的流体质点是否具有加速度?

4.13　均匀流动中的流体质点运动速度是否能够随时间变化?

4.14　用以下三种形式表达伯努利方程:(1)水头;(2)压强;(3)能量。

4.15　沿流线运动的流体质点,压强为 p,流速为 V,位置高度为 z,写出该流体质点的压强水头、速度水头和位置水头。

4.16　什么是动能修正系数?该系数值是否很大?

4.17　什么是动量修正系数?该系数值是否很大?

4.18　简述牛顿第一、第二和第三定律。

4.19　简述动量守恒定律。

4.20　直径 40mm 的水管内水流速为 15m/s,若管径扩大至 100mm,计算水流速和流量。

4.21　向底面尺为 4m×5m 的水池内注水,水池下方直径 40mm 的出口同时以 5m/s 的流速排水。若池内水面以 10mm/min 的速度上升,则向池内注水的流量是多少?

4.22　管内输送液体($S_g = 0.9$)流量为 270kg/s,管道的两个方向有直径 100mm 和直径 60mm 的两个分支,如图 E4.22 所示。若大直径分支管内的流速为 20m/s,则小直径分支管内的流速是多少?

4.23　如图 E4.23 所示,不可压缩且无黏的水流经渐缩截面时产生了 20kPa 压降,估算水刚刚流过渐缩截面后的流速。

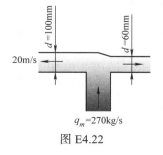

图 E4.22

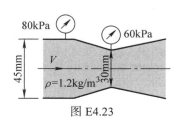

图 E4.23

ter stream from the nozzle is observed to rise 18m above the ground. Determine the water level in the tank.

4.25   An 89%-efficient pump is inserted in a 40mm-diameter line transporting 40L/s of water. A pressure rise of 400kPa is desired. Determine the power required by the pump.

4.26   Air flows through a pipe at a flow rate of 157L/s. The pipe consists of two sections of diameters 200mm and 120mm with a smooth reducing section. The pressure difference between the two sections is measured by a water manometer. Neglecting frictional effects, determine the differential height of water between the two pipe sections. Take the air density to be 1.20kg/m^3.

4.27   Determine the flow rate of air through a Venturi meter shown in Fig.E4.27 if the differential height of water between the two pipe sections reads $h = 150$mm.

4.28   A pressurized tank of water has a 10mm-diameter orifice at the bottom shown in Fig.E4.28, where water discharges to the atmosphere. Neglecting the frictional effects, determine the initial discharge rate of water from the tank.

4.29   A tank supplies water to a fountain, air above the water is pressurized of 400kPa, the nozzle is connected to a pipe at the bottom of the tank as shown in Fig. E4.29. Determine the maximum height to which the water stream could rise.

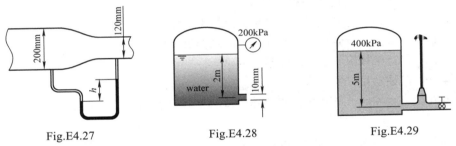

Fig.E4.27   Fig.E4.28   Fig.E4.29

4.30   Consider the water tank in Exercise 4.29 is open to the atmosphere. A pump is attached to the nozzle to increase the pressure of water. If the water jet rises to a height of 15m, determine the minimum pressure rise supplied by the pump to the water.

4.31   The velocity profile for turbulent flow in a circular pipe is approximated as $u(r) = u_{\max}(1 - r/R)^{1/n}$ where $n = 5$. Determine the kinetic energy correction factor for this flow.

4.32   Water is flowing through a Venturi meter whose diameter is 70mm at the entrance part and 40mm at the throat. The pressure difference between the entrance and the throat is measured to be 230kPa. Neglecting frictional effects, determine the flow rate of water.

4.33   A 4m-high tank filled with water has two discharge orifices, one near the bottom and another near the top. (a) Find the locations of this two orifices that the jets will meet at $x = 2$m away from the tank, as it is shown in Fig.E4.33. (b) Find a location of an orifice that the water jet will make the maximum distance away from the tank and determine this maximum distance.

4.34   A contraction of $A_1/A_2 = 4$ as shown in Fig.E4.34. The manometer is used to determine the velocity of the fluid providing the viscous effects are negligible. If water is flowing steadily, determine the velocity V_1 if $H = 100$mm. Losses through the contraction are negligible.

4.35   What is the velocity of the water in the pipe if the manometer shown in Fig. E4.35 reads: (a) 60mm, (b) 120mm.

4.36   Water at 15 ℃ flows steadily through the contraction shown in Fig.E4.36 such

4.24　地面上放置的开口容器底部有出水口，在出水口用软管连接喷嘴竖直向上喷水。观察到水柱高约 18m。计算容器内水位高度。

4.25　水泵效率 89%，出水管直径 40mm，出流量 40L/s，出口压强较进口升高 400kPa。计算泵的驱动功率。

4.26　某输气管道有一段变径区，直径从 200mm 缓慢过渡到 120mm，管道内空气流量 157L/s。用工作液体为水的测压管测量两截面间的压强差，试估算测压管两端水柱高度差。忽略黏性，空气密度取 $1.20kg/m^3$。

4.27　如图 E4.27 所示，用文丘里流量计测量管内空气流量，若水柱高度差 $h = 150mm$，则管内空气流量是多少？

4.28　内部加压容器下方有一直径 10mm 的出口向空气中排水，如图 E4.28 所示，忽略摩擦，计算出水口的初始流量。

4.29　一储水罐为喷泉供水，罐内上方空气加压 400kPa，如图 E4.29 所示喷嘴连接于罐底出口，试计算喷泉水柱可能达到的最大高度。

4.30　若习题 4.29 中的水罐上方通大气压，用水泵为喷泉加压。若要求水柱高度达到 15m，计算泵的最小输出压强。

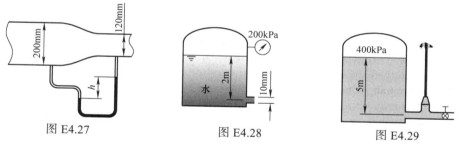

图 E4.27　　　　　　　　图 E4.28　　　　　　　　图 E4.29

4.31　某涡轮机管道内流体的速度分布规律为 $u(r) = u_{max}(1 - r/R)^{1/n}$，式中 $n = 5$。计算该流动的动能修正系数。

4.32　文丘里流量计主管直径 70mm、喉部直径 40mm，现测得两不同直径截面处的压强差为 230kPa，忽略黏性，计算管内水的流量。

4.33　水位高度 4m 的储水罐有上、下两个出水口，如图 E4.33 所示。（1）若两出口喷出的水流均落在距罐壁 2m 处，求两出口的位置高度；（2）找出一个出口位置使水流喷出的距离 x 能够达到最大，且计算该最大出流距离 x 值。

4.34　图 E4.34 所示渐缩管两截面面积比 $A_1/A_2 = 4$，用测压管测流速。管内水流稳定，测压管读数 $H = 100mm$，忽略黏性损失计算管内流速。

4.35　如图 E4.35 所示测流量，计算测压管读数 H 分别为 60mm 和 120mm 时水的流量。

4.36　15℃水稳定流过渐缩管，如图 E4.36 所示，两截面速度比 $V_2 = 9V_1$。当压

that $V_2 = 9V_1$. If the gage reading is maintained at 200kPa, determine the maximum velocity V_1 possible before cavitation occurs.

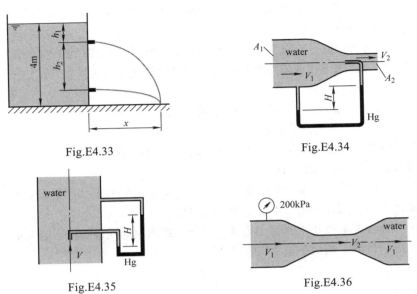

Fig.E4.33   Fig.E4.34

Fig.E4.35   Fig.E4.36

4.37   In the pipe contraction shown in Fig.E4.37, water flows steadily with a velocity of $V_1 = 0.5$m/s. Two Piezometer tubes are attached to the pipe at sections 1 and 2. Determine the velocity at section 2. Neglect any losses through the contraction.

4.38   A fireman reduces the exit area on a nozzle so that the velocity inside the hose is quite small relative to the velocity at the exiting (Fig.E4.38). What is the maximum exiting velocity and what is the maximum height the water can reach if the pressure inside the hose is 1.2MPa. Neglect any losses.

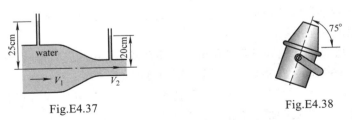

Fig.E4.37   Fig.E4.38

4.39   Air at 20℃ is drawn into the hose of a vacuum cleaner through a head. Estimate the velocity of air in the hose if the vacuum in the hose measures 6cm water. Assume the flow id inviscid and the density of air at 20℃ is about 1.2kg/m³.

4.40   A 5kW pump is used to raise water to a 15m higher elevation. If the mechanical efficiency of the pump is 82%, determine the maximum volume flow rate of water.

4.41   The flow rate of water in a 30mm-diameter horizontal pipeline at a pressure of 500kPa is 30L/min. If the pipeline increases to 45mm diameter, calculate the pressure after the expansion, neglect the losses.

力表读数保持在 200kPa 时，计算不发生气穴现象的最大流速 V_1。

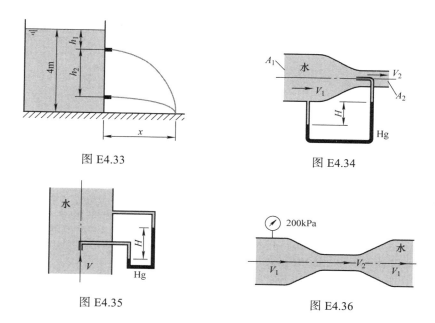

图 E4.33

图 E4.34

图 E4.35

图 E4.36

4.37 渐缩管内水以 $V_1 = 0.5$m/s 的速度稳定流动，两静压管读数如图 E4.37 所示，忽略各项损失，计算截面 2 处的流速。

4.38 消防员为了提高喷水射程将消防栓喷水口调至很小，使得水管内水的流速较出口小很多（见图 E4.38）。若水管内水压达到 1.2MPa，忽略各项损失，估算出口水流速和水流喷射的最大高度。

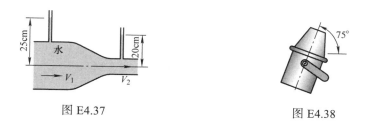

图 E4.37

图 E4.38

4.39 吸尘器通过软管吸入 20℃的空气，管内真空压强为 6cm 水柱，计算管内空气流速。假设空气无黏性，且 20℃空气密度为 1.2kg/m³。

4.40 用一台功率为 5kW 的水泵可以将水位提升 15m，若水泵的机械效率为 82%，试计算输水的最大流量。

4.41 直径 30mm 的水平管道内水压 500kPa 时流量为 30L/min，若管道直径增加至 45mm，忽略损失，计算扩张后管道内部的压强。

4.42 The velocity downstream of a gate is assumed to be uniform (Fig.E4.42). Express V in terms of H and h. Use a streamline along the top downstream and a stream along the bottom downstream respectively. The flow can be assumed to be inviscid.

4.43 Water exits from a pressurized tank as shown in Fig.E4.43. Calculate the flow rate at the exit near the bottom of the tank.

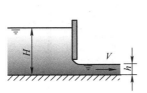

Fig.E4.42

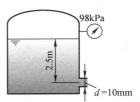

Fig.E4.43

4.44 A siphon of 10mm diameter as shown in Fig.E4.44, where $h_1 = 0.2$m and $h_2 = 1.2$m, find the flow rate of siphon and the pressure head at the highest point A.

4.45 A horizontal pipe has a 20mm diameter inlet and a 10mm diameter outlet. Pressure measured at inlet and outlet are 80kPa and 50kPa respectively. Neglect all the friction and find the flow rate in the pipe.

4.46 Water of 40℃ flowing in the blocked pipe as shown in Fig.E4.46. The pressure gage reads 35kPa and the local atmospheric pressure reads 95kPa. Find the minimum flow rate at which the cavitation will not occur. Neglect all the friction.

Fig.E4.44

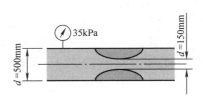

Fig.E4.46

4.47 In the Venturi meter of Fig.E4.47, calculate the flow rate of water if $d_1 = 3d_2 = 150$mm, $h = 250$mm. Assume no losses.

4.48 In Fig.E4.48, determine the minimum possible opening percentage of throttle if cavitation is to be avoided at 20℃ . Neglect all losses and assume $p_{atm} = 100$kPa.

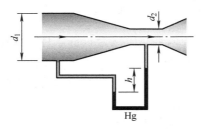

Fig.E4.47

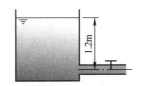

Fig.E4.48

4.42　假设水流通过闸门后流速均匀分布（见图 E4.42），忽略黏性，分别对自由液面和水底平面建立等式，列出闸门后流速 V 与闸门前、后水位 H 及 h 的关系式。

4.43　如图 E4.43 所示水从加压的容器内流出，计算罐底部出口处流量。

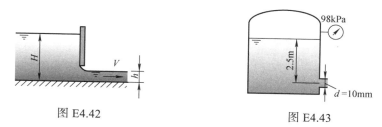

图 E4.42　　　　　　　　　图 E4.43

4.44　如图 E4.44 所示，$h_1 = 0.2m$，$h_2 = 1.2m$，水管直径 10mm，假设虹吸管中无摩擦。求吸水量和管最高点 A 点处的压强水头。

4.45　一段水平直管入口处直径 20mm，出口直径 10mm。入口处测得水压为 80kPa，出口处水压为 50kPa。不计摩擦损失，求管内水流量。

4.46　40℃的水在与图 E4.46 类似的发生堵塞的管道内流动，压力计读数为 35kPa，当地大气压 95kPa，不计摩擦，计算可能发生气穴现象时的流量。

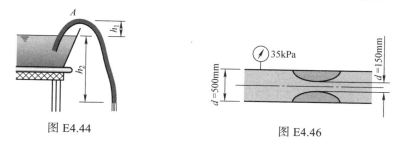

图 E4.44　　　　　　　　　图 E4.46

4.47　如图 E4.47 所示文丘里流量计，已知 $d_1 = 3d_2 = 150mm$，$h = 250mm$，不计损失，计算管内流量。

4.48　如图 E4.48 所示，需在 20℃条件下避免气穴现象，计算水罐底部出口阀门的最小开度。不计损失，假设大气压 $p_{atm} = 100kPa$。

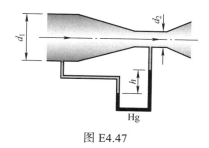
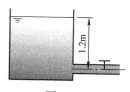

图 E4.47　　　　　　　　　图 E4.48

4.49   Find the water pressure at position A in Fig.E4.49.

4.50   Find the horizontal force of the water on the horizontal bend shown in Fig.E4.50.

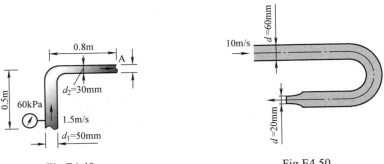

Fig.E4.49   Fig.E4.50

4.51   A nozzle with an exit diameter of 50mm is attached to a 100mm-diameter pipe transporting $0.1m^3/s$ of water. Calculate the force that the water exerts on the nozzle.

4.52   Water flows in a pipe of diameter 100mm at a pressure 300kPa. It flows out from a nozzle of diameter 50mm to the atmosphere. Calculate the force of the water on the nozzle.

4.53   Find the horizontal force components of the water on the horizontal bend shown in Fig.E4.53.

4.54   Assuming uniform velocity profiles, find the weight of the plug needed to hold itself in the pipe shown in Fig.E4.54. Neglect viscous effects.

Fig.E4.53   Fig.E4.54

4.55   Assuming uniform velocity profiles, and neglect viscous effects, find the vertical force needed to rise the gate shown in Fig.E4.55. The coefficient of friction between the gate and the support is 0.5.

4.56   Water flows at 20m/s into a 100mm-diameter horizontal T-section that branches into 50mm-diameter pipes as shown in Fig.E4.56. Find the force of the water on the T-section if the branches exit to the atmosphere. Neglect viscous effects.

4.57   A horizontal 60mm-diameter jet of water with $q_m = 200kg/s$ strikes a vertical plate. Calculate the force needed to hold the plate stationary.

4.58   What is the force needed to hold the plate in Exercise 4.57 if it moves toward the water at 2m/s?

4.59   A nozzle and hose are attached to a support. What force is needed to hold a nozzle supplied by an 80mm-diameter hose with a pressure of 510kPa? The nozzle outlet

4.49　水流经图 E4.49 所示的弯管，计算 A 点处的压强。

4.50　图 E4.50 所示水平放置的弯管，计算水流在水平方向对其形成的冲击力。

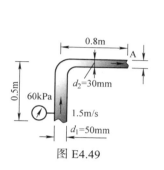

图 E4.49

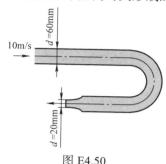

图 E4.50

4.51　出口直径 50mm 的喷嘴安装在直径 100mm 的管路上，已知出水量为 0.1m³/s，计算水对喷嘴的冲击力。

4.52　直径 100mm 的管道内水压为 300kPa，水从直径 50mm 的喷嘴流出到空气中，计算水对喷嘴的作用力。

4.53　如图 E4.53 所示水平面内放置的弯管，计算水在水平方向对弯管的作用力。

4.54　假设图 E4.54 中管道内水均匀流动，不计黏性影响，欲维持图示状态，计算塞子的质量。

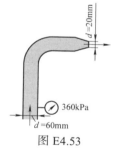

图 E4.53

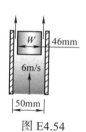

图 E4.54

4.55　图 E4.55 中水的流速均匀分布，不计黏性影响，计算竖直向上提起闸门所需的力。闸门与壁面间摩擦因数为 0.5。

4.56　放置在水平面的三通如图 E4.56 所示，水从直径 100mm 的入口以 20m/s 流速流入，从两个直径 50mm 出口流出至大气中。忽略黏性，计算水对三通的作用力。

4.57　质量流量为 q_m = 200kg/s 的水自直径 60mm 的水管沿水平方向流出，冲击到一块竖直放置的平板上，计算保持平板固定所需的力。

4.58　若习题 4.57 中的平板以 2m/s 速度迎向水流移动，此时持住平板所需的力是多少？

4.59　出口直径 20mm 的喷嘴与直径 80mm 的水管连接，已知管内水流压力为

diameter is 20mm.

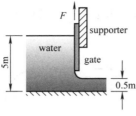

Fig.E4.55

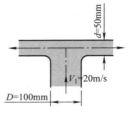

Fig.E4.56

4.60   What is the force needed to hold the orifice shown in Fig.E4.60 onto the pipe?

4.61   A horizontal water jet of constant velocity 20m/s impinges normally on a vertical flat plate and splashes off the sides in the vertical plane (Fig.E4.61). Calculate the force required to maintain the plate stationary if: (a) the plate is initially stationary; (b) the plate is moving toward the oncoming water jet with velocity 5m/s; (c) the plate is moving away from the oncoming water jet with velocity 5m/s.

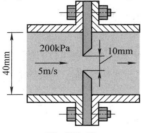

Fig.E4.60

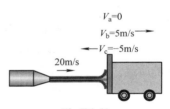

Fig.E4.61

4.62   A 100mm diameter 90° elbow in a horizontal pipe as shown in Fig.E4.62 is used to discharge water flow upward at a flow rate of 20kg/s. The pressure at the exit is atmospheric. The elevation difference between the centers of the exit and the inlet of the elbow is 500mm. Neglect the weight of the elbow and the water in it. Determine (a) the gage pressure at the center of the inlet of the elbow and (b) the anchoring force needed to hold the elbow in place.

4.63   A 100mm diameter U-turn is used to discharge water flow at a mass flow rate of 30kg/s. The pressure at the exit is atmospheric. The elevation difference between the centers of the exit and the inlet of the elbow is 400mm. Neglect the weight of the U-turn and the water in it. Determine (a) the gage pressure at the center of the inlet of the U-turn and (b) the anchoring force needed to hold the U-turn in place.

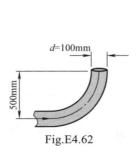

Fig.E4.62

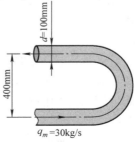

Fig.E4.63

510kPa，计算持住喷嘴所需的力。

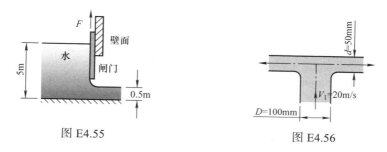

图 E4.55　　　　　　　　　　图 E4.56

4.60　计算图 E4.60 中孔板固定在管道上所需的力。

4.61　水平方向水的流速保持在 20m/s，并垂直冲击一块竖直放置的平板，如图 E4.61 所示。计算以下情况维持平板状态所需的力：（1）平板最初保持固定；（2）平板以 5m/s 的速度迎向水流；（3）平板以 5m/s 的速度与水流同向移动。

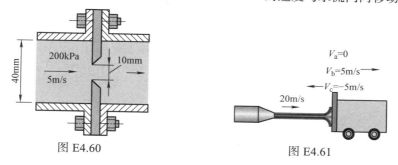

图 E4.60　　　　　　　　　　图 E4.61

4.62　直径 100mm 的直角弯与水平水管连接使水流竖直向上流至空气中，流量为 20kg/s。如图 E4.62 所示，弯头出口至水平管中心线高度为 500mm。不计弯头及水的重量，计算（1）弯头入口的水压；（2）固定弯头所需的力。

4.63　如图 E4.63 所示，直径 100mm 的 U 形弯头入口处流量 30kg/s，出口为大气压，进出口间高度差为 400mm。不计弯头和水的重量，计算（1）弯头入口的水压；（2）固定弯头所需的力。

图 E4.62

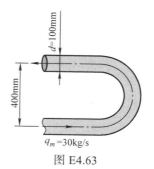

图 E4.63

4.64   A tripod holding a nozzle at the end of a hose in Fig.E4.64. If the nozzle exit diameter is 60mm and the water flow rate is 5m³/min, determine the horizontal force required of the tripod to hold the nozzle.

4.65   Water enters a two-armed lawn sprinkler along the vertical axis at a flow rate of 3L/s, and leaves the sprinkler nozzles as 4mm diameter jets at an angle of 45° from the tangential direction, as shown in Fig.E4.65. The length of each sprinkler arm is 0.5m. Disregarding any frictional effects, determine (a) the flow rate of rotation of the sprinkler in rev/min, (b) the torque required to prevent the sprinkler from rotating.

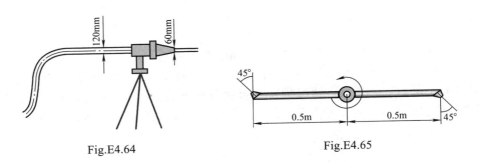

Fig.E4.64   Fig.E4.65

4.64　如图 E4.64 所示，用三脚架固定水管末端的喷嘴。已知喷嘴出口直径 60mm 流量 5m³/min，计算三脚架固定喷嘴所需的水平方向的力。

4.65　3L/s 的水流从中心竖直管道流入草坪喷淋器的双臂，如图 E4.65 所示，臂长为 0.5m，喷嘴出口直径为 4mm，与水平臂的切向成 45°。不计损失，计算（1）喷淋器每分钟的转数；（2）强迫喷淋器停止转动所需的力矩。

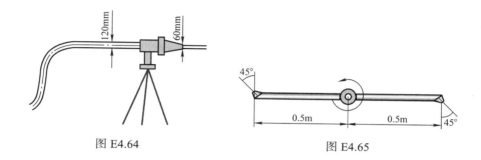

图 E4.64　　　　　　　　　　　图 E4.65

CHAPTER 5
DIMENSIONAL ANALYSIS AND SIMILARITY

In this chapter we introduce the fundamental principle of dimensional homogeneity, and a powerful tool for engineers and scientists called dimensional analysis is described. The concept of similarity between a prototype and a model is also discussed, its application to experimental modeling is presented.

5.1   Introduction

There are three basic research methods of fluid mechanics, the derivative method, the control volume method and the experimental method. The derivative method has definite physical significance but always following the fluid in motion and involving large amount of computation, complex programing algorithm are required. The control volume method has easier model while a transformation of relationship of physical quantities are necessary. The experimental method has a high degree of accuracy but usually ask for a reasonable cost, it is necessary to keep the required experimentation to a minimum.

In the study of fluid flows experimentally, there are invariably many flow and geometric parameters involved. In the interest of saving time and money, the fewest possible combinations of parameters should be utilized. We can greatly reduce the experimental time and costs needed by using *dimensional analysis* and the laws of *similarity*.

Dimensional analysis is useful when it is necessary to design and conduct experiments. Similarity is the study of predicting prototype conditions from model observations, and enable us to experiment with a model and then apply the results to a prototype.

5.2   Dimensional Analysis

The primary purposes of dimensional analysis are: (1) To the design of experiments, (2) To obtain scaling laws, (3) To predict trends in the relationship between parameters.

Before we illustrate the basic principle of dimensional analysis, let us review the concept of dimension.

5.2.1   Dimensions and units

A *dimension* is a measure of a physical quantity, all of the quantities have some combination of dimensions of length, time, mass, and other physical quantities. A *unit* is a way to assign a number to that dimension. For example, length is a dimension that is measured in units such as meters (m), kilometers (km), centimeters (cm), microns (μm), etc.

We may express physical quantities in the mass-length-time [M-L-T] system, because the system is interrelated through Newton's second law, which states that $F = ma$. In terms of dimensions, it is written as

$$[F] = \frac{[ML]}{[T]^2} \tag{5.2.1}$$

where [F], [M], [L], and [T] are the dimensions of force, mass, length, and time, respec-

第5章
量纲分析和相似原理

本章首先介绍量纲和谐原理和量纲分析法，量纲分析法对科学研究和工程应用都是一个非常有效的方法。另一个主要内容是相似原理，主要通过对原型和模型相似关系的分析指导实验建模。

5.1　引言

流体力学常用的研究方法包括微元法、控制体法和实验法。微元法跟随流体质点运动，具有明确的物理意义，但通常需要大量复杂的计算。控制体法模型简单，但需要对物理量进行必要的关系转换。实验法精度高但往往费用也高，经常面临着降低实验成本的压力。

流体力学的研究总会涉及很多流动和流场几何参数，出于节约实验成本的考量，我们希望尽量涉及最少的参数来获得满意的实验结果。应用量纲分析和相似原理能够有效地降低实验研究成本、节约时间。

量纲分析法在设计和简化实验方面作用显著，相似原理着眼于从模型中获得原型的物理规律，使实验结果能够正确地指导原型的应用。

5.2　量纲分析

量纲分析的主要意义包括：（1）指导设计实验系统；（2）明确相似准则；（3）预测变量关系和规律。

介绍量纲分析方法之前，我们先回顾一下量纲的概念。

5.2.1　量纲和物理单位

量纲就是物理量的类型，任何物理变量都可以由包括长度、时间、质量等的物理量构成；而物理**单位**则是量纲的大小和数量的标准。例如，长度的量纲可以用米（m）、千米（km）、厘米（cm）、微米（μm）等单位表示。

我们常用质量 - 长度 - 时间，即 [M-L-T] 系统来表示物理量，这几种量纲的关系通过牛顿第二定律相统一，牛顿第二定律 $F = ma$ 写成量纲形式

$$[F] = \frac{[ML]}{[T]^2} \qquad (5.2.1)$$

式中，[F]、[M]、[L] 和 [T] 分别是力、质量、长度和时间的量纲，括在中括号内表

tively, and the brackets indicate that are dimensions.

We see that it is sufficient to use [M-L-T] system because the force dimension can be eliminated with Eq.5.2.1. We call this three [M], [L] and [T] *fundamental dimensions* (also called primary or basic dimensions) in study of incompressible fluid flow. The quantities of interest in this book are listed with their respective dimensions in Tab.5.2.1. For the dimensions of quantities shown in Tab.5.2.1 can be expressed with the fundamental dimensions, called *derived dimensions*.

Tab.5.2.1   The derived dimensions used in Fluid Mechanics

Quantity	Symbol	Dimensions
Velocity	V	$[LT^{-1}]$
Acceleration	a	$[LT^{-2}]$
Volume flow rate	q	$[L^3T^{-1}]$
Mass flow rate	q_m	$[MT^{-1}]$
Force	F	$[MLT^{-2}]$
Pressure	p	$[ML^{-1}T^{-2}]$
Shearing stress	τ	$[ML^{-1}T^{-2}]$
Density	ρ	$[ML^{-3}]$
Viscosity	μ	$[ML^{-1}T^{-1}]$
Kinematic viscosity	ν	$[L^2T^{-1}]$
Surface tension	σ	$[MT^{-2}]$

5.2.2   Dimensional homogeneity

The technique of dimensional analysis based on the notion of *dimensional homogeneity*, which was first formalized by Baron Joseph Fourier (1768—1830), a French mathematician and physicist, states that: all terms in an equation must have the same dimensions.

We note from Fig.5.2.1 that the dimension of each term in Bernoulli Equation is length.

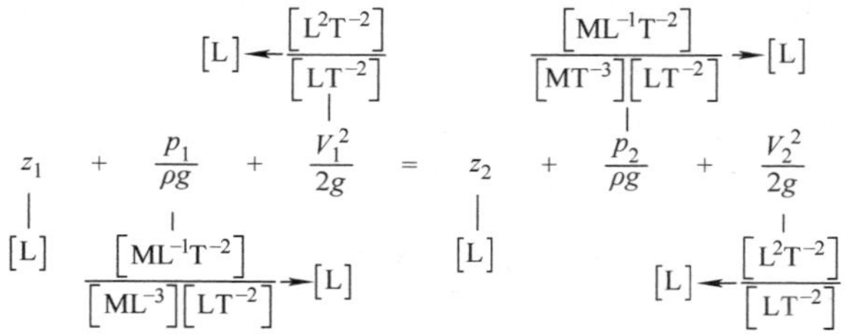

Fig.5.2.1   The dimension of each term in Bernoulli Equation is length

明其为量纲。

由式 5.2.1 可以看到，力的量纲完全可以由 [M-L-T] 三个量纲表达，我们称 [M]、[L] 和 [T] 为**基础量纲**。表 5.2.1 列出了本书中一些常用物理量的量纲。表中物理量均可由基本量纲构成，称为**导出量纲**。

表 5.2.1　流体力学中常用的导出量纲

物理量	符号	量纲
速度	V	$[LT^{-1}]$
加速度	a	$[LT^{-2}]$
体积流量	q	$[L^3T^{-1}]$
质量流量	q_m	$[MT^{-1}]$
力	F	$[MLT^{-2}]$
压强	p	$[ML^{-1}T^{-2}]$
剪切应力	τ	$[ML^{-1}T^{-2}]$
密度	ρ	$[ML^{-3}]$
黏度	μ	$[ML^{-1}T^{-1}]$
运动黏度	v	$[L^2T^{-1}]$
表面张力	σ	$[MT^{-2}]$

5.2.2　量纲和谐原理

量纲分析法的基础是**量纲和谐原理**，最初由法国数学家、物理学家 Baron Joseph Fourier（1768—1830）提出，内容为：正确反映客观规律的物理方程中各项的量纲必定是一致的。

图 5.2.1 表明伯努利方程中各项的量纲都是长度。

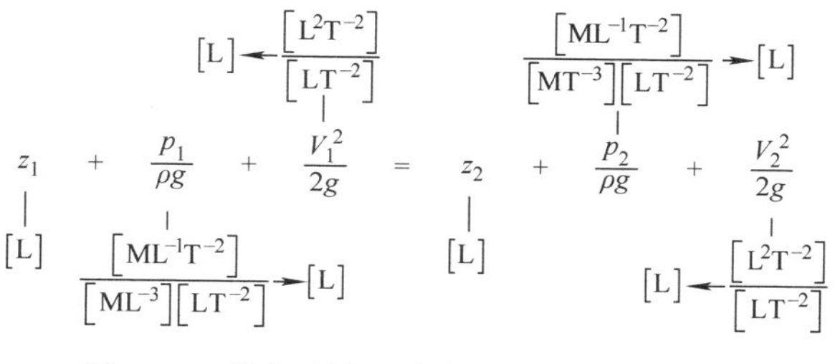

图 5.2.1　伯努利方程中各项的量纲均匀长度

To illustrate the basic principles of dimensional analysis, let us explore the equation for the critical velocity that of the determination of laminar flow and turbulent flow. We recognized that the physical factors probably influence the critical velocity are density ρ and viscosity μ of fluid, and the diameter d of the flowing pipe.

The dimensions of these quantities are

$$V_c = \left[LT^{-1}\right], \quad \rho = \left[ML^{-3}\right], \quad \mu = \left[ML^{-1}T^{-1}\right], \quad d = [L]$$

We multiply these factors in a way that their dimensions balance,

$$V_c = k\rho^{a_1}\mu^{a_2}d^{a_3} \tag{5.2.2}$$

where k is a dimensionless constant, and let us solve for the exponents a_1, a_2, and a_3. Substituting the dimensions, we get

$$\left[LT^{-1}\right] = \left[L^{-3}M\right]^{a_1}\left[L^{-1}MT^{-1}\right]^{a_2}\left[L\right]^{a_3} \tag{5.2.3}$$

To satisfy dimensional homogeneity, the exponents of each dimension must be identical on both sides of this equation. Thus

For [L]: $1 = -3a_1 - a_2 + a_3$

For [M]: $0 = a_1 + a_2$

For [T]: $-1 = -a_2$

Solving these three equations, we get

$$a_1 = -1, \quad a_2 = 1, \quad a_3 = -1$$

So that

$$V_c = k\frac{\mu}{\rho d} \tag{5.2.4}$$

We see that the critical velocity V_c varies with the diameter of the experimental pipe, type of fluid and even the temperature which effects the viscosity of fluid. In engineering practice, we prefer to use the nondimensional constant k, which is called Reynolds number and presented by Re_c (Eq.5.2.5) named after Osborne Reynold (1842—1912), an English engineer, as the parameter to distinguish the flow type. You will find that many problems in fluid mechanics involve a Reynolds number, we will discuss the Re in section 6.2 furtherly.

$$Re_c = \frac{\rho d V_c}{\mu} \tag{5.2.5}$$

The method used in the uppercase was developed by Lord Rayleigh (1842—1919), an English physicist. The *Rayleigh's method* shows the basic idea in dimensional analysis, however can only solve the relationship not more than three factors involved because we have only three fundamental dimension, has been superseded.

我们以判定流体流动状态为层流还是紊流时依据的临界速度为例，说明量纲分析法的基本原理。研究人员通过实验和观察总结出，与临界速度相关的物理量包括流体的密度 ρ、黏度 μ 以及实验中所使用管路的直径 d。

这些变量的量纲构成为

$$V_c = \left[LT^{-1}\right], \qquad \rho = \left[ML^{-3}\right], \qquad \mu = \left[ML^{-1}T^{-1}\right], \qquad d = [L]$$

把这些变量按下式方式相乘，以使得等式两边量纲平衡，

$$V_c = k\rho^{a_1}\mu^{a_2}d^{a_3} \tag{5.2.2}$$

式中，k 为无量纲常系数，接下来求解各项指数 a_1，a_2 和 a_3。代入各项的量纲，得

$$\left[LT^{-1}\right] = \left[L^{-3}M\right]^{a_1}\left[L^{-1}MT^{-1}\right]^{a_2}[L]^{a_3} \tag{5.2.3}$$

为符合量纲和谐原理，同一量纲等式两边的指数一定是相等的，所以

对于 [L]：　　$1 = -3a_1 - a_2 + a_3$

对于 [M]：　　$0 = a_1 + a_2$

对于 [T]：　　$-1 = -a_2$

解这三个方程，得

$$a_1 = -1, \qquad a_2 = 1, \qquad a_3 = -1$$

因此

$$V_c = k\frac{\mu}{\rho d} \tag{5.2.4}$$

我们看到，临界速度 V_c 的变化与实验管路直径有关，受流体种类甚至温度（影响流体的黏度）影响。工程中，我们常用符号 Re_c 代替式 5.2.5 中的无量纲数 k 作为判定流态的标准，Re 是英国工程师 Osborne Reynold（1842—1912）名字的首写字母，称为雷诺数。流体力学问题的研究中经常会用到雷诺数，本书 6.2 节会详细介绍雷诺数的相关知识。

$$Re_c = \frac{\rho d V_c}{\mu} \tag{5.2.5}$$

上述方法称为瑞利法，由英国物理学家 Lord Rayleigh（1842—1919）提出。瑞利法对三个基本量纲建立方程，因此不适用于求解超过三个变量的情况，但这一方法清楚地说明了量纲分析法的基本思想。

5.2.3 Buckingham π-theorem

The law of dimensional homogeneity guarantees that every term in an equation has the same dimensions. Therefore if we divide them all by a quantity that has the same dimensions then all the terms will become dimensionless.

If we factored out z_1 and z_2 from the left- and right-hand side respectively, the Bernoulli Equation can be written as

$$1+\frac{p_1}{\rho g z_1}+\frac{V_1^2}{2g z_1}=\left(1+\frac{p_2}{\rho g z_2}+\frac{V_2^2}{2g z_2}\right)\frac{z_2}{z_1} \tag{5.2.6}$$

The terms in Eq.5.2.6 are all dimensionless and we have written the equation as a combination of dimensionless parameters, this shows the idea of *nondimensionalization*.

For example, consider the pressure drop across inlet of the spool valve of Fig.A4.5.1a. We may suspect that the pressure drop depends on such parameters as mean velocity V, the density ρ of the fluid, the fluid viscosity μ, the seat diameter d, and the width of opening x_T. This could be expressed as

$$\Delta p = f\left(V,\rho,\mu,d,x_T\right) \tag{5.2.7}$$

Now, if we attempt an experimental study of this problem, consider a way for finding the dependence of the pressure drop on the parameters involved. We could fix all parameters except the velocity and investigate the dependence of the pressure drop on the average velocity. Then the diameter could be changed and the experiment repeated. Following that set of experiments the width of opening could be changed, different fluids could be studied. Extensive experiments imply significant cost of time and money.

Consider the notion that any equation that relates a certain set of variables can be written in terms of dimensionless parameters, we can organize the variables of Eq.5.2.7 into dimensionless parameters as follows:

$$\frac{\Delta p}{\rho V^2}=f\left(\frac{\rho d V}{\mu},\frac{x_T}{d}\right) \tag{5.2.8}$$

Obviously, this is a much simpler relationship. We could perform a set of experiments with a fixed x_T/d by varying $\rho d V/\mu$, this has greatly reduced the effort and the cost.

We can express many equations more simply as relationships between dimensionless groups or numbers. However, the selection of these parameters requires a detailed understanding of the physics involved.

A more generalized method of dimensional analysis developed by Edgar Buckingham (1867—1940), called *Buckingham π-theorem*, states that $(n-m)$ dimensionless groups of variables, called π-terms.

We will explain the steps of Buckingham π-theorem in further detail by example.

Step 1, List the parameters in the problem, including dependent variable and independent variables, and count their numbers n.

Step 2, List the fundamental dimensions for each of the n parameters.

Step 3, Set the reduction m as the number of fundamental dimensions. The expected number of π's is equal to $(n-m)$.

Step 4, Choose $(n-m)$ parameters to build each π.

Step 5, Build $(n-m)$ π's, forcing the product to be dimensionless. By convention, designated π_1 as the dependent π.

5.2.3　布金汉 -π 原理法

量纲和谐原理要求保证方程中各项的量纲一致。所以，如果我们对等式除以一个具有相同量纲的量，则方程中的各项都将成为无量纲数。

例如，对伯努利方程两边分别提取 z_1 和 z_2，则方程可改写为

$$1+\frac{p_1}{\rho g z_1}+\frac{V_1^2}{2 g z_1}=\left(1+\frac{p_2}{\rho g z_2}+\frac{V_2^2}{2 g z_2}\right)\frac{z_2}{z_1} \tag{5.2.6}$$

方程 5.2.6 中的各项都是无量纲量，把方程化为无量纲量形式的过程称为**无量纲化**。

回顾图 A4.5.1a 中的滑阀，我们分析滑阀的压降可能与平均流速 V、油液的密度 ρ、油液的黏度 μ、阀座的直径 d 以及阀口开度有关 x_{T}，那么这些变量的关系可以写为

$$\Delta p=f\left(V,\rho,\mu,d,x_{\mathrm{T}}\right) \tag{5.2.7}$$

如果希望通过实验来分析以上因素与压差间的关系，我们可以首先仅改变流速而保持其他实验条件不变，研究油液平均流速对压强差的影响；然后对阀座直径和其他，如，阀口开度、油液种类等各项参数重复同样的工作。显然，这一系列实验将耗费大量的时间和费用。

既然可以用无量纲量构成物理方程，那么式 5.2.7 可以组织为无量纲式：

$$\frac{\Delta p}{\rho V^2}=f\left(\frac{\rho d V}{\mu},\frac{x_{\mathrm{T}}}{d}\right) \tag{5.2.8}$$

显然，这种形式的函数关系更加简单。我们仅需对 x_{T}/d 和 $\rho d V/\mu$ 两个值，改变一个、固定另一个重复进行实验，大大减少了实验次数、节约了人力物力。

很多方程都可以由无量纲量和无量纲数构成更为简单的形式，当然，这种化简工作需要在充分理解物理关系、掌握一定背景知识的前提下进行。

应用更广泛的量纲分析方法是由 Edgar Buckingham（1867—1940）提出的，使用了（$n-m$）个无量纲式，即 π- 式，因此被称为**布金汉 -π 原理法**。

我们先列出布金汉 -π 原理法的各个步骤，再通过具体实例来详细说明。

步骤 1，列出待研究问题相关的全部变量，包括自变量和因变量，统计变量的个数 n。

步骤 2，列出 n 个变量各自的量纲。

步骤 3，选出 m 各变量作为基本变量，把 π- 式的个数减少至（$n-m$）个。

步骤 4，用（$n-m$）个变量构建各个 π- 式。

步骤 5，构建（$n-m$）个幂指数乘积形式的无量纲式，因变量记为 π_1，自变量记为各 π- 式。

Step 6, To satisfy dimensional homogeneity, solve for the exponents and the forms of the dimensionless groups.

Step 7, Write the final functional relationship in the form of

$$\pi_1 = f\left(\pi_2, \pi_3 \cdots \pi_{n-m}\right) \tag{5.2.9}$$

Example 5.2.1 Consider the incompressible steady turbulent flow of viscous fluid in pipes, suppose that all we know is the frictional force on the wall of pipe must be a function of length of pipe l, diameter of pipe d, density of fluid ρ, fluid viscosity μ, average velocity V, and roughness of pipe wall ε. As we go through each step of the method of Buckingham π-theorem, we explain the technique in more detail using the frictional force as an example.

Solution

Step 1 List the relevant parameters, $F_\tau = f\left(l, d, \rho, \mu, V, \varepsilon\right)$, that is, $n = 7$.

Step 2 The fundamental dimensions of each parameter are listed here.

$$\begin{array}{ccccccc} F_\tau & l & d & \rho & \mu & V & \varepsilon \\ [MLT^{-2}] & [L] & [L] & [MLT^{-3}] & [ML^{-1}T^{-1}] & [LT^{-1}] & [L] \end{array}$$

Step 3 Set $m = 3$, the number of fundamental dimensions represented in the problem $[L, M, T]$.

Number of expected π's is $n - m = 7 - 3 = 4$.

Make sure that any one of the m parameters is independent of the others, i.e., it cannot be expressed in terms of them. And the m parameters including all of the fundamental dimensions. The selection of the proper parameters is the first crucial step in the application of dimensional analysis. Here we select ρ, d and V as the fundamental variables.

Step 4 Use the other 4 parameters F_τ, l, ρ and ε to build each π.

Step 5 Now we combine the m variables into products with each of the remaining parameters, one at a time, to create the π's. Each variable is written with exponents since this is helpful with later algebra. The first π is always formed with the dependent variable.

$$\pi_1 = F_\tau \rho^{a_1} d^{a_2} V^{a_3}$$
$$\pi_2 = l \rho^{b_1} d^{b_2} V^{b_3}$$
$$\pi_3 = \mu \rho^{c_1} d^{c_2} V^{c_3}$$
$$\pi_4 = \varepsilon \rho^{d_1} d^{d_2} V^{d_3}$$

where a_i, b_i, c_i and d_i are constant exponents that need to be determined.

Step 6 Using the Rayleigh's method to solve each equation, for example:

$$\pi_1 = F_\tau \rho^{a_1} d^{a_2} V^{a_3} \quad \rightarrow \quad [0] = \left[MLT^{-2}\right]\left[ML^{-3}\right]^{a_1}\left[L\right]^{a_2}\left[LT^{-1}\right]^{a_3}$$

M: $\quad 0 = 1 + a_1$
L: $\quad 0 = 1 - 3a_1 + a_2 + a_3$
T: $\quad 0 = -2 - a_3$
$a_1 = -1, \quad a_2 = -2, \quad a_3 = -2$

$$\rightarrow \quad \pi_1 = \frac{F_\tau}{\rho V^2 d^2}$$

步骤 6，根据量纲和谐原理求解各指数和无量纲式；

步骤 7，写出各 π- 式间的函数关系式，如

$$\pi_1 = f\left(\pi_2, \pi_3, \cdots, \pi_{n-m}\right) \tag{5.2.9}$$

例 5.2.1　对于圆管内不可压缩流体稳定紊流，已知管道内壁上的摩擦阻力 F_τ 与管道长度 l、管径 d、流体密度 ρ 和黏度 μ、平均流速 V 以及管壁粗糙度 ε 有关。我们以管壁摩擦力的函数关系为例，解释应用布金汉 -π 原理法进行量纲分析的步骤。

解

步骤 1：列出相关变量，$F_\tau = f\left(l, d, \rho, \mu, V, \varepsilon\right)$，变量的个数 $n = 7$。

步骤 2：列出各变量的量纲构成。

$$
\begin{array}{ccccccc}
F_\tau & l & d & \rho & \mu & V & \varepsilon \\
[\mathrm{MLT^{-2}}] & [\mathrm{L}] & [\mathrm{L}] & [\mathrm{MLT^{-3}}] & [\mathrm{ML^{-1}T^{-1}}] & [\mathrm{LT^{-1}}] & [\mathrm{L}]
\end{array}
$$

步骤 3：取 $m = 3$ 个基本变量，组成基本变量的量纲包括 $[\mathrm{L, M, T}]$，待建立 π- 式的个数为 $n - m = 7 - 3 = 4$。

一定要保证 m 个变量互相独立，即，彼此不能互相表示，且 m 个变量中应该包含全部基础量纲。合理选取基本变量是量纲分析的第一个关键步骤，本例中，我们选取 ρ、d 和 V 作为基本变量。

步骤 4：用剩余的 4 个变量 F_τ、l、ρ 和 ε 建立无量纲式。

步骤 5：每次选择一个变量与基本变量构建无量纲 π- 式，共（$n - m$）个，为方便求解，式中变量都写为幂指数乘积的形式。通常用第一个 π- 式表示因变量：

$$\pi_1 = F_\tau \rho^{a_1} d^{a_2} V^{a_3}$$
$$\pi_2 = l \rho^{b_1} d^{b_2} V^{b_3}$$
$$\pi_3 = \mu \rho^{c_1} d^{c_2} V^{c_3}$$
$$\pi_4 = \varepsilon \rho^{d_1} d^{d_2} V^{d_3}$$

式中，a_i、b_i、c_i 和 d_i 就是待求解的指数。

步骤 6：用瑞利法求解每个指数，如

$$\pi_1 = F_\tau \rho^{a_1} d^{a_2} V^{a_3} \quad \rightarrow \quad [0] = \left[\mathrm{MLT^{-2}}\right]\left[\mathrm{ML^{-3}}\right]^{a_1}\left[\mathrm{L}\right]^{a_2}\left[\mathrm{LT^{-1}}\right]^{a_3}$$

$$\mathrm{M}: \quad 0 = 1 + a_1$$
$$\mathrm{L}: \quad 0 = 1 - 3a_1 + a_2 + a_3$$
$$\mathrm{T}: \quad 0 = -2 - a_3$$
$$a_1 = -1, \quad a_2 = -2, \quad a_3 = -2$$
$$\rightarrow \quad \pi_1 = \frac{F_\tau}{\rho V^2 d^2}$$

By the same way, we can solve each π as

$$\pi_2 = \frac{l}{d}, \; \pi_3 = \frac{\mu}{\rho dV} \text{ and } \pi_4 = \frac{\varepsilon}{d}$$

Step 7   Write the final functional relationship between the dependent variable and the independent variables as

$$\frac{F_\tau}{\rho V^2 d^2} = f\left(\frac{l}{d}, \frac{\mu}{\rho Vd}, \frac{\varepsilon}{d}\right) \; \text{or,} \; F_\tau = k\rho V^2 d^2 \left(\frac{\mu}{\rho dV}\right)^{k_1} \left(\frac{\varepsilon}{d}\right)^{k_2} \left(\frac{l}{d}\right)^{k_3}$$

and rearrange the π groups as desired.

We use the Buckingham π-theorem requires some knowledge of the phenomenon being studied, we see that $\frac{\mu}{\rho dV}$ is the revers of $\frac{\rho dV}{\mu}$, which is the famous Reynolds number we will discuss in section 6.2. And we know that the frictional force is proportional to the length of the pipe, then the exponent number k_3 can be taken as $k_3 = 1$.

Then the final function is manipulated to

$$F_\tau = k\rho V^2 l d Re^{k_1} \left(\frac{\varepsilon}{d}\right)^{k_2}$$

Rather than the original relationship of seven variables we have reduced the relationship to one involving four π terms, a much simpler expression. The method of Buckingham π-theorem cannot predict the exact mathematical form of the equation. This would result in an expression that includes an arbitrary constant such as k and exponents such as k_1 and k_2 in the above equation, that could be determined through analysis or experimentation. This is often the case in fluid mechanics.

The use of Buckingham π-theorem requires some knowledge of the problem being studied in order that the appropriate quantities of interest are included. We extract the dimensionless parameters that influence a particular flow situation. Some established non-dimensional parameters, most of which are named after a notable scientist or engineer (sec.5.3), are very helpful and it is necessary to manipulate your π's into them. You are encouraged to use this powerful tool of dimensional analysis in other subjects as well, not just in fluid mechanics.

Example 5.2.2   The drag force F_D on a cylinder of diameter d and length l is to be studied. What functional form relates the dimensionless variables if a fluid with velocity V flows normal to the cylinder?

Solution   The involved factors F_D, d, l, V and density of fluid ρ, fluid viscosity μ, resulting in $n = 6$. This is written as

$$F_D = f(d, l, V, \rho, \mu)$$

The dimension of these six factors are

$$F_D = \left[MLT^{-2}\right], \quad d = [L], \quad l = [L], \quad V = \left[LT^{-1}\right], \quad \rho = \left[ML^{-3}\right], \quad \mu = \left[ML^{-1}T^{-1}\right]$$

The variables are observed to include three fundamental dimensions. Then $m = 3$. Consequently, we can expect $n - m = 6 - 3 = 3\pi$-terms.

应用同样的方法，解各个无量纲式，得

$$\pi_2 = \frac{l}{d}, \quad \pi_3 = \frac{\mu}{\rho d V} \quad \text{及} \quad \pi_4 = \frac{\varepsilon}{d}$$

步骤 7：写出因变量和自变量间的函数关系式，如

$$\frac{F_\tau}{\rho V^2 d^2} = f\left(\frac{l}{d}, \frac{\mu}{\rho V d}, \frac{\varepsilon}{d}\right) \quad \text{或,} \quad F_\tau = k \rho V^2 d^2 \left(\frac{\mu}{\rho d V}\right)^{k_1} \left(\frac{\varepsilon}{d}\right)^{k_2} \left(\frac{l}{d}\right)^{k_3}$$

并适当调整方程形式。

应用布金汉 $-\pi$ 原理需要对所研究的问题有充分的了解，如例中 $\frac{\mu}{\rho d V}$ 项是 $\frac{\rho d V}{\mu}$ 的倒数，而后者是著名的雷诺数（将在 6.2 节详细讨论）。并且，我们知道摩擦阻力与流动路径的长度成正比，所以指数 k_3 取值应为 $k_3 = 1$。

调整后方程的最终形式为

$$F_\tau = k \rho V^2 l d Re^{k_1} \left(\frac{\varepsilon}{d}\right)^{k_2}$$

通过应用布金汉 $-\pi$ 原理法，方程从最初七个变量的函数关系化简为四个无量纲量的函数关系。该方法无法精准预测方程的数学模型，例如例题中的常系数 k 和指数 k_1、k_2，还需要通过进一步的实验和分析求解。这就是流体力学问题研究的普遍过程。

了解待研究的问题、合理选取变量对应用布金汉 $-\pi$ 原理法至关重要。有一些流体力学研究中常用的无量纲量，其中大部分都是以提出该变量的著名科学家或工程师命名（详见 5.3 节），可以直接选用为 π- 式。当然，量纲分析方法的应用不仅仅局限于流体力学问题，对其他科学问题的研究同样有极大的助力。

例 5.2.2　直径 d、长度 l 的缸内，流体平均流速为 V，求缸内阻力 F_D 的无量纲函数表达式。

解: 与该问题相关的变量包括 F_D、d、l、V 和液体密度 ρ 及黏度 μ，变量数 $n = 6$。写为

$$F_D = f(d, l, V, \rho, \mu)$$

六个变量的量纲为

$$F_D = \left[MLT^{-2}\right], \quad d = [L], \quad l = [L], \quad V = \left[LT^{-1}\right], \quad \rho = \left[ML^{-3}\right], \quad \mu = \left[ML^{-1}T^{-1}\right]$$

选择三个包含基础量纲的变量，即 $m = 3$.
接下来需要建立 $n - m = 6 - 3 = 3$ 个 π- 式。

The fundamental variables are chosen to be d, V, and ρ. We could not combine d and l as fundamental parameters because they have the same dimension. Also we cannot combine ρ and μ, they are non-independent since we have $v = \mu/\rho$.

To form the 3π-terms effectively, we need to do some inspection rather than the Rayleigh's method.

$$\pi_1 = F_D \rho^a V^b d^c$$

When the fundamental variables are combined with F_D we observe that only F_D and ρ have the mass dimension; hence F_D must be divided by ρ, then we have $a = -1$. Only F_D and V have the time dimension, hence, F_D must be divided by V^2, that is $b = -2$. Thus F_D divided by ρ has L^4 in the numerator; when divided by V^2 this results in L^2 remaining in the numerator. Hence we must have d^2 in the denominator resulting in $c = -2$. The form of the first π is determined,

$$\pi_1 = \frac{F_D}{\rho V^2 d^2}$$

The next π-term results from combining μ, with d, V, and ρ, which remind us of the Reynold Number

$$\pi_2 = Re = \frac{\rho d V}{\mu}$$

The last one

$$\pi_3 = \frac{l}{d}$$

The dimensionless, functional relationship relating the π-terms is

$$\frac{F_D}{\rho V^2 d^2} = f\left(\frac{\rho d V}{\mu}, \frac{l}{d}\right), \quad \text{or} \quad F_D = k\rho V^2 d^2 \left(\frac{\rho d V}{\mu}\right)^m \left(\frac{l}{d}\right)^n$$

Example 5.2.3   The rise of liquid in a capillary tube is to be studied. It is anticipated that the rise h will depend on surface tension σ, tube diameter d, liquid specific weight $\gamma = \rho g$, and angle θ of contact between the liquid and tube. Write the functional form of the dimensionless variables.

Solution   The expression relating the variables is

$$h = f(\sigma, d, \gamma, \theta)$$

The number of factors $n=5$
The dimensions of the variables are

$$h = [L], \quad \sigma = [MT^{-2}], \quad d = [L], \quad \gamma = [ML^{-2}T^{-2}], \quad \theta(\text{dimensionless})$$

We choose σ and d as the fundamental variables, then $m = 2$, $n - m = 3$.

选取 d、V 和 ρ 作为基本变量。注意，d 和 l 量纲相同所以不能同时选为基本变量；同理，ρ 和 μ 具有函数关系 $\nu=\mu/\rho$，不互相独立，因此也不能同时出现在基本变量中。

本例中，我们不使用瑞利法，而是通过推理求解 3 个 π- 式。

$$\pi_1 = F_D \rho^a V^b d^c$$

当 F_D 和基本变量同时出现在一个等式中，我们注意到，只有 F_D 和 ρ 含有质量的量纲，因此一定有 $a = -1$；时间量纲只出现在 F_D 和 V，所以一定是 F_D 和 V^2 相除，得出 $b = -2$；F_D 除以 ρ 得到 L^4 项，除以 V^2 时余下 L^2 项，所以函数式中一定含有 d^2 项，即 $c = -2$。推理得出第一个无量纲量为

$$\pi_1 = \frac{F_D}{\rho V^2 d^2}$$

下一个无量纲量由 μ 和 d、V 及 ρ 构成，显然为雷诺数

$$\pi_2 = Re = \frac{\rho d V}{\mu}$$

最后一个

$$\pi_3 = \frac{l}{d}$$

所以，各无量纲项的函数关系写作

$$\frac{F_D}{\rho V^2 d^2} = f\left(\frac{\rho d V}{\mu}, \frac{l}{d}\right), \quad 或，\quad F_D = k\rho V^2 d^2 \left(\frac{\rho d V}{\mu}\right)^m \left(\frac{l}{d}\right)^n$$

例 5.2.3　分析毛细管内液体上升高度的影响因素。观察发现浸润高度 h 与表面张力 σ、毛细管直径 d、液体重度 $\gamma = \rho g$ 以及液体和管壁间接触角 θ 有关。写出无量纲函数关系式。

解　变量间函数关系为

$$h = f(\sigma, d, \gamma, \theta)$$

变量的个数 $n = 5$

各变量的量纲构成

$$h = [L], \quad \sigma = \left[MT^{-2}\right], \quad d = [L], \quad \gamma = \left[ML^{-2}T^{-2}\right], \quad \theta(无量纲)$$

选取 σ 和 d 作为基本变量，则 $m = 2$，$n - m = 3$。

When combined with h, the first π-term is

$$\pi_1 = \frac{h}{d}$$

The dimensionless contact angle θ forms a π-term by itself

$$\pi_2 = \theta$$

When σ and d are combined with γ, the last π-term is

$$\pi_3 = \frac{\sigma}{\gamma d^2}$$

The final functional form relating the π-terms is

$$\frac{h}{d} = f\left(\theta, \frac{\sigma}{\gamma d^2}\right) \quad \text{or} \quad h = k\theta^a (\rho g)^{-b} d^{-2b+1} \sigma^b$$

Compare to the Eq.1.4.20, $h = \dfrac{4\sigma \cos\theta}{\rho g d}$, we know that $k=4$, and $b=1$.

5.3   Similarity

Similarity is the study of predicting prototype conditions from model observations. The similarity laws enable us to perform tests on a model if testing is not practical on a full-scale prototype, be it too large or too small. It also enable us to experiment with a convenient fluid and then apply the results to a fluid that is less convenient to work with.

5.3.1   Geometric similarity, Kinematic similarity and Dynamic similarity

When a *dynamic similarity* between model and prototype is developed, the forces which act on corresponding masses in model flow and the prototype flow are in the same ratio throughout the entire flow fields. This makes that a quantity measured on the model can be used to predict the associated quantity on the prototype.

We signified the model with a subscript m, and prototype by a subscript p. Suppose that inertial forces, pressure forces, viscous forces, and gravity forces are present; then at corresponding points in the model and prototype flow fields, dynamic similarity requires that

$$\frac{(F_I)_m}{(F_I)_p} = \frac{(F_p)_m}{(F_p)_p} = \frac{(F_\tau)_m}{(F_\tau)_p} = \frac{(F_g)_m}{(F_g)_p} = \text{Constant} \tag{5.3.1}$$

Let us take the inertial forces in steady flow of incompressible fluid along a streamline, the forces can be expressed in the simplest terms as

$$F_I = ma_S = mV\frac{dV}{dS} \quad \rightarrow \quad \rho l^3 V \frac{V}{l} = \rho l^2 V^2 \tag{5.3.2}$$

We can write the inertial force ratio as

与 h 组合，第一个无量纲式为

$$\pi_1 = \frac{h}{d}$$

接触角 θ 单独组成无量纲式

$$\pi_2 = \theta$$

σ 和 d 与 γ 组合成最后一个无量纲式

$$\pi_3 = \frac{\sigma}{\gamma d^2}$$

各无量纲式间的函数关系为

$$\frac{h}{d} = f\left(\theta, \frac{\sigma}{\gamma d^2}\right) \quad \text{或} \quad h = k\theta^a \left(\rho g\right)^{-b} d^{-2b+1}\sigma^b$$

与式 1.4.20 式 $h = \dfrac{4\sigma\cos\theta}{\rho g d}$ 比较，可知 $k = 4$，$b = 1$。

5.3　相似原理

相似性就是通过对模型的观察预测原型的流动规律。依据相似原理，我们可以把不方便在原型上操作的实验（如原型太大或太小）转移至模型上进行实验研究；或者使用方便取得的流体进行实验，并将实验结果用于指导难控制流体的研究。

5.3.1　几何相似、运动相似和动力相似

原型与模型间**动力相似**是指原型与模型流场内各种力的比值相同。有了动力相似就可以根据在模型上测得的变量预测其在原型流场的数值。

我们用下标 m 代表模型，下标 p 代表原型。假设原型和模型的流场动力相似，则两个流场中的惯性力、压力、黏性力和重力等具有以下相似关系

$$\frac{(F_I)_m}{(F_I)_p} = \frac{(F_p)_m}{(F_p)_p} = \frac{(F_\tau)_m}{(F_\tau)_p} = \frac{(F_g)_m}{(F_g)_p} = \text{常数} \tag{5.3.1}$$

以沿流线运动的不可压缩流体稳定流动时的惯性力为例，其物理量构成可表示为

$$F_I = ma_s = mV\frac{dV}{dS} \quad \rightarrow \quad \rho l^3 V \frac{V}{l} = \rho l^2 V^2 \tag{5.3.2}$$

则模型和原型间惯性力之比为

$$\frac{(F_I)_m}{(F_I)_p} = \frac{m_m a_m}{m_p a_p} = \frac{m_m}{m_p}\frac{a_m}{a_p} = \text{Constant} \tag{5.3.3}$$

Illustrating that the acceleration ratio between corresponding points on the model and prototype is a constant provided that the mass ratio of corresponding fluid elements is also a constant.

We can write the acceleration ratio as

$$\frac{a_m}{a_p} = \frac{V_m^2/l_m}{V_p^2/l_p} = \left(\frac{V_m}{V_p}\right)^2 \bigg/ \left(\frac{l_m}{l_p}\right) = \text{Constant} \tag{5.3.4}$$

Illustrating that the velocity ratio between corresponding points is a constant if the length ratio is a constant.

The length ratio being constant between all corresponding points in the flow fields is the demand of *geometric similarity*, which results in the model having the same shape as the prototype.

The velocity ratio being a constant between all corresponding points in the flow fields is the statement of *kinematic similarity*.

Hence, there are three necessary conditions for complete similarity between a model and a prototype, the geometric similarity, the kinematic similarity and the dynamic similarity.

The first condition, the geometric similarity, the model must be the same shape as the prototype, but may be scaled by some constant scale factor. The second condition, the kinematic similarity, which means that the velocity at any point in the model flow must be proportional to the velocity at the corresponding point in the prototype flow by a constant scale factor. Specifically, for kinematic similarity the velocity at corresponding points must scale in magnitude and must point in the same relative direction. The third and most restrictive similarity condition, the dynamic similarity, is achieved when all forces in the model flow scale by a constant factor to corresponding forces in the prototype flow.

Just as the geometric scale factor can be less than, equal to, or greater than one, so can the scale factor for the velocity and the forces. You may think of geometric similarity as *length-scale* equivalence, kinematic similarity as *time-scale* equivalence, and dynamic similarity as *force-scale* equivalence.

Geometric similarity is a prerequisite for kinematic similarity. Kinematic similarity is a necessary but insufficient condition for dynamic similarity. It is thus possible for a model flow and a prototype flow to achieve both geometric and kinematic similarity, yet not dynamic similarity. In a general flow field, complete similarity between a model and prototype is achieved only when there is geometric, kinematic, and dynamic similarity, also including the similarity of initial condition and boundary condition. The streamline pattern around the model is the same as that around the prototype. When using models the fluid velocity should never be so low that flow is laminar when the flow in the prototype is turbulent, for example.

5.3.2 Common nondimensional numbers of dynamic similitude

Let us assume that two geometrically similar flow systems also possess kinematic similarity, and that the forces acting on any fluid element can be pressure forces F_p, viscous forces F_τ, and gravity forces F_g, and elasticity F_E, also, if the element of fluid is at a

$$\frac{(F_{\mathrm{I}})_{\mathrm{m}}}{(F_{\mathrm{I}})_{\mathrm{p}}} = \frac{m_{\mathrm{m}} a_{\mathrm{m}}}{m_{\mathrm{p}} a_{\mathrm{p}}} = \frac{m_{\mathrm{m}}}{m_{\mathrm{p}}} \frac{a_{\mathrm{m}}}{a_{\mathrm{p}}} = 常数 \tag{5.3.3}$$

可见，模型和原型间加速度和质量之比也同样为常数。

把两个流场的加速度之比写为

$$\frac{a_{\mathrm{m}}}{a_{\mathrm{p}}} = \frac{V_{\mathrm{m}}^2 / l_{\mathrm{m}}}{V_{\mathrm{p}}^2 / l_{\mathrm{p}}} = \left(\frac{V_{\mathrm{m}}}{V_{\mathrm{p}}}\right)^2 \Big/ \left(\frac{l_{\mathrm{m}}}{l_{\mathrm{p}}}\right) = 常数 \tag{5.3.4}$$

可见，模型和原型间速度之比和长度之比也为常数。

模型和原型流场各处长度之比为常数称为**几何相似**，几何相似的流场具有相同的几何形状。

模型和原型流场各处速度之比为常数称为**运动相似**。

可见，完全相似的流场必须同时满足三个相似条件，即几何相似、运动相似和动力相似。

第一个条件几何相似，模型和原型须具有相同的形状且各处比例尺相同；第二个条件运动相似，要求模型流场各处的流体质点不仅运动速度与原型保持相同比值，且相对位置处的速度方向也须一致；第三个条件也是最严格的相似条件，动力相似，要求模型流场中各种力与原型流场力比值恒定。

我们可以这样理解相似，即几何相似为流场各处长度比例尺处处相等，运动相似的流场时间比例尺处处相等，动力相似的流场力的比例尺处处相等。几何、运动以及力的比例尺都可以小于、等于或大于 1。

几何相似是运动相似的前提，运动相似是动力相似的必要但不充分条件。模型和原型流场实现了几何相似和运动相似，但不一定满足动力相似。模型和原型流场完全相似不仅仅指几何相似、运动相似和动力相似，还包括初始条件和边界条件相似以及流动类型相同。例如，原型为紊流，则模型流场内的流速也一定不能处于层流的范围内。

5.3.2　常用动力相似准则数

假设有两个几何和运动相似的流场，流场中流体质点所受的力包括：压力 F_p、黏性力 F_τ、重力 F_g、弹性力 F_E，如果存在气 - 液分界面，则还存在表面张力 F_σ。构成各种力的物理量如下式

liquid-gas interface, there are forces due to surface tension F_σ. The forces in the equation can be expressed in terms of Eq.5.3.2 as

$$F_p = \Delta pA \rightarrow \Delta pL^2$$

$$F_\tau = \mu \frac{\mathrm{d}u}{\mathrm{d}y}A \rightarrow \mu \frac{V}{L}L^2 = \mu VL$$

$$F_g = mg \rightarrow \rho g L^3 \qquad (5.3.5)$$

$$F_E = EA \rightarrow EL^2$$

$$F_\sigma = \sigma L$$

The variation of fluid flow is the consequence of interaction between the inertial force and active forces. The resultant of these active forces can be balanced by inertial force F_I. Then Eq.5.3.1 is rearranged to express the relations as,

$$\left(\frac{F_I}{F_p}\right)_m = \left(\frac{F_I}{F_p}\right)_p, \quad \left(\frac{F_I}{F_\tau}\right)_m = \left(\frac{F_I}{F_\tau}\right)_p, \quad \left(\frac{F_I}{F_g}\right)_m = \left(\frac{F_I}{F_g}\right)_p, \quad \left(\frac{F_I}{F_E}\right)_m = \left(\frac{F_I}{F_E}\right)_p, \quad \left(\frac{F_I}{F_\sigma}\right)_m = \left(\frac{F_I}{F_\sigma}\right)_p$$

$$(5.3.6)$$

Each of the quantities in Eq.5.3.6 is dimensionless. Then each ratio of two forces can be written as the dimensionless number. We may use the established numbers frequently without thinking of them as dimensionless numbers. These commonly used numbers have been given names for convenience, and to honor persons who have contributed in the development, such as Euler number (Eu), Reynolds number (Re), Froude number(Fr), Mach number (Ma) and Weber number (We),

$$Eu \propto \frac{\text{Pressure force}}{\text{Inertial force}} = \frac{\Delta p}{\rho V^2}$$

$$Re \propto \frac{\text{Inertial force}}{\text{Viscous force}} = \frac{\rho LV}{\mu}$$

$$Fr \propto \frac{\text{Inertial force}}{\text{Gravity force}} = \frac{V}{\sqrt{gL}} \qquad (5.3.7)$$

$$Ma \propto \frac{\text{Inertial force}}{\text{Elastic force}} = \frac{V}{\sqrt{E/\rho}} = \frac{\text{Flow speed}}{\text{Speed of sound}} = \frac{V}{c}$$

$$We \propto \frac{\text{Inertial force}}{\text{Surface tension force}} = \frac{\rho LV^2}{\sigma}$$

For flow with an unsteady component that repeats itself periodically, Strauhal number (Sr) is identified as

$$Sr = \frac{\text{Centrifugal force}}{\text{Inertial force}} = \frac{L\omega}{V} \qquad (5.3.8)$$

$$F_p = \Delta p A \to \Delta p L^2$$

$$F_\tau = \mu \frac{du}{dy} A \to \mu \frac{V}{L} L^2 = \mu V L$$

$$F_g = mg \to \rho g L^3 \qquad (5.3.5)$$

$$F_E = EA \to E L^2$$

$$F_\sigma = \sigma L$$

流体的流动都是惯性力与各种主动力较量的结果，惯性力 F_I 与各种主动力平衡时，式 5.3.1 中的关系可改写为

$$\left(\frac{F_I}{F_p}\right)_m = \left(\frac{F_I}{F_p}\right)_p, \quad \left(\frac{F_I}{F_\tau}\right)_m = \left(\frac{F_I}{F_\tau}\right)_p, \quad \left(\frac{F_I}{F_g}\right)_m = \left(\frac{F_I}{F_g}\right)_p, \quad \left(\frac{F_I}{F_E}\right)_m = \left(\frac{F_I}{F_E}\right)_p, \quad \left(\frac{F_I}{F_\sigma}\right)_m = \left(\frac{F_I}{F_\sigma}\right)_p$$

$$(5.3.6)$$

上式各项都是无量纲量。所以式 5.3.6 中每一对力的比值都可以写为无量纲数。这些常用的无量纲数大都以其最初提出者或参与研究者命名的，如下式中依次为：欧拉数（Eu）、雷诺数（Re）、弗劳德数（Fr）、马赫数（Ma）和韦伯数（We）等，遇到动力相似问题时，我们可以不需思考地直接使用这些无量纲数，大大方便了流体力学问题的分析和研究。

$$Eu \propto \frac{压力}{惯性力} = \frac{\Delta p}{\rho V^2}$$

$$Re \propto \frac{惯性力}{黏性力} = \frac{\rho L V}{\mu}$$

$$Fr \propto \frac{惯性力}{重力} = \frac{V}{\sqrt{gL}} \qquad (5.3.7)$$

$$Ma \propto \frac{惯性力}{弹性力} = \frac{V}{\sqrt{E/\rho}} = \frac{流速}{声速} = \frac{V}{c}$$

$$We \propto \frac{惯性力}{表面张力} = \frac{\rho L V^2}{\sigma}$$

对于非稳定流动，可用斯特劳哈数（Sr）表达

$$Sr = \frac{离心力}{惯性力} = \frac{L\omega}{V} \qquad (5.3.8)$$

The Reynolds number is the most well-known and useful dimensionless parameter in all of numbers mentioned above, which turns out for comparing different flows. The linear dimension L may be any length that is significant in the flow pattern. Thus, for a pipe completely filled, we use the pipe diameter d for L most commonly. Thus, for a pipe flowing full, the Reynolds number is written in form of

$$Re = \frac{\rho d V}{\mu} \tag{5.3.9}$$

where d is the diameter of the pipe.

The numbers in Eq.5.3.7 and 5.3.8 are by no means exhaustive. We have introduced the more common dimensionless flow parameters of interest in fluid mechanics. You should, as far as possible, convert your π-terms into the established nondimensional parameters in the process of dimensional analysis.

The nondimensional number groups are easily obtained, however, the problem is that it is not always possible to match all the nondimensional parameters of a model to the corresponding numbers of the prototype, even if we are careful to achieve geometric similarity. In other words, there are no two systems are exactly similar. In fact, it is not always necessary or desirable to adhere to these various dimensionless ratios in every case. All of the effects included in fluid mechanics would not be of interest in any one situation. It would be very unlikely that both viscous effects and compressibility effects would influence a flow simultaneously. In experiments, similarity of one main force would be obtained, and the influence of other minor forces are ignored. One curtain type of nondimensional number is introduced to each active force of interest. That is the rule of dynamic similarity.

Thus, we will have viscous dynamic similarity when

$$\left(\frac{\rho d V}{\mu}\right)_m = Re_m = Re_p = \left(\frac{\rho d V}{\mu}\right)_p \tag{5.3.10}$$

For the same fluid in the model and the prototype, Eq.5.3.10 shows that for dynamic similarity, we must have a high velocity with a model of small linear dimensions. It may be the same device, however, the fluid used in the model need not be the same as that in the prototype, provided d and V are chosen so that they give the same value of Re in model and prototype.

Example 5.3.1   The aerodynamic drag of a car is to be predicted at a speed of 80.0km/h at an air temperature of 20℃. Automotive engineers build a 1:3 scale model of the car to test in a wind tunnel. The temperature of the wind tunnel air is about 30℃. Determine how fast the engineers should run the wind tunnel in order to achieve similarity between the model and the prototype.

Solution

Compressibility of the air is negligible.

For air at atmospheric pressure and 20℃, $\rho = 1.204 \text{kg/m}^3$ and $\mu = 1.81 \times 10^{-5} \text{Pa·s}$. Similarly, at 30℃, $\rho = 1.164 \text{kg/m}^3$ and $\mu = 1.86 \times 10^{-5} \text{Pa·s}$.

The Reynolds numbers of a model and its prototypes must be the same, by Eq.5.3.10

以上无量纲数中最常用的是雷诺数，常用于判定流态。式中长度尺寸 L 可以是流场中的任何特征尺寸，如，对于完全充满流体的管路，该特征尺寸为管路直径 d，所以管道流动中雷诺数的常用形式为

$$Re = \frac{\rho d V}{\mu} \qquad (5.3.9)$$

式中，d 是管路直径。

　　式 5.3.7 和式 5.3.8 列举了部分常用的无量纲数。量纲分析中，可以根据所分析的具体问题，选取适当的无量纲数构建 π- 式。

　　建立无量纲数并非难事，难的是在模型和原型间实现各种无量纲数的完全一致，即使是达到完美的几何相似条件也几乎不可能实现各种无量纲数的完全一致。也就是说，不会有两个完全相似的系统。事实上，我们也没必要做到完全相似。流体力学的研究不会同时关注多种问题，例如不会同时研究黏性和可压缩性。实验中仅需实现主要力相似，其他次要力的影响可以忽略，因此只需保证研究的主要力无量纲数，即该种力的相似准则数相同。

　　所以，当黏性力动力相似时

$$\left(\frac{\rho d V}{\mu}\right)_{m} = Re_{m} = Re_{p} = \left(\frac{\rho d V}{\mu}\right)_{p} \qquad (5.3.10)$$

　　从式 5.3.10 可以看出，原型和模型为同种流体时，动力相似条件下模型尺寸虽小但流速却要求很高。如果适当选择实验装置直径 d 和流速 V，即使模型与原型使用不同的流体，也能够保证两个流场具有相同的雷诺数 Re。

　　例 5.3.1　为研究小车在时速 80.0km/h、气温 20℃ 条件下的空气阻力，汽车工程师搭建了一个 1∶3 的实验风洞，风洞内气温约为 30℃。求为实现模型和原型动力相似，风洞内的风速应该是多少？

　　解　忽略空气的可压缩性，20℃ 大气压条件下，空气密度 $\rho = 1.204\text{kg/m}^3$，黏度 $\mu = 1.81 \times 10^{-5}\text{Pa·s}$。30℃ 条件下密度 $\rho = 1.164\text{kg/m}^3$，黏度 $\mu = 1.86 \times 10^{-5}\text{Pa·s}$。

　　模型和原型雷诺数相同，根据式 5.3.10，有

$$\left(\frac{\rho L V}{\mu}\right)_m = \left(\frac{\rho L V}{\mu}\right)_p \quad \rightarrow \quad \left(\frac{1.164 L_m V}{1.86}\right)_m = \left(\frac{1.204 \times 3 L_m 80}{1.81}\right)_p$$

That results in, $V_m = 255 \text{km/h} = 70.86 \text{m/s}$

Example 5.3.2 A test is to be performed on a proposed design for a system that is to deliver water from a 150mm-diameter pipe with average velocity of 3m/s at 20℃. A pressure drop of 20kPa is measured between two locations have a distance of 12m along the stream. A model with a 50mm-diameter kerosene system is to be used. The kinematic viscosity and the density of the kerosene are known as $3.6 \times 10^{-6} \text{m}^2/\text{s}$ and 830kg/m^3 respectively. What flow average velocity should be used and what pressure drop is to be expected on a 4m length of model pipe? The model fluid is kerosene at the same temperature as the water in the prototype.

Solution For similarity to exist in this confined incompressible flow problem, the Reynolds numbers must be equal; by Eq.5.3.10, and the kinematic viscosity of water at 20℃ is $1.01 \times 10^{-6} \text{m}^2/\text{s}$, then we have

$$\left(\frac{dV}{\nu}\right)_m = Re_m = Re_p = \left(\frac{dV}{\nu}\right)_p \quad \rightarrow \quad \left(\frac{50V}{3.6}\right)_m = \left(\frac{150 \times 3}{1.01}\right)_p$$

The result is $V_m = 32.08 \text{m/s}$.

The average velocity of kerosene is to be taken as 32.08m/s.
The dimensionless pressure drop is found using the Euler number:

$$\left(\frac{\Delta p}{\rho V^2}\right)_m = \left(\frac{\Delta p}{\rho V^2}\right)_p \quad \rightarrow \quad \left(\frac{\Delta p}{0.83 \times 32.08^2}\right)_m = \left(\frac{20}{3^2}\right)_p$$

Hence the pressure drop for the model is

$$\Delta p = 1898 \text{kPa} = 1.898 \text{MPa}$$

Exercises 5

5.1 Combine power W, diameter d, pressure drop Δp, and average velocity V into a dimensionless group.

5.2 Use dimensional analysis to arrange the following groups into dimensionless parameters: (a) τ, V, p; (b) V, L, p, σ.

5.3 Use dimensional analysis to arrange the following groups into dimensionless parameters: (a) Δp, V, ρ, g; (b) F, ρ, L, V.

5.4 Write Bernoulli's equation in dimensionless form.

5.5 If the [M]-[L]-[T] dimensions were used, what would the dimensions be on each of the following? (a) Mass flow rate (b) Pressure gradient (c) Specific weight (d) Work (e) Power.

5.6 Assume that the velocity V of fall of an object depends on the height H through

$$\left(\frac{\rho L V}{\mu}\right)_{\mathrm{m}} = \left(\frac{\rho L V}{\mu}\right)_{\mathrm{p}} \quad \rightarrow \quad \left(\frac{1.164 L_{\mathrm{m}} V}{1.86}\right)_{\mathrm{m}} = \left(\frac{1.204 \times 3 L_{\mathrm{m}} 80}{1.81}\right)_{\mathrm{p}}$$

解得　$V_{\mathrm{m}} = 255 \mathrm{km/h} = 70.86 \mathrm{m/s}$。

例 5.3.2　欲设计一输水系统，系统管道直径 150mm，20℃时平均水的流速为 3m/s。沿流动路径相距 12m 的两点间压降为 20kPa。现使用管道直径 50mm 以煤油为流体的实验模型。已知煤油的黏度和密度分别为 $3.6 \times 10^{-6} \mathrm{m^2/s}$ 和 $830 \mathrm{kg/m^3}$。求（1）实验中煤油的平均流速；（2）长度 4m 的管道上产生的压降为多少？模型中煤油与原型中水的温度相同。

解　（1）两个相似的不可压缩流体流场，雷诺数一定相等，根据式 5.3.10，取 20℃水的运动黏度为 $1.01 \times 10^{-6} \mathrm{m^2/s}$，可知

$$\left(\frac{dV}{\nu}\right)_{\mathrm{m}} = Re_{\mathrm{m}} = Re_{\mathrm{p}} = \left(\frac{dV}{\nu}\right)_{\mathrm{p}} \quad \rightarrow \quad \left(\frac{50 V}{3.6}\right)_{\mathrm{m}} = \left(\frac{150 \times 3}{1.01}\right)_{\mathrm{p}}$$

解得　$V_{\mathrm{m}} = 32.08 \mathrm{m/s}$。

煤油的平均流速应为 32.08m/s。

（2）压强差问题的无量纲数应选欧拉数：

$$\left(\frac{\Delta p}{\rho V^2}\right)_{\mathrm{m}} = \left(\frac{\Delta p}{\rho V^2}\right)_{\mathrm{p}} \quad \rightarrow \quad \left(\frac{\Delta p}{0.83 \times 32.08^2}\right)_{\mathrm{m}} = \left(\frac{20}{3^2}\right)_{\mathrm{p}}$$

所以模型中压降为

$$\Delta p = 1898 \mathrm{kPa} = 1.898 \mathrm{MPa}$$

习题五

5.1　用功率 W、直径 d、压差 Δp 和平均流速 V 组成一个无量纲量。

5.2　用量纲分析法将以下两组变量分别组成无量纲量：（1）τ，V，p；（2）V，L，p，σ。

5.3　用量纲分析法将以下两组变量分别组成无量纲量：（1）Δp，V，ρ，g；（2）F，ρ，L，V。

5.4　把伯努利方程改写为无量纲形式。

5.5　用 [M]-[L]-[T] 量纲表示以下物理量：（1）质量流量；（2）压力梯度；（3）重度；（4）功；（5）功率。

5.6　假设物体坠落的速度 V 与其坠落前高度 H、重力加速度 g、质量 m 有关，

which it falls, the gravity g, and the mass m of the object. Find an expression for V.

5.7　Consider the resistance force F of the fluid in Exercise 5.6 including the density ρ and viscosity μ of the surrounding fluid.

5.8　Find an expression for the centrifugal force F_C if it depends on the mass m, the angular velocity ω, and the radius R of an impeller.

5.9　Find an expression for the average velocity V in a smooth pipe if it depends on the viscosity μ, the diameter d, and the pressure gradient $\partial p/\partial x$.

5.10　It is suggested that the velocity of the water flowing from an exit of an open tank depends on the height H of the exit from the free surface, the gravity g, and the density of the water ρ. What expression relates the variables?

5.11　Derive an expression for the velocity of liquid issuing from an exit in the side of an open tank if the velocity V depends on the height H of the exit from the free surface, the fluid viscosity μ and density ρ, gravity g, and the exit diameter d.

5.12　The pressure drop Δp in a relatively rough pipe depends on the average velocity V, the pipe diameter d, the kinematic viscosity v, the length L of pipe, the wall roughness height ε, and the fluid density ρ. Find an expression for Δp.

5.13　A new design of a valve is to be tested. Which dimensionless parameter is the most important if hydraulic oil flows through the valve?

5.14　A new design of an airplane is to be tested. Which dimensionless parameter is the most important if hydraulic oil flows through the valve?

5.15　Derive an expression for the torque T needed to rotate a cylinder shown in Fig. E5.15 of diameter d at the angular velocity ω in a fluid with density ρ and viscosity μ, if the cylinder is a distance t from a wall and h from the bottom.

5.16　Derive an expression for the torque T necessary to rotate the shaft with installation error as show in Fig.E5.16.

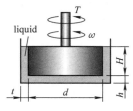 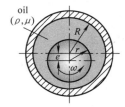

Fig.E5.15　　　　　　　　　　　　Fig.E5.16

5.17　The flow rate q in an open channel depends on the hydraulic radius R, the cross-sectional area A, the wall roughness height ε, gravity g, and the slope S. Relate q to the other variables.

5.18　The drag force F_D acting on a ship is considered to be a function of the fluid density ρ, viscosity μ, gravity g, ship velocity V, and a characteristic length l. Determine a set of suitable dimensionless numbers to describe the relationship $F_D = f(\rho, \mu, g, V, l)$.

5.19　Water at 15℃ in a 25mm-diameter pipe flows with a velocity of 2kg/s. What is the Reynolds number?

5.20　What velocity should be selected in a wind tunnel where a 4:1 scale model of an automobile is to simulate a speed of 10m/s? Neglect compressibility effects.

用这些变量表示速度 V。

5.7　如果习题 5.6 中物体坠落的速度还与周围流体的密度 ρ 和黏度 μ 有关，则推导其速度表达式。

5.8　若叶轮离心力 F_C 与其质量 m、角速度 ω 和旋转半径 R 有关，写出离心力的表达式。

5.9　光滑管道内流体的平均流速 V 受流体黏度 μ、管道直径 d 和压力梯度 $\partial p/\partial x$ 影响，写出流速的表达式。

5.10　观察发现，顶部开放的水池底部出流速度 V 取决于出口至池内自由液面的高度 H、重力加速度 g 和水的密度 ρ，试写出这些变量间的关系。

5.11　写出顶部开放的水池底部出流速度 V 与出口至池内自由液面高度 H、重力加速度 g、液体的黏度 μ 和密度 ρ 以及出口直径 d 之间的关系。

5.12　相对粗糙的管道内流体压强差 Δp 与平均流速 V、管道直径 d、流体运动黏度 ν 和密度 ρ 以及管道长度 L 和管壁粗糙高度 ε 有关，写出压强差 Δp 的函数关系。

5.13　如果要对一个新设计的液压阀进行测试，应选用哪种相似准则数？

5.14　如果要对一新型飞机在空气中的飞行情况进行实验，应选用哪种相似准则数？

5.15　如图 E5.15 所示，写出转动图中圆柱体所需力矩 T 与圆柱体直径 d、转动角速度 ω、液体密度 ρ 和黏度 μ 间的函数关系。圆柱体与壁面和容器底的距离分别为 t 和 h。

5.16　如图 E5.16 所示存在安装误差的轴，写出其转动力矩 T 的函数关系。

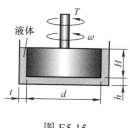

图 E5.15

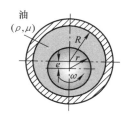

图 E5.16

5.17　水力半径为 R、截面面积为 A 的明渠内水流量为 q，已知壁面粗糙度平均高度为 ε，坡度 S，重力加速度 g，写出流量 q 与这些变量的关系式。

5.18　船航行时所受阻力 F_D 与流体密度 ρ、黏度 μ、重力加速度 g 以及船速 V 及船的特征长度 l 存在函数关系 $F_D = f(\rho, \mu, g, V, l)$，写出其表达式。

5.19　15℃水在直径 25mm 的管道内流动，流量为 2kg/s，计算水流的雷诺数。

5.20　在 4∶1 的风洞模型中模拟汽车车速为 10m/s 时的情况，忽略空气的可压缩性，计算应选择的风速。

5.21　A 180m-long ship is to operate at a speed of 30km/h. If a model of this ship is 3m-long, what should its speed be to give the same Froude number?

5.22　A 1:3 scale model of a large pump is used for tests. The prototype pump produces a pressure rise of 500kPa at a flow rate of 600m³/s. Determine the flow rate to be used in the model pump and the expected pressure rise.

5.23　A test is to be performed on a proposed design for a piping system that is to deliver 2m³/s of water from a 500mm-diameter pipe with a pressure drop of 400kPa. A model with a 100mm-diameter pipe is to be used. What flow rate should be used and what pressure drop is to be expected? The water in the model and prototype are at the same temperature.

5.24　A 1:16 scale model of a surface boat is used to test the influence of a proposed design on the wave drag. A wave drag of 30N is measured at a model speed of 2.5m/s. What speed does this correspond to on the prototype, and what wave drag is predicted for the prototype? Neglect viscous effects, and assume the same fluid for model and prototype.

5.25　A model study of oil (S_g=0.9, ν=50mm²/s) flowing in a 0.8m-diameter pipe is performed by using water at 20℃. Pipe of what diameter should be used if the average velocities are to be the same? What pressure drop ratio is expected?

5.26　The flow rate over a weir is 10m³/s of water. A 1:20 scale model of the weir is tested in a water channel. What flow rate should be used?

5.21　长度 180m 的船，航行速度为 30km/h，若以长度 3m 的模型进行实验，两流场弗劳德数（Fr）相同，实验模型的速度应为多少？

5.22　用 1∶3 的模型模拟大型水泵进行实验，水泵原型在流量为 600m³/s 时的压强可升高 500kPa。计算模型应选择的流量和可能的压强增加值。

5.23　欲设计一输水管网，预期管道直径为 500mm、流量为 2m³/s 时，压力损失为 400kPa。现使用直径为 100mm 的管路进行实验。实验模型应使用的流量是多少？模型上的压力损失为多少？假设模型和原型水温相同。

5.24　用 1∶16 的模型测量船行驶过程中所受的波浪阻力，当模型的速度为 2.5m/s 时，测得阻力值为 30N。此实验对应的原型船速应该为多少？原型会遇到的阻力将是多少？不计黏性，假设原型船与模型在同种流体中运动。

5.25　用 20℃水模拟直径 0.8m 管道内油液（$S_g = 0.9$，$\nu = 50$mm²/s）的流动情况。如果模型和原型中流体的平均流速相同，计算应选的实验管道直径。两个流场压力损失的比值为多少？

5.26　鱼梁坝表面过水量为 10m³/s，在水渠内以 1∶20 的模型进行实验，应该选取的流量为多少？

CHAPTER 6
INTERNAL FLOW FOR VISCOUS
INCOMPRESSIBLE FLUID

The fluids we discussed in previous chapters are all ideal fluid, also called non-viscous fluid, which is assumed have no viscosity or the viscous effects do not significantly influence the flow and are thus neglected. Inviscid flows are of primary importance in derivation of basic formulas. But in fact, considerable analysis must be taken to justify the inviscid flow assumption since all fluids have viscosity.

Viscous flows include the broad class of internal flows, such as flows in pipes and in fluid transmission devices. In such flows viscous effects cause substantial"losses" and account for the huge amounts of energy that must be used to transport oil and gas in pipelines. The no-slip condition resulting in zero velocity at the wall and the resulting shear stresses lead directly to these losses.

In this chapter, Bernoulli equation of viscous fluid and the energy losses will be studied, two types of viscous flow, laminar and turbulent are discussed. In this chapter we also consider internal flow where the conduit is completely filled with the fluid, and the flow is driven primarily by a pressure difference in the orifice and gaps.

6.1 Bernoulli Equation for Viscous Incompressible Flow

Let us review the Bernoulli equation as in the previous section, except that now we shall consider a real fluid. The real fluid along a stream is similar to that depicted in section 4.3, except that now with the real fluid there is an additional shear stress acting because of fluid friction. The friction causing a loss of mechanical energy occurs over the boundary or internal surfaces between the fluid particles. The justified Bernoulli equation can be expressed as

$$z_1 + \frac{p_1}{\rho g} + \frac{u_1^2}{2g} = z_2 + \frac{p_2}{\rho g} + \frac{u_2^2}{2g} + h_L \qquad (6.1.1)$$

If we compare Eq.6.1.1 with Bernoulli Eq.4.3.14 for ideal flow we see the only difference is the additional term h_L, which represents the irreversible loss of energy per unit weight due to fluid friction between sections 1 and 2. The dimension of this energy loss term is length either, which agrees with all the other terms, and so this term is a form of head, we commonly refer to as total loss head, denoted by h_L. The loss head is often written in terms of equivalent fluid column height as

$$h_L = k \frac{u^2}{2g} \qquad (6.1.2)$$

where k is the loss coefficient, a dimensionless factor that could be different forms for different type of losses.

The loss terms in Eq.6.1.1 include the losses between the inlet and the respective exits,

第6章
黏性不可压缩流体的内部流动

前面各章节我们分析的都是理想流体，也称为无黏流体，是假设其不存在黏性或其黏性影响可以忽略的流体。无黏流体对推导基本方程非常重要，但实际的流体都是具有黏性的，有时黏性的影响不能忽略，因此需要对无黏假设加以修正。

相当多的黏性流动发生在内部流情况下，如管道流动或流体传动装置等。壁面不滑移效应使得壁面处的流体速度为零，因此流体流动过程为克服剪切应力会导致大量的能量损失，这也是为什么我们传输液体和气体时要消耗很多能量。

本章首先介绍黏性流体的伯努利方程和能量损失；介绍黏性流体的层流和紊流以及如何区分这两种流态，还将介绍黏性流体在管道流动及孔口和缝隙流动。

6.1 黏性不可压缩流动的伯努利方程

我们首先把前面章节介绍的伯努利方程与实际流体联系起来。实际流体沿流线运动情况与 4.3 节理想流体沿流线运动的不同之处在于由于内摩擦的存在，也就是说实际流体还会受到剪切应力的作用。内摩擦阻力会在流体质点与周围流体以及外界的界面处导致机械能损失，所以，修正后的伯努利方程为

$$z_1 + \frac{p_1}{\rho g} + \frac{u_1^2}{2g} = z_2 + \frac{p_2}{\rho g} + \frac{u_2^2}{2g} + h_L \qquad (6.1.1)$$

把式 6.1.1 中的伯努利方程与理想流体的式 4.3.14 对比，我们看到，不同之处在于前者增加了能量损失 h_L 项，这一项代表截面 1 和 2 之间单位重量流体上产生的不可逆能量损失。这一损失能量的量纲与方程中其他项一致，均为长度量纲，也就是说它也是某种水头，通常我们称之为总损失水头，用 h_L 表示。损失水头用液柱高度表示为

$$h_L = k \frac{u^2}{2g} \qquad (6.1.2)$$

式中，k 是损失系数，为无量纲数，对不同形式的能量损失系数 k 有对应的具体形式。

式 6.1.1 中的损失发生在进、出截面间，如果这些损失可忽略，则式 6.1.1 化为式 4.3.14，所以式 6.1.1 是伯努利方程的一般形式。

if the losses are negligible, the equivalent reduce to the form of Eq.4.3.14, the Bernoulli equation for ideal fluid. So, Eq.6.1.1 is the general form of the Bernoulli equation.

Note that in fluid flow, it is convenient to work with an average velocity V over the cross section, a velocity distribution can be accounted for by introducing the kinetic energy correction factor α, and then the Bernoulli equation can be written as

$$z_1 + \frac{p_1}{\rho g} + \alpha_1 \frac{V_1^2}{2g} = z_2 + \frac{p_2}{\rho g} + \alpha_2 \frac{V_2^2}{2g} + h_L \qquad (6.1.3)$$

For most internal turbulent flows, we simply let $\alpha=1$ since it is so close to unity, this will always be done unless otherwise stated. And then we have the Bernoulli equation for real fluid, it is

$$z_1 + \frac{p_1}{\rho g} + \frac{V_1^2}{2g} = z_2 + \frac{p_2}{\rho g} + \frac{V_2^2}{2g} + h_L \qquad (6.1.4)$$

The basic assumptions involved in Eq.6.1.1 and Eq.6.1.4, that we need always to bear in mind, are still (1) steady flow, (2) of incompressible fluid, (3) along a streamline, (4) with no energy added or removed.

In a real Bernoulli-type flow, the various forms of mechanical energy are converted to each other, their sum decreases along the stream because of the mechanical energy losses due to fluid viscosity. The total head line of the stream flow shown in Fig.6.1.1 is declined.

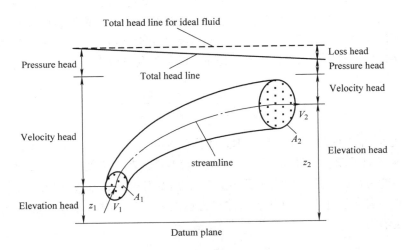

Fig.6.1.1   The total head of a real fluid flow declines along the stream.

The mechanical losses of real fluid are due to two primary effects:

1. Viscosity causes internal friction that results in increased internal energy (temperature increase) or heat transfer.

2. Changes in geometry result in separated flows that require useful energy to maintain the resulting secondary motions in which viscous dissipation occurs.

　　处理流动问题时，我们常常方便地取截面上的平均流速 V，实际速度分布的不均匀可以用动能修正系数 α 来加以修正，这样，伯努利方程也可以写为

$$z_1 + \frac{p_1}{\rho g} + \alpha_1 \frac{V_1^2}{2g} = z_2 + \frac{p_2}{\rho g} + \alpha_2 \frac{V_2^2}{2g} + h_L \qquad (6.1.3)$$

　　对大多数内部流动的紊流，流速分布是相当均匀的，动量修正系数可以取为 1，以后如果没有特别说明的情况，我们都取 $\alpha = 1$。所以，实际流体伯努利方程的常用形式为

$$z_1 + \frac{p_1}{\rho g} + \frac{V_1^2}{2g} = z_2 + \frac{p_2}{\rho g} + \frac{V_2^2}{2g} + h_L \qquad (6.1.4)$$

　　时刻需要记得，式 6.1.1 和式 6.1.4 成立的条件与理想流体伯努利方程相同：（1）稳定流动；（2）不可压缩流体；（3）沿流线成立；（4）与外界没有能量交换。

　　满足伯努利条件的实际流体流动的能量守恒和转化关系如图 6.1.1 所示，可以看到，由于流体黏性导致机械能的损失，所以其总机械能或称为总水头线沿流线总是下降的。

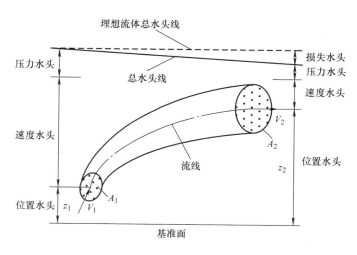

图 6.1.1　实际流体的总水头沿流线下降

导致实际流体机械能损失的原因主要有两个：

1. 克服黏性内摩擦阻力的机械能转化为内能导致流体温升或以热的形式传递出去。

2. 流场的几何形状改变导致的机械能损耗。

In a conduit, the losses due to viscous effects are distributed over the entire length and directly proportional to the length of the conduit, always directly related to the wall shear stress, is called *frictional losses*, represented by h_f. Whereas the loss due to a geometry change (a valve, an elbow, an enlargement) is concentrated in the vicinity of the geometry change, called *minor losses*, represented by h_j.

Bernoulli equation 6.1.4 can also take the form of

$$z_1 + \frac{p_1}{\rho g} + \frac{V_1^2}{2g} = z_2 + \frac{p_2}{\rho g} + \frac{V_2^2}{2g} + \sum h_f + \sum h_j \qquad (6.1.5)$$

The frictional losses always happen in gradually varied fluid flow, and the primary consequence is pressure drop, temperature rise due to frictional heating is usually too small to warrant any consideration and thus is disregarded. The frictional loss for all types of fully developed internal flows, valid for laminar or turbulent flows, circular or noncircular pipes, and pipes with smooth or rough surfaces, can be expressed as

$$\Delta p = \lambda \frac{l}{d} \frac{\rho V^2}{2} \qquad (6.1.6)$$

In practice, it is convenient to express the frictional loss as

$$h_f = \lambda \frac{l}{d} \frac{V^2}{2g} \qquad (6.1.7)$$

The equation with form of Eq.6.1.6 or Eq.6.1.7 is called the *Darcy's formula* or *Darcy-Weisbach formula*, and the dimensionless coefficient λ is called *Darcy friction factor* or just *friction factor*.

Note from Eq.6.1.6 and Eq.6.1.7, the friction loss is proportional to the length of flow stream since friction occurs as long as the flow process over the boundaries or surfaces of fluid elements.

The friction loss is also inversely proportional to the diameter of pipe. As we noted on two circular cross section of diameter d_1, d_2 respectively and $d_1 < d_2$, with the same average velocity V, shown in Fig.6.1.2. Recall from chapter one that $\tau = \mu(du/dr)$, we assume the velocity is linear distribution in r-direction, the maximum velocity $2V$ occurs at the center of the pipe and velocity at wall is zero due to noslip condition, then we have

$$\frac{du}{dr} = \frac{2V}{d/2} = \frac{4V}{d} \qquad (6.1.8)$$

Comparing two pipe of different diameters and solving for shear stress

$$\tau_1 = \mu \frac{4V}{d_1} \quad > \quad \tau_2 = \mu \frac{4V}{d_2} \qquad (6.1.9)$$

Therefore, the shear stress (the internal friction), at the wall of pipe and stream tubes, is inversely proportional to the diameter of pipe.

在管渠流动中，由于固体壁面的阻滞而产生的黏性剪切应力造成的能量损失称为**沿程损失**，用 h_f 表示，沿程损失沿流动路径分布且与路径长度成正比。由于流场界面的几何形状改变（如阀口、弯头和截面变化等）导致的机械能损耗称为局部损失，用 h_j 表示。

把两种能量损失写入式 6.1.4，伯努利方程形式为

$$z_1 + \frac{p_1}{\rho g} + \frac{V_1^2}{2g} = z_2 + \frac{p_2}{\rho g} + \frac{V_2^2}{2g} + \sum h_f + \sum h_j \qquad (6.1.5)$$

沿程损失往往发生在缓变流并导致压强下降，摩擦热导致的温升由于不很明显，常常可以忽略。对于充分发展的内部流动（无论是层流还是紊流、圆形或非圆形管道、粗糙或光滑管道），沿程损失导致的压强差可以表示为

$$\Delta p = \lambda \frac{l}{d} \frac{\rho V^2}{2} \qquad (6.1.6)$$

实践中常用沿程损失水头形式

$$h_f = \lambda \frac{l}{d} \frac{V^2}{2g} \qquad (6.1.7)$$

式 6.1.6 和式 6.1.7 称为**达西公式**，式中无量纲系数 λ 称为**沿程损失系数**。

从式 6.1.6 和式 6.1.7 中可以看出，沿程损失与流动路径的长度成正比，这是因为流动的整个过程中流体始终与边界面发生摩擦，流体质点间的摩擦也始终存在。

沿程损失与管道直径成反比。我们以图 6.1.2 中两个不同直径管道为例来解释，两个管道直径分别为 d_1 和 d_2，且 $d_1 < d_2$，假设两管道内流体的平均流速相同，均为 V。由第 1 章的内容可知，单位面积上的黏性内摩擦阻力 $\tau = \mu(\mathrm{d}u/\mathrm{d}r)$，假设流速沿半径 r 方向线性分布，管壁处由于壁面不滑移，所以流速为 0；管道中心处流速最大，等于 $2V$，所以

$$\frac{\mathrm{d}u}{\mathrm{d}r} = \frac{2V}{d/2} = \frac{4V}{d} \qquad (6.1.8)$$

对比两个不同半径管道内的剪切应力，得

$$\tau_1 = \mu \frac{4V}{d_1} \qquad > \qquad \tau_2 = \mu \frac{4V}{d_2} \qquad (6.1.9)$$

所以，管壁处的剪切应力（内摩擦阻力）与管道直径成反比。

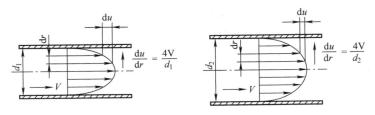

Fig.6.1.2   Velocity gradient in pipe is inversely proportional to the diameter

As for the minor losses, happen in rapid varied fluid flow where the geometric changes are significant, can also be expressed as the general form of Eq.6.1.2, especially substitute the minor loss factor ζ for k

$$h_j = \zeta \frac{V^2}{2g} \tag{6.1.10}$$

We shall discuss minor loss in pipe further in Sec.6.4.

Many fluid systems are designed to transport a fluid from one location to another at a specified flow rate, velocity, and elevation difference, and the system may generate mechanical work in a turbine or it may consume mechanical work in a pump during this process. A pump transforms mechanical energy to a fluid by raising its pressure, and a turbine extracts mechanical energy from a fluid by dropping its pressure. Therefore, the pressure of a flowing fluid is also associated with its mechanical energy. Energy conservation principle expressed by Bernoulli equation still provides a sound basis for studying such systems. The systems can be analyzed conveniently by considering only the mechanical forms of energy and the frictional effects that cause the mechanical energy to be lost.

In this form we have equated the energy at the inlet plus added energy (energy per unit weight, which has the unit of length) to the energy at the exit plus extracted energy. The energy equation, for an incompressible flow involving pump and turbine, takes the form

$$h_p + z_1 + \frac{p_1}{\rho g} + \frac{V_1^2}{2g} = z_2 + \frac{p_2}{\rho g} + \frac{V_2^2}{2g} + h_T + h_L \tag{6.1.11}$$

the terms h_p and h_T represent the energy that is transferred to and from the fluid, respectively. Where h_L is the irreversible head loss between inlet and exit due to all components of the piping system other than the pump or turbine.

If there is no pump or no turbine, the appropriate term is simply omitted. These special cases take the basis form of Bernoulli equation as Eq.6.1.4.

If the energy delivered by the turbine or required by the pump is desired, the efficiency of each device must be used.

Example 6.1.1   The pump of a water distribution system is powered by a 15-kW electric motor whose efficiency is 0.92 (Fig.X6.1.1). The water flow rate through the pump is 50L/s. The diameters of the inlet and outlet pipes are the same, and the exit area is 200mm above the inlet area. If the absolute pressures at the inlet and outlet of the pump are measured to be 90kPa and 300kPa, respectively, determine the mechanical efficiency of the pump. Neglect other losses.

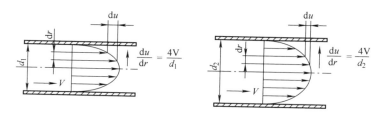

图 6.1.2　速度梯度与管道直径成反比

局部损失通常发生在几何特征急剧变化的急变流，用局部损失系数 ζ 代替式 6.1.2 中系数 k，得

$$h_j = \zeta \frac{V^2}{2g} \qquad (6.1.10)$$

本章 6.4 节将详细讨论管流的局部损失。

某些流体系统会利用泵的机械能将流体从一处输送至另一处，同时满足对流量、流速或高度的要求。同样，也可能把泵提供的压力能用于涡轮或马达而产生机械能做功。机械能通过泵转化成压力能，而马达消耗压力能产生机械能。可见，流体流动的压力与机械能相关。伯努利方程给出的能量守恒原理，也是这类机械装置的应用基础。为分析方便，我们仅考虑系统的机械能，假设黏性摩擦也仅产生机械能损失。

我们在进口端新增附加能量项，在出口端新增消耗能量项（均为单位重量流体内的能量，量纲同样为长度）。这样，不可压缩流体含泵和涡轮系统的能量方程可写作

$$h_p + z_1 + \frac{p_1}{\rho g} + \frac{V_1^2}{2g} = z_2 + \frac{p_2}{\rho g} + \frac{V_2^2}{2g} + h_T + h_L \qquad (6.1.11)$$

式中，h_p 和 h_T 分别表示注入系统和从系统中提取的能量。h_L 表示进、出口间包含泵和涡轮等所有环节产生的不可逆能量损失。

如果系统无泵或涡轮，相关项可以省略，而这种特例正是如式 6.1.4 所示的基本形式的伯努利方程。

考虑涡轮消耗能量或泵输入能量时，还经常需要计算其工作效率。

例 6.1.1　使用 15kW 效率 0.92 的电动机驱动水泵抽水，输水量 50L/s（见图 X6.1.1）。假设泵进、出口管道直径相同，出口中心高于进口中心 200mm。测得水泵进、出口绝对压强分别为 90kPa 和 300kPa，忽略其他损失，试计算泵的机械效率。

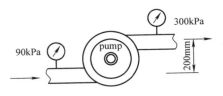

Fig.X6.1.1

Solution   Equation 6.1.11 across the pump provides

$$h_p + z_1 + \frac{p_1}{\rho g} + \frac{V_1^2}{2g} = z_2 + \frac{p_2}{\rho g} + \frac{V_2^2}{2g} + h_T + h_L$$

The inlet and outlet diameters are the same and thus the average inlet and outlet velocities are equal, $V_1 = V_2$. And we take the center of inlet section as datum, it is $z_1 = 0$. There is no turbine in the system. And then the above equation reduces to

$$h_p + \frac{p_1}{\rho g} = z_2 + \frac{p_2}{\rho g} + h_L$$

Each term has the unit of meter, the pressure head of pump represent the mechanical energy per unit weight of fluid

$$h_p = \frac{W_p}{\rho g q}$$

The motor draws 15kW of power and is 0.92 efficient, thus the mechanical power it delivers to the pump is

$$W_p = 15 \times 0.92 \text{kW} = 13.8 \text{kW}$$

The loss head of the fluid as it flows through the pump

$$h_L = \left(\frac{13.8 \times 10^3}{9.81 \times 50} + \frac{90}{9.81} - 0.2 - \frac{300}{9.81} \right) \text{m} = 6.528 \text{m}$$

Then the mechanical efficiency of the pump becomes $\eta_p = \dfrac{h_p - h_L}{h_p} = 0.768$.

Example 6.1.2 $100 \text{m}^3/\text{s}$ of water flows from a reservoir through to a turbine-generator unit where electric power is generated and exits to a river that is 120m below the reservoir surface (Fig.X6.1.2). The total irreversible head loss in the piping system from the surface of reservoir to the river (excluding the turbine unit) is determined to be 30m. If the overall efficiency of the turbine-generator is 0.8, estimate the electric power output.

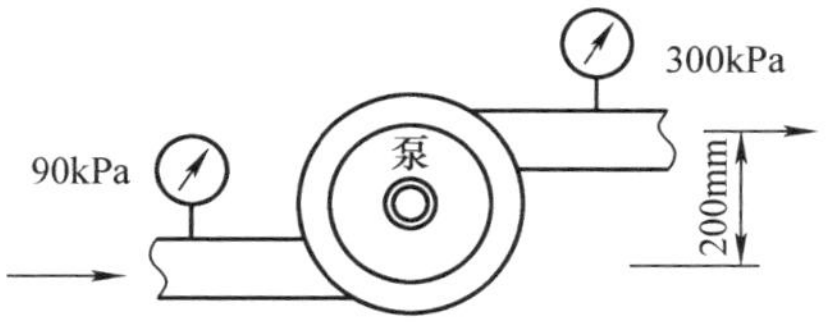

图 X6.1.1

解　将式 6.1.11 用于水泵系统

$$h_{\mathrm{p}} + z_1 + \frac{p_1}{\rho g} + \frac{V_1^2}{2g} = z_2 + \frac{p_2}{\rho g} + \frac{V_2^2}{2g} + h_{\mathrm{T}} + h_L$$

进、出口管路直径相同，所以进、出口截面平均流速相等，$V_1=V_2$。取进口中心为基准面，所以 $z_1=0$。系统不含涡轮。上式简化为

$$h_{\mathrm{p}} + \frac{p_1}{\rho g} = z_2 + \frac{p_2}{\rho g} + h_L$$

上式各项的单位均为 m，泵的压强水头代表每单位重量水所含的机械能

$$h_{\mathrm{p}} = \frac{W_{\mathrm{p}}}{\rho g q}$$

驱动电动机功率 15kW 效率 0.92，所以电动机输出至水泵的功率为

$$W_{\mathrm{p}} = 15 \times 0.92\mathrm{kW} = 13.8\mathrm{kW}$$

经过水泵而损失的水头为

$$h_L = \left(\frac{13.8 \times 10^3}{9.81 \times 50} + \frac{90}{9.81} - 0.2 - \frac{300}{9.81} \right)\mathrm{m} = 6.528\mathrm{m}$$

所以，水泵的机械效率为 $\eta_{\mathrm{p}} = \dfrac{h_{\mathrm{p}} - h_L}{h_{\mathrm{p}}} = 0.768$。

例 6.1.2　从河流上方 120m 高处的水库向涡轮发电机输水并排放至河流中，涡轮输入流量 100m³/s（见图 X6.1.2）。从水库至河流，包括管道系统，但不包括涡轮机，全部不可逆水头损失为 30m。如果涡轮机的整体效率为 0.8，试估算其最大发电量。

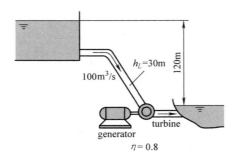

Fig.X6.1.2

Solution The river surface is taken as datum plane, thus $z_2 = 0$. Both the surface of reservoir and river are open to the atmosphere, it is $p_1 = p_2 = 0$. The flow velocities at this two surface are negligible, so we have $V_1 = V_2 = 0$.

The energy equation for steady, incompressible flow reduces to

$$\cancel{h_p} + z_1 + \frac{\cancel{P_1}}{\rho g} + \frac{\cancel{V_1^2}}{2g} = \cancel{z_2} + \frac{\cancel{P_2}}{\rho g} + \frac{\cancel{V_2^2}}{2g} + h_T + h_L \quad \rightarrow \quad h_T = z_1 - h_L = 120\text{m} - 30\text{m} = 90\text{m}$$

$$W_T = \rho g q h_T = 9810 \times 100 \times 90\text{kW} = 88290\text{kW} = 88.29\text{MW}$$

Therefore, a turbine would generate 88.29MW of electricity without any loss.
The electric power generated by the actual turbine-generator unit is

$$W_{out} = W_T \eta = 88.29 \times 0.8\text{MW} = 70.632\text{MW}$$

Example 6.1.3 A hydraulic motor is used to drive a rotary load of 200N · m and 120rpm rotational speed. Hydraulic oil is pumping to the motor through a pump with 5MPa output pressure and 30L/min outlet flow rate. Efficiency of pump and hydraulic motor are 0.85 and 0.8 respectively. Pressure at outlet of hydraulic motor is measured at 1.5MPa. Assuming that the diameter of pipes is unchanged, and the loss of hydraulic oil is negligible. Also neglect the friction. Find the power of the electric motor need to drive the pump.

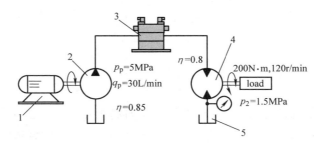

Fig.X6.1.3
1—electric motor   2—hydraulic pump
3—control valves   4—hydraulic motor   5—oil tank

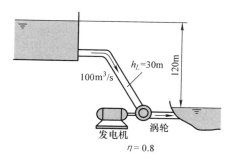

图 X6.1.2

解　取河流水面为基准面，所以 $z_2=0$。水库和河流的水面都是大气压，因此有 $p_1=p_2=0$。忽略两处相对巨大水面的流速，即 $V_1=V_2=0$。

对稳定、不可压缩流体的能量方程简化为

$$\frac{\cancel{p_1}}{\rho g} + z_1 + \frac{\cancel{p_1}}{\rho g} + \frac{\cancel{V_1}^2}{2g} = \cancel{z_2} + \frac{\cancel{p_2}}{\rho g} + \frac{\cancel{V_2}^2}{2g} + h_{\mathrm{T}} + h_L \quad\rightarrow\quad h_{\mathrm{T}} = z_1 - h_L = 120\mathrm{m} - 30\mathrm{m} = 90\mathrm{m}$$

$$W_{\mathrm{T}} = \rho g q h_{\mathrm{T}} = 9810\times100\times90\mathrm{kW} = 88290\mathrm{kW} = 88.29\mathrm{MW}$$

所以，在无任何损失的情况下，涡轮可产生 88.29MW 的电力。

涡轮发电机组可实际产生电力

$$W_{\mathrm{out}} = W_{\mathrm{T}}\eta = 88.29\times0.8\mathrm{MW} = 70.632\mathrm{MW}$$

例 6.1.3　如图 X6.1.3 所示，液压马达需以 120r/min 转速驱动 200N·m 转矩的负载。现使用一台输出压力 5MPa、输出流量 30L/min 的液压泵供油。泵和马达的效率分别为 0.85 和 0.8。液压马达出口压力为 1.5MPa。假设管路直径一致，忽略液压油的流量损失和摩擦损失，计算驱动液压泵所需的电动机功率。

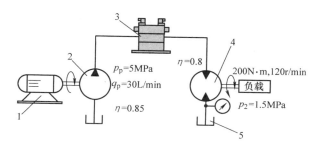

图 X6.1.3

1—电动机　2—液压泵
3—控制阀　4—液压马达　5—油箱

Solution Oil is absorbed from the tank by a pump and return back to the tank through a variety of control valves and the hydraulic motor. Pressure in the tank is almost atmospheric. Velocity of oil in the pipes is constant since the diameter of pipe is uniform. The height differential of pump and hydraulic motor is taken to be zero.

The energy equation between the pump and hydraulic motor is written as

$$\cancel{z_1} + \frac{\cancel{p_1}}{\rho g} + \frac{\cancel{V_1^2}}{2g} + h_p = \cancel{z_2} + \frac{p_2}{\rho g} + \frac{\cancel{V_2^2}}{2g} + h_T + h_L$$

where h_L is the energy losses due to the efficiency of the hydraulic motor, which is 0.8, then it is, $h_L = 0.25 h_T$, thus

$$h_p = \frac{p_2}{\rho g} + 1.25 h_T$$

The head of hydraulic motor is the mechanic energy per unit weight, that is

$$h_T = \frac{W_T}{\rho q g} = \frac{2\pi n T}{\rho q g} = \frac{2 \times 3.142 \times 120 \times 200}{30 \times 9.8} \text{m} = 512.5 \text{m}$$

$$h_p = 665.4 \text{m} \quad \rightarrow \quad W_{\text{pout}} = \rho q g h_p = 3.26 \text{kW}$$

The power of the electric motor needed to drive this system is

$$W_{\text{motor}} = W_{\text{pin}} = \frac{W_{\text{pout}}}{\eta_p} = 3.84 \text{kW}$$

A minimum of 4kW electric motor is needed to drive this system.

6.2 Laminar and Turbulent Flows of Viscous incompressible Fluid

A viscous flow can be classified as either a laminar flow or a turbulent flow. These two different types of flow are so important because of the strongly different effects they have on a variety of flow features, including on energy losses, velocity profiles and even mixing of the materials.

British physicist Osborne Reynolds (1842—1912) demonstrated this two distinctly different types of fluid flow in 1883. He designed a device as shown in Fig.6.2.1a. Water flowing through a large glass tube from a constant head tank. The temperature of the water keeps constant, gross variations of the water velocity are eliminated. Uniform and low velocity water enter the bell mouth glass tube, which is designed to accelerate the water uniformly without any spurious inertial effects. A faucet at the discharge end permit us to vary the flow rate, i.e. the velocity of water. A fine, threadlike stream of dye having the same density as water is injected at the entrance of the tube.

解　液压泵从油箱内吸取液压油，经控制阀和马达后返回油箱，油箱内压强接近大气压。由于管道直径一致，所以管道各处液压油流速相等。忽略液压泵和马达中心线间的高度差。

泵和马达间可列能量方程

$$z_1 + \frac{p_1}{\rho g} + \frac{v_1^2}{2g} + h_p = z_2 + \frac{p_2}{\rho g} + \frac{v_2^2}{2g} + h_T + h_L$$

式中，h_L 是马达效率导致的能量损失，马达效率等于 0.8，所以可得出 $h_L = 0.25 h_T$，由此

$$h_p = \frac{p_2}{\rho g} + 1.25 h_T$$

液压马达的水头是单位重量液压油所携带的机械能，所以

$$h_T = \frac{W_T}{\rho q g} = \frac{2\pi n T}{\rho q g} = \frac{2 \times 3.142 \times 120 \times 200}{30 \times 9.8}\text{m} = 512.5\text{m}$$

$$h_p = 665.4\text{m} \quad \rightarrow \quad W_{pout} = \rho q g h_p = 3.26\text{kW}$$

驱动该系统所需的电动机功率为

$$W_{motor} = W_{pin} = \frac{W_{pout}}{\eta_p} = 3.84\text{kW}$$

所以，该系统需要一台功率最小为 4kW 的电动机。

6.2　黏性不可压缩流体的层流和紊流

黏性不可压缩流体的流动状态可以分为层流和紊流。这两种流态在能量损失、速度分布甚至是流体成分掺混等方面特征区别显著，因此区分两种流态十分重要。

英国物理学家雷诺于 1883 年首次证实了这两种流态的存在和区别。他的实验装置如图 6.2.1a 所示，盛水容器内由隔板保证液面高度不变，且水温保持恒定（内能一定），水通过阀口玻璃管从容器中均匀、稳定流出，水管末端的阀门可调节出流速度，通过滴管可将与水密度相同的有色颜料从水管入口滴入。

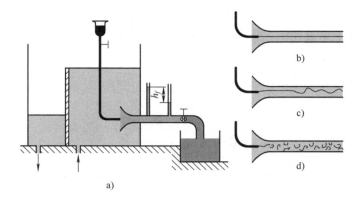

Fig.6.2.1   Reynolds experiment

a) The experimental device   b) Laminar flow   c) Transition flow   d) Turbulent flow

Turn the faucet on so the water flows out very slowly, we may observe that the dye streak forms a straight and smooth line at low velocities. The streamline would not mix with the neighboring fluid, and would retain its identity for a relatively long period of time. This flow regime characterized by smooth streamlines and highly ordered motion as the (b) case in Fig.6.2.1 is said to be *laminar*. Viscous shear stresses always influence a laminar flow, the fluid appears to move by the sliding of laminations of infinitesimal thickness over adjacent layers.

Open the faucet slowly and observe streamline zigzags rapidly and disorderly. The dye mix immediately by the action of the randomly moving fluid particles and would quickly lose its identity when the flow becomes fully turbulent. The flow regime characterized by velocity fluctuations and highly disordered motion, is said to be *turbulent* as shown in Fig.6.2.1d, where the flow varies irregularly so that flow quantities show random variation with time and space coordinates. However, we can define a "steady" turbulent flow in which the time-average physical quantities described by statistical averages do not change with time. Inertial forces plays significant part in turbulent flow.

Note that the *transition* from laminar to turbulent flow as shown in the (c) case, occurs over some region in which the flow fluctuates between laminar and turbulent flows before it becomes fully turbulent.

If we measure the head loss of laminar flow in a given length of uniform pipe at different velocities, the head loss which is represented by h_f as shown in Fig.6.2.1a, due to friction, is directly proportional to the velocity, as shown in Fig.6.2.2. But with increasing velocity, at some point B, where visual observation of dye show that the flow changes from laminar to turbulent, there will be an abrupt increase in the flow rate at which the head loss varies. If we plot the logarithms of these two variables on linear scales, we will find that, after passing a certain transition region, like BC in Fig.6.2.2, the lines will have slopes ranging from about 1.75

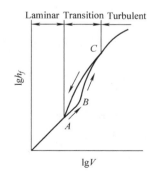

Fig.6.2.2   The plot of V-h_f for flow in a uniform pipe

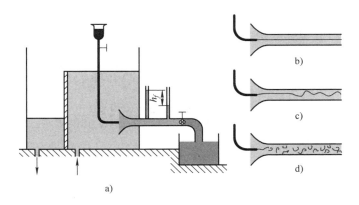

图 6.2.1　雷诺实验

a）实验装置　b）层流　c）过渡流　d）紊流

　　阀门开口较小流速很慢时，可以看到颜料线平直而光滑，没有与周围流体掺混，且可以较持久的保持这种状态，如图 6.2.1b 所示。我们把这种流体质点沿平滑直线高度有序的流动称为**层流**。层流中黏性剪切力主导流动，流体像是在分层地光滑运动。

　　随着阀门开度逐渐增大，可以观察到流线开始弯曲且变得无序。当流体运动变得非常紊乱时，颜料在随机运动的流体质点带动下迅速扩散至周围流体并无法辨识踪迹，如图 6.2.1d 所示。流体这种高度无序、散乱的流动称为**紊流**（或湍流、乱流），紊流中流体质点运动呈现高度无序，各项流动参数随时间和空间位置无规律地变化。但是注意，紊流也存在稳定流动，如果各项物理量的时间平均值不随时间变化，这时的紊流也为稳定流动。惯性力对紊流影响显著。

　　层流充分过渡到紊流之前的一段状态，如图 6.2.1c 所示，称为**过渡流**，在这段区间，流体运动状态介于层流和紊流之间。

　　图 6.2.1a 中的装置还能够测量管内两点间沿程损失 h_f 随流速的变化关系。由图 6.2.2 可以看出，摩擦阻力导致的沿程损失随流速增加而增大。当流速增加到 B 点值附近时，通过颜料的变化能够观察到流态从层流变为紊流，此时损失水头有一个激增。取流速和沿程损失的对数作曲线，可以发现经过一段过渡流之后，曲线的斜率明显增加至 1.75 ~ 2，对应图 6.2.2 中的 BC 段。可见，层流状态流速与沿程损失为线性关系，而紊流状态摩擦损失随 V^n 变化，n 值在 1.75 ~ 2

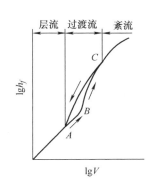

图 6.2.2　均匀管流的
V-h_f 曲线

to 2.00. Thus we see that for laminar flow the drop in energy due to friction varies as V, while for turbulent flow the friction varies as V^n where n ranges from about 1.75 to 2. If we gradually reduce the velocity from a high value, the points will not return along line CB. Instead, the points will lie along curve CA. We call the velocity at dividing point of laminar and turbulent flow the critical velocity V_c, velocity at point B the higher critical velocity, and velocity at A the lower critical velocity.

The critical velocity depends on three or two physical parameters describing the flow conditions, the diameter of a pipe, the density and viscosity of fluid (Eq.5.2.4), or the latter two are represented by the kinematic viscosity.

Osborne Reynolds developed the criterion number to determine whether a flow is laminar or turbulent. This number called the *Reynolds number* is the ratio of inertial forces to viscous forces in the fluid, which we introduced in 5.2 and 5.3, and is expressed for internal flow in a circular pipe as

$$Re = \frac{\rho dV}{\mu} = \frac{dV}{\nu} \qquad (6.2.1)$$

here Re is a dimensionless number.

The upper critical Reynolds number corresponding to point B of Fig.6.2.2, is really indeterminate. Its value is normally about 4000, but experimenters up to values as high as 50000. Flow under that critical condition is inherently unstable because that initial disturbance affecting the flow, and the least disturbance will transform it instantly into turbulent flow. However, the lower value is much more definite to persist at 2300, and is the real dividing point between laminar and turbulent flows. So we define this lower value as the true critical Reynolds number, and take the critical value as

$$Re_c = 2300 \qquad (6.2.2)$$

Under most practical conditions, the flow in a circular pipe is laminar for $Re \leq 2300$, turbulent for $Re > 2300$, that is

$$Re \leq 2300 \quad \text{Laminar flow}$$
$$Re > 2300 \quad \text{Turbulent flow} \qquad (6.2.3)$$

At large Reynolds numbers region, the inertial forces which are proportional to ρ and V^2, are large relative to the viscous forces, and thus the viscous forces cannot prevent the random and rapid fluctuations of the fluid. At small or moderate Reynolds numbers region, however, the viscous forces are large enough to suppress these fluctuations and to keep the fluid in "laminations". At transitional flow region in a circular pipe, the flow switches between laminar and turbulent in a disorderly and not typical fashion, can be considered as turbulent flow.

Practically most cases of engineering importance encountered such fluid as water and air, are turbulent flow. Laminar flow is encountered when highly viscous fluids such as oils flow in small pipes or narrow passages.

Example 6.2.1   The 50mm-diameter pipe is used to transport water at 20℃. What is the maximum flow rate that may exist in the pipe for which laminar flow is guaranteed?

Solution   The kinematic viscosity of water at 20℃ is $\nu = 1 \times 10^{-6} \, \text{m}^2/\text{s}$.

区间。如果从紊流状态逐渐关小阀门降低流速，我们会发现曲线没有沿 CB 返回层流，而是沿 CA 变化至层流状态。我们把划分层流和紊流状态的流速称为临界速度 V_c，称 B 点的流速为上临界速度，A 点流速为下临界速度。

临界速度值取决于描述流动特征的三个或两个物理量：管道直径、流体密度和黏度，见式 5.2.4，或者用运动黏度代替后面两个变量。

雷诺开发了判定流态的准则数，称为雷诺数 Re。雷诺数是取惯性力与黏性力之比，在 5.2 节和 5.3 节中已有介绍，对于管流其常用形式为

$$Re = \frac{\rho dV}{\mu} = \frac{dV}{\nu} \tag{6.2.1}$$

式中，Re 是无量纲数。

上临界速度的雷诺值，即图 6.2.2 中 B 点的雷诺数值，很不稳定，通常大于 4000，由于状态的不稳定，有些实验条件下瞬间产生的巨大扰动会使雷诺数高达 50000。而下临界速度的雷诺数则相当稳定，基本都在 2300 左右，适合作为分界点。所以我们常用下临界速度的雷诺数作为判定层流和紊流的标准，其值为

$$Re_c = 2300 \tag{6.2.2}$$

大多数实践条件下，对于圆管层流都有 $Re \leqslant 2300$，对紊流则 $Re > 2300$，所以

$$\begin{matrix} 层流 & Re \leqslant 2300 \\ 紊流 & Re > 2300 \end{matrix} \tag{6.2.3}$$

雷诺数大的流动，惯性力与 ρ 和 V^2 成正比，远远大于黏性力，所以黏性力无法阻止流体迅速而随机的扩散。而雷诺数小的流动，黏性力足以压制流体扩散而维持分层流动状态。管流中的过渡流总是在层流和紊流间切换，不具有典型特征，所以常被归为紊流。

工程实践中的大多数流体，如水和空气的流动都是紊流。层流仅出现在高黏度流体如油液，在小管或狭缝流动中。

例 6.2.1　用直径 50mm 的管道输送 20℃的水，求能够维持管内层流状态的最大流量。

解　水在 20℃时的运动黏度为 $\nu = 1 \times 10^{-6} \mathrm{m^2/s}$。

Using a Reynolds number of 2300 so that a laminar flow is guaranteed, we find that

$$Re = \frac{dV}{v} \leqslant 2300 \quad \rightarrow \quad V \leqslant 0.046 \text{m/s}$$

The maximum flow rate is

$$q = \frac{\pi d^2}{4} V = 90.275 \times 10^{-6} \text{m}^3 / \text{s}$$

Example 6.2.2   In a refinery oil ($\rho = 850 \text{kg/m}^3$, $\mu = 15.3 \times 10^{-3}$ Pa · s) flows through a 100mm-diameter pipe at 90L/min. Is the flow laminar or turbulent?
Solution
The average velocity of oil flow is

$$V = \frac{4q}{\pi d^2} = \frac{4 \times 90 \times 10^{-3}}{60 \times 3.142 \times 0.01} \text{m/s} = 0.19 \text{m/s}$$

The Reynolds number

$$Re = \frac{\rho dV}{\mu} = \frac{850 \times 0.1 \times 0.19}{15.3 \times 10^{-3}} = 1056 < 2300$$

The flow is laminar.

6.3   Laminar Flow in Pipes

Internal flows are of particular importance to engineers. Flow in a circular pipe is undoubtedly the most common internal fluid flow. When considering internal flows we are interested primarily in developed flows in pipes. In laminar flow, fluid particles flow in an orderly manner along pathlines, and momentum and energy are transferred across streamlines by molecular diffusion. A developed laminar flow results where the velocity profile does not change in the flow direction and viscous stresses dominate the entire cross section. In this section we investigate incompressible, steady, developed laminar flow in a pipe, using an elemental approach.

6.3.1   Velocity and shearing stress distribution

A cylindrical elemental volume of the fluid as sketched in Fig.6.3.1 can be considered as an infinitesimal control volume into which and from which fluid flows, or it can be considered an infinitesimal fluid mass upon which forces are acting. The fluid mass is very small so that the action of gravity is negligible. Since the velocity profile does not change in the y-direction, and there is no acceleration of the mass element, the resultant force in the y-direction must be zero. Consequently, a force balance in the y-direction yields

$$p \pi r^2 - \left(p + \frac{dp}{dy} \Delta y \right) \pi r^2 - 2 \pi r \Delta y \tau = 0 \tag{6.3.1}$$

which can be simplified to

雷诺数值不超过 2300 时为层流，所以

$$Re = \frac{dV}{\nu} \leqslant 2300 \quad \rightarrow \quad V \leqslant 0.046 \text{m/s}$$

此时最大流量为

$$q = \frac{\pi d^2}{4} V = 90.275 \times 10^{-6} \text{m}^3/\text{s}$$

例 6.2.2　精炼油（ρ=850kg/m³, μ=15.3 × 10⁻³ Pa·s）在直径 100mm 的管道内流量为 90L/min，判断其流态为层流还是紊流。

解　油液平均流速为

$$V = \frac{4q}{\pi d^2} = \frac{4 \times 90 \times 10^{-3}}{60 \times 3.142 \times 0.01} \text{m/s} = 0.19 \text{m/s}$$

雷诺数值

$$Re = \frac{\rho dV}{\mu} = \frac{850 \times 0.1 \times 0.19}{15.3 \times 10^{-3}} = 1056 < 2300$$

流态为层流。

6.3　圆管层流

内部流动对工程应用非常重要，管流则为最常见的内流，通常我们关注最多的是充分发展管流。层流的特点是流体质点沿特定轨迹线有序运动，动能和动量通过分子扩散在流线间传递。充分发展层流的速度分布不沿流动方向变化，黏性剪切力作用显著。本节我们用微元法分析不可压缩黏性流体在圆管内的稳定充分发展层流。

6.3.1　流速和剪切应力分布

取图 6.3.1 所示的圆柱形体积微元，流体可以流入或流出该体积；或者，也可以把该体积看作具有质量的流体微元，力作用于该微元内流体。流体微元质量很小，所以可忽略重力作用。因为流体速度不随 y 方向改变，且流体微元无加速度，因此在 y 方向的外力为零。所以，流体微元在 y 方向的受力平衡方程为

$$p\pi r^2 - \left(p + \frac{dp}{dy} \Delta y \right) \pi r^2 - 2\pi r \Delta y \tau = 0 \qquad (6.3.1)$$

化简为

$$\tau = -\frac{r}{2}\frac{dp}{dy} \tag{6.3.2}$$

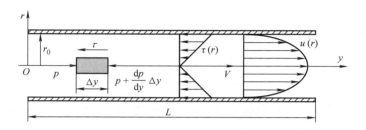

Fig.6.3.1 Developed viscous flow in a circular pipe

Since $u = u(r)$, we can conclude that the pressure drop in the y-direction due to friction loss is considered to be linear variation, the pressure gradient must be at most a constant, can be expressed by

$$-\frac{dp}{dy} = \frac{p_{in} - p_{out}}{L} = \frac{\Delta p}{L} \tag{6.3.3}$$

We may find the shearing stress change in linearly fashion

$$\tau = \frac{r}{2}\frac{\Delta p}{L} \tag{6.3.4}$$

The maximum shear stress τ_0 occurs on the wall of the pipe where the radius $r = r_0$

$$\tau_0 = \frac{r_0}{2}\frac{\Delta p}{L} \tag{6.3.5}$$

Recall from Eq.1.4.10 that $\tau = \mu\frac{du}{dr}$, the shear stress in this laminar flow giving

$$\mu\frac{du}{dr} = \frac{r}{2}\frac{\Delta p}{L} \tag{6.3.6}$$

Eq.6.3.6 can then be integrated by $\int du = \frac{1}{2\mu}\frac{\Delta p}{L}\int r dr$, to give the velocity distribution

$$u = \frac{1}{4\mu}\frac{\Delta p}{L}\left(r^2 + C\right) \tag{6.3.7}$$

$$\tau = -\frac{r}{2}\frac{\mathrm{d}p}{\mathrm{d}y} \tag{6.3.2}$$

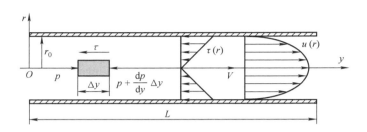

图 6.3.1   圆管发展黏性层流

因为 $u=u(r)$，可以认为压强在 y 方向变化是由黏性力引起，且为线性变化规律。所以 y 方向压强梯度为常值，可以写为

$$-\frac{\mathrm{d}p}{\mathrm{d}y} = \frac{p_{\text{in}} - p_{\text{out}}}{L} = \frac{\Delta p}{L} \tag{6.3.3}$$

因此剪切应力沿半径方向为线性分布

$$\tau = \frac{r}{2}\frac{\Delta p}{L} \tag{6.3.4}$$

最大剪切应力 τ_0 发生在管壁 $r = r_0$ 处

$$\tau_0 = \frac{r_0}{2}\frac{\Delta p}{L} \tag{6.3.5}$$

由式 1.4.10 中剪切应力的定义式 $\tau = \mu\dfrac{\mathrm{d}u}{\mathrm{d}r}$ 可知，圆管内层流的剪切应力与压力间的平衡关系为

$$\mu\frac{\mathrm{d}u}{\mathrm{d}r} = \frac{r}{2}\frac{\Delta p}{L} \tag{6.3.6}$$

对式 6.3.6 积分，有 $\displaystyle\int \mathrm{d}u = \frac{1}{2\mu}\frac{\Delta p}{L}\int r\,\mathrm{d}r$，可得速度分布

$$u = \frac{1}{4\mu}\frac{\Delta p}{L}\left(r^2 + C\right) \tag{6.3.7}$$

where C is the constant of integration can be determined by using $u = 0$ at $r = r_0$ due to noslip condition, then the velocity distribution is presented by

$$u = \frac{1}{4\mu} \frac{\Delta p}{L}\left(r^2 - r_0^2\right) = \frac{1}{4\mu}\frac{dp}{dy}\left(r_0^2 - r^2\right) \tag{6.3.8}$$

The maximum velocity at $r = 0$ from Eq.6.3.8 is

$$u_{max} = \frac{1}{4\mu}\frac{\Delta p}{L}r_0^2 \tag{6.3.9}$$

The velocity distribution of a parabolic profile as shown in Fig.6.3.1 is often referred to as *Poiseuille flow*, a laminar flow with a parabolic profile in a pipe or between parallel plates.

6.3.2 Flow rate and average velocity

For steady, laminar, developed flow in a circular pipe, the flow rate q is found to be

$$
\begin{aligned}
q &= \int_0^{r_0} 2\pi r u \, dr = -\frac{\Delta p}{L}\frac{\pi}{2\mu}\int_0^{r_0} r\left(r_0^2 - r^2\right) dr \\
&= -\frac{\Delta p}{L}\frac{\pi}{2\mu}\int_0^{r_0}\left(r_0^2 r - r^3\right) dr \\
&= -\frac{\Delta p}{L}\frac{\pi r_0^4}{8\mu} \\
&= \frac{dp}{dy}\frac{\pi r_0^4}{8\mu}
\end{aligned} \tag{6.3.10}
$$

Note that the pressure drop is a positive quantity, whereas the pressure gradient is negative.

The value of average velocity on the cross section is found to be

$$V = \frac{q}{A} = \frac{\Delta p}{L}\frac{r_0^2}{8\mu} \tag{6.3.11}$$

Refer to the maximum velocity on the cross section shown in Eq.6.3.9, we have

$$V = \frac{1}{2}u_{max} \tag{6.3.12}$$

The velocity distribution can also be expressed by the average velocity as

$$u = 2V\left(1 - \frac{r^2}{r_0^2}\right) \tag{6.3.13}$$

Then the kinetic energy correction factor for a fully laminar flow in pipe can be evaluated

式中，C 为积分常数，可应用边界条件求得。代入壁面不滑移条件：$r=r_0$ 处 $u=0$，解得流速分布规律为

$$u = \frac{1}{4\mu}\frac{\Delta p}{L}\left(r^2 - r_0^{\,2}\right) = \frac{1}{4\mu}\frac{\mathrm{d}p}{\mathrm{d}y}\left(r_0^{\,2} - r^2\right) \tag{6.3.8}$$

由式 6.3.8 可知管道中心，即 $r=0$ 处，流速最大

$$u_{\max} = \frac{1}{4\mu}\frac{\Delta p}{L}r_0^{\,2} \tag{6.3.9}$$

如图 6.3.1 所示，速度分布为旋转抛物面形，圆管或平行平板缝隙中的这种具有抛物面速度分布的层流，也称 Poiseuille 流动。

6.3.2　流量和平均流速

稳定、充分发展圆管层流的流量 q 可由速度积分而得

$$
\begin{aligned}
q &= \int_0^{r_0} 2\pi r u\,\mathrm{d}r = -\frac{\Delta p}{L}\frac{\pi}{2\mu}\int_0^{r_0} r\left(r_0^{\,2} - r^2\right)\mathrm{d}r \\
&= -\frac{\Delta p}{L}\frac{\pi}{2\mu}\int_0^{r_0}\left(r_0^{\,2}r - r^3\right)\mathrm{d}r \\
&= -\frac{\Delta p}{L}\frac{\pi r_0^{\,4}}{8\mu} \\
&= \frac{\mathrm{d}p}{\mathrm{d}y}\frac{\pi r_0^{\,4}}{8\mu}
\end{aligned}
\tag{6.3.10}
$$

上式压力差取为正值，所以压力梯度值为负。

截面上平均流速的值为

$$V = \frac{q}{A} = \frac{\Delta p}{L}\frac{r_0^{\,2}}{8\mu} \tag{6.3.11}$$

与式 6.3.11 中最大流速比较，可知

$$V = \frac{1}{2}u_{\max} \tag{6.3.12}$$

用平均流速表示实际流速分布为

$$u = 2V\left(1 - \frac{r^2}{r_0^{\,2}}\right) \tag{6.3.13}$$

计算完全层流的动能修正系数，有

$$\alpha = \frac{1}{A}\int_A \left(\frac{u}{V}\right)^3 dA = \frac{8}{r_0^2}\int_0^{r_0}\left(1-\frac{r^2}{r_0^2}\right)^3 rdr = 2 \tag{6.3.14}$$

As we introduced in section 4.3, the value of kinetic energy correction factor is always more than 1 and less than 2. Under most common conditions, it is customary to assume that $\alpha = 1$.

The momentum correction factor for a parabolic profile in a circular pipe can be calculated

$$\beta = \frac{1}{A}\int_A \left(\frac{u}{V}\right)^2 dA = \frac{4}{r_0^2}\int_0^{r_0}\left(1-\frac{r^2}{r_0^2}\right)^2 rdr = \frac{4}{3} \tag{6.3.15}$$

6.3.3 Frictional loss

We now discuss the pressure drop due to the frictional loss in circular pipe for laminar flow. Substitute diameter of the pipe d for radius r_0 and the pressure drop can be express by the average velocity form Eq.6.3.11 as

$$\frac{\Delta p}{L} = \frac{32\mu}{d^2}V \tag{6.3.16}$$

Rewrite the pressure drop in term of pressure loss head, we have, for a horizontal pipe

$$\frac{\Delta p}{\rho g} = \frac{32\mu}{\rho g}\frac{L}{d^2}V \tag{6.3.17}$$

Combining the Darcy's formula Eq.6.1.7, the friction loss has been shown to be

$$h_f = \lambda\frac{L}{d}\frac{V^2}{2g} = \frac{32\mu}{\rho g}\frac{L}{d^2}V \tag{6.3.18}$$

And then the friction factor λ, a dimensionless wall shear valid for laminar flow can be written as

$$\lambda = \frac{64\mu}{\rho dV} \tag{6.3.19}$$

Recalling the definition of Reynolds number, we find that

$$\lambda = \frac{64}{Re} \tag{6.3.20}$$

This is the friction loss factor for laminar flow that generally is applied to developed, laminar flows in circular pipe.

Example 6.3.1 Consider the fully developed flow of glycerin at 40℃ through a 80m-long, inclined 15° upward, 50mm-diameter, circular pipe. If the flow velocity at the centerline is measured to be 5m/s, determine the velocity profile and the pressure difference

$$\alpha = \frac{1}{A}\int_A \left(\frac{u}{V}\right)^3 \mathrm{d}A = \frac{8}{r_0^2}\int_0^{r_0}\left(1-\frac{r^2}{r_0^2}\right)^3 r\mathrm{d}r = 2 \qquad (6.3.14)$$

如 4.3 节所述，动能修正系数取值总是大于 1 而小于 2。通常习惯取 $\alpha=1$。

圆管流抛物面形速度分布的动量修正系数值为

$$\beta = \frac{1}{A}\int_A \left(\frac{u}{V}\right)^2 \mathrm{d}A = \frac{4}{r_0^2}\int_0^{r_0}\left(1-\frac{r^2}{r_0^2}\right)^2 r\mathrm{d}r = \frac{4}{3} \qquad (6.3.15)$$

6.3.3　沿程损失

接下来讨论圆管层流的黏性压力损失。用管路直径 d 替换式 6.3.11 中的管壁半径 r_0，并用平均流速表示压强差，有

$$\frac{\Delta p}{L} = \frac{32\mu}{d^2}V \qquad (6.3.16)$$

把压强差写为损失水头形式，则对水平放置的圆管，有

$$\frac{\Delta p}{\rho g} = \frac{32\mu}{\rho g}\frac{L}{d^2}V \qquad (6.3.17)$$

结合式 6.1.7 的达西公式，圆管层流沿程损失为

$$h_f = \lambda\frac{L}{d}\frac{V^2}{2g} = \frac{32\mu}{\rho g}\frac{L}{d^2}V \qquad (6.3.18)$$

进一步整理可得出层流的沿程损失系数表达式

$$\lambda = \frac{64\mu}{\rho dV} \qquad (6.3.19)$$

回顾雷诺数定义式，可以发现

$$\lambda = \frac{64}{Re} \qquad (6.3.20)$$

这是圆管层流常用的沿程损失系数表达式。

例 6.3.1　如图 X6.3.1 所示，40 ℃的甘油在直径 50mm，且向上倾斜 15°的 80m 长圆管内充分发展流动。若在管道中心测得流速为 5m/s，求（1）速度分布和

across this 80m-long section of the pipe, and the useful pumping power required to maintain this flow. The pump is located outside this pipe section. The density and dynamic viscosity of glycerin at 40℃ are $\rho = 1252\text{kg/m}^3$ and $\mu = 0.3073\text{Pa} \cdot \text{s}$, respectively.

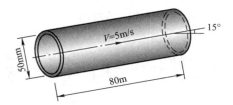

Fig.X6.3.1

Solution   The velocity at the centerline is the maximum velocity, from Eq.6.3.13, average velocity is determined to be

$$V = \frac{1}{2}u_{\max} = 2.5\text{m/s}$$

The Reynolds number is

$$Re = \frac{\rho d V}{\mu} = \frac{1252 \times 0.05 \times 2.5}{0.3073} = 509 < 2300$$

Therefore, the flow is laminar. Then the friction factor and the head loss become

$$\lambda = \frac{64}{Re}, \quad h_f = \frac{64}{Re} \frac{L}{d} \frac{V^2}{2g} = 64.1\text{m}$$

The pumping power is required to overcome the frictional loss and the upward inclination of the pipe. The elevation difference is

$$z_2 - z_1 = L \sin 15° = 20.7\text{m}$$

The pressure drop due to the inclination and frictional loss is

$$\Delta p = \rho g \left[h_f + (z_2 - z_1) \right] = 1041.5\text{kPa}$$

Then useful pumping power required is

$$W_p = \Delta p q = \Delta p V \frac{\pi d^2}{4} = 5.1\text{kW}$$

Example 6.3.2   Find the radius on the cross section of a circular pipe at where the elemental velocity is the same with average velocity for an incompressible, steady and fully developed laminar flow.

Solution   The velocity profile in a circular pipe from Eq.6.3.14 is

80m 长管道两端截面的压强差；（2）维持流动所需的液压泵功率。泵安装于该段管道外部。甘油 40℃时的密度和动力黏度分别为 $\rho=1252\text{kg/m}^3$ 和 $\mu=0.3073\text{Pa·s}$。

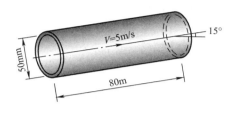

图 X6.3.1

解　管道中心处的流速为最大流速，由式 6.3.13 可知，平均流速为

$$V = \frac{1}{2}u_{\max} = 2.5\text{m/s}$$

雷诺数为

$$Re = \frac{\rho d V}{\mu} = \frac{1252 \times 0.05 \times 2.5}{0.3073} = 509 < 2300$$

流动为层流，沿程损失系数和损失水头为

$$\lambda = \frac{64}{Re}, \quad h_f = \frac{64}{Re}\frac{L}{d}\frac{V^2}{2g} = 64.1\text{m}$$

泵的功率用于克服黏性摩擦损失和管道向上的倾斜角。管道两截面高度差为

$$z_2 - z_1 = L\sin 15° = 20.7\text{m}$$

由于摩擦损失和管路倾斜而产生的压强差为

$$\Delta p = \rho g\left[h_f + (z_2 - z_1)\right] = 1041.5\text{kPa}$$

泵需提供的功率为

$$W_p = \Delta p q = \Delta p V \frac{\pi d^2}{4} = 5.1\text{kW}$$

例 6.3.2　对于圆管内不可压缩流体稳定、充分发展层流，找出截面上实际速度与平均流速相等处的半径。

解　由式 6.3.14 可知圆管内速度分布

$$u = 2V\left(1 - \frac{r^2}{r_0^2}\right) = V$$

where r_0 is the internal radius of the pipe.

Then it must have

$$\frac{1}{2} = \frac{r^2}{r_0^2} \quad \rightarrow \quad r = \frac{\sqrt{2}}{2}r_0$$

The velocity profile on the circle with a radius of $r = \frac{\sqrt{2}}{2}r_0$, has the same value with the average velocity.

Example 6.3.3   Water flows in an open rectangular channel at a depth of $h_1 = 1\text{m}$ with a velocity of 5m/s, as shown in Fig.X6.3.3. The bottom of the channel drops over a short length at a distance of $h = 1.2\text{m}$. The head loss across the channel drop h_L is 0.2m. Calculate the two possible depths of flow after the drop.

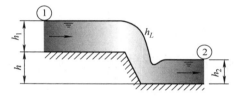

Fig.X6.3.3

Solution   Take the control volume from section ① to ②, the Bernoulli equivalent between center of section ① and ② gives that

$$\frac{h_1}{2} + h + \frac{p_1}{\rho g} + \frac{V_1^2}{2g} = \frac{h_2}{2} + \frac{p_2}{\rho g} + \frac{V_2^2}{2g} + h_L$$

Pressures at center of section 1 and 2 are

$$p_1 = \rho g \frac{h_1}{2}, \quad p_2 = \rho g \frac{h_2}{2}$$

By continuity, $V_1 h_1 B = V_2 h_2 B$, where B is the width of the channel.

That is, $V_2 = \frac{h_1}{h_2}V_1$

Substituting and solving a cubic equation, the results are

$$h_2 = 3.147, \quad 0.704 \quad \text{and} \quad -0.576$$

$$u = 2V\left(1 - \frac{r^2}{r_0^{\,2}}\right) = V$$

式中，r_0 为圆管内壁半径，由此可知

$$\frac{1}{2} = \frac{r^2}{r_0^{\,2}} \rightarrow r = \frac{\sqrt{2}}{2}r_0$$

管道截面上 $r = \dfrac{\sqrt{2}}{2}r_0$ 处，实际流速与平均流速相等。

例 6.3.3　如图 X6.3.3 所示，开放水渠内水位高度 $h_1 = 1\text{m}$，水流平均流速为 5m/s。水渠底部在短距离内降低了 $h = 1.2\text{m}$，并在水渠截面上产生 0.2m 水头损失 h_L。求落差发生后，水渠内水位高度（两种可能）。

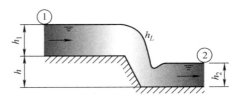

图 X6.3.3

解　取截面①和②间为控制体，两截面间能量平衡方程为

$$\frac{h_1}{2} + h + \frac{p_1}{\rho g} + \frac{V_1^2}{2g} = \frac{h_2}{2} + \frac{p_2}{\rho g} + \frac{V_2^2}{2g} + h_L$$

截面①和②中心处的静压强为

$$p_1 = \rho g\frac{h_1}{2}, \quad p_2 = \rho g\frac{h_2}{2}$$

根据连续性方程：$V_1 h_1 B = V_2 h_2 B$，式中 B 为水渠宽度

得出 $V_2 = \dfrac{h_1}{h_2}V_1$

代入已知量并解一元三次方程，得三个解

$$h_2 = 3.147, \quad 0.704 \quad 和 \quad -0.576$$

Obviously, the two possible depths of flow after the drop are 3.147m and 0.704m respectively.

Example 6.3.4 A water tank is placed at 50m above the ground and supplying water to a building through a 100mm-diameter pipe with flow rate of 0.012m³/s, as shown in Fig.X6.3.4. If the head loss up to an apartment at the third level is 4mH₂O, where the elevation is 12m. Find the water pressure at the apartment.

Solution Average velocity at section 2 is

$$V_2 = \frac{4q}{\pi d^2} = 1.53 \text{m/s}$$

The energy equation from section 1 to 2 is given as

Fig.X6.3.4

$$z_1 + \frac{\cancel{p_1}}{\rho g} + \frac{\cancel{V_1^2}}{2g} = z_2 + \frac{p_2}{\rho g} + \frac{V_2^2}{2g} + h_L$$

The result is $p_2 = 311$kPa.

Example 6.3.5 Oil of density 900kg/m³ flow through a 6mm-diameter 2m-long pipe at a flow rate of 7.3×10^{-6}m³/s. Height difference of manometer as shown in Fig.X6.3.5 is 120mm. Determine the viscosity of the oil.

Solution The pressure difference between section 1 and 2 is

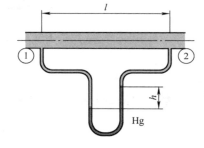

Fig.X6.3.5

$$\Delta p = (\rho' - \rho) gh = 14.9 \text{kPa}$$

The average velocity in the pipe and the Reynolds number are

$$V = \frac{4q}{\pi d^2}, \quad \text{and,} \quad Re = \frac{\rho dV}{\mu}$$

The head loss due to the viscosity of oil is

$$h_f = \frac{\Delta p}{\rho g} = \frac{64}{Re} \frac{l}{d} \frac{V^2}{2g}$$

Substituting and calculating, then gives the result

显然两个可能的深度分别为 3.147m 和 0.704m。

例 6.3.4 图 X6.3.4 所示的水箱安装在距地面 50m 高处，并通过直径 100mm 的水管向建筑供水，已知管内流量为 $0.012\text{m}^3/\text{s}$。若从水箱至距地面 12m 高的三层某公寓水头损失为 $4\text{mH}_2\text{O}$，求该公寓内的水压。

图 X6.3.4

解 截面 2 的平均流速为

$$V_2 = \frac{4q}{\pi d^2} = 1.53\text{m/s}$$

截面①—②间的伯努利方程为

$$z_1 + \frac{\cancel{p_1}}{\rho g} + \frac{\cancel{V_1^2}}{2g} = z_2 + \frac{p_2}{\rho g} + \frac{V_2^2}{2g} + h_L$$

解得 $p_2 = 311\text{kPa}$。

例 6.3.5 密度为 900kg/m^3 的油液流经直径 6mm 长度 2m 的管道，流量 $7.3 \times 10^{-6}\text{m}^3/\text{s}$。如图 X6.3.5 所示，测压管高度差为 120mm，求油液黏度。

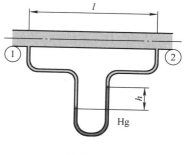

图 X6.3.5

解 截面①—②间的压强差为

$$\Delta p = (\rho' - \rho) gh = 14.9\text{kPa}$$

管内平均流速和雷诺数为

$$V = \frac{4q}{\pi d^2} \quad \text{和} \quad Re = \frac{\rho d V}{\mu}$$

黏性水头损失为

$$h_f = \frac{\Delta p}{\rho g} = \frac{64}{Re} \frac{l}{d} \frac{V^2}{2g}$$

代入已知条件并解得

319

$$\mu = \frac{\Delta p \pi d^4}{128 q l} = 0.0325 \text{Pa} \cdot \text{s}$$

Example 6.3.6 The oil flow in example 6.3.5, find (a) the friction factor, (b) the shear stress at the pipe wall, and (c) the head loss per meter of pipe length.

Solution

(a) The frictional loss between section 1 and 2 is

$$h_f = \lambda \frac{l}{d} \frac{V^2}{2g} = \frac{\Delta p}{\rho g}$$

The frictional factor is given then

$$\lambda = \frac{2 d \Delta p}{\rho l V^2}$$

Substituting the pressure differential and the average velocity, and have the result

$$\lambda = 1.49$$

(b) The shear stress at the pipe wall is calculated from Eq.6.3.6

$$\tau_0 = \frac{r_0}{2} \frac{\Delta p}{L} = \frac{3}{2} \times \frac{14.9}{2} \text{N/m}^2 = 11.175 \text{N/m}^2$$

(c) The head loss per meter of pipe length is

$$\frac{h_f}{L} = \frac{\Delta p}{\rho g L} = 0.76 \text{m/m}$$

Example 6.3.7 A downward inclined oil pipeline of 500mm-diameter and 5000m-long with a flow rate of 15489kg/min. The entrance is 10m above the exit and pressure at entrance is measured of 490kPa. Density of oil and the friction factor of the pipeline are known as 895.4kg/m³ and 0.003 respectively. Calculate the pressure at exit of the pipeline.

Solution

The energy equation from the entrance to the exit is written as

$$h + \frac{p_1}{\rho g} + \frac{V_1^2}{2g} = \frac{p_2}{\rho g} + \frac{V_2^2}{2g} + h_L$$

The head loss is assumed the frictional loss only, and that is, $h_L = h_f = \lambda \dfrac{l}{d} \dfrac{V^2}{2g}$

Velocity in the pipeline is constant, then we have the result

$$p_2 = 555.6 \text{kPa}$$

$$\mu = \frac{\Delta p \pi d^4}{128 q l} = 0.0325 \text{Pa} \cdot \text{s}$$

例 6.3.6　例 6.3.5 中的油管，求（1）沿程损失系数；（2）管壁处的剪切应力；（3）每米长度管道上的沿程损失。

解　（1）截面①—②间的沿程损失为

$$h_f = \lambda \frac{l}{d} \frac{V^2}{2g} = \frac{\Delta p}{\rho g}$$

沿程损失系数

$$\lambda = \frac{2 d \Delta p}{\rho l V^2}$$

代入压强差和平均流速，解得

$$\lambda = 1.49$$

（2）根据式 6.3.6 计算管壁处剪切应力为

$$\tau_0 = \frac{r_0}{2} \frac{\Delta p}{L} = \frac{3}{2} \times \frac{14.9}{2} \text{N/m}^2 = 11.175 \text{N/m}^2$$

（3）每米长度管道上的损失水头为

$$\frac{h_f}{L} = \frac{\Delta p}{\rho g L} = 0.76 \text{m/m}$$

例 6.3.7　一段向下倾斜，长度 5000m 直径 500mm 的输油管，流量为 15489kg/min。入口高于出口 10m，且入口处测得的压强为 490kPa。管内输送的油液密度为 895.4kg/m³，沿程损失系数为 0.003，计算管道出口处压强。

解　进出口间的能量平衡方程可写为

$$h + \frac{p_1}{\rho g} + \frac{V_1^2}{2g} = \frac{p_2}{\rho g} + \frac{V_2^2}{2g} + h_L$$

假设能量损失仅为沿程损失，且 $h_L = h_f = \lambda \frac{l}{d} \frac{V^2}{2g}$

管道内油液流速恒定，则出口压强为

$$p_2 = 555.6 \text{kPa}$$

Example 6.3.8 Water is spraying out from a fountain nozzle of 0.5m-long, the diameter of entrance and exit are 40mm and 20mm respectively, as shown in Fig.X6.3.8. The pressure at the entrance is measured of 98kPa, and head loss at the nozzle is known as 1.6m. What is the maximum height that the jet could achieve?

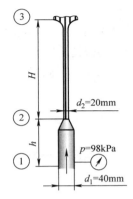

Fig.X6.3.8

Solution The Bernoulli equation between sections ① — ② can be given as

$$\frac{p_1}{\rho g} + \frac{V_1^2}{2g} = z_2 + \frac{V_2^2}{2g} + h_L, \text{ from continuum } V_2 = 4V_1$$

Then we have $\dfrac{p_1}{\rho g} + \dfrac{V_1^2}{2g} = h + \dfrac{16V_1^2}{2g} + h_L$

it gives $V_1 = 3.21\text{m/s}, \quad V_2 = 12.85\text{m/s}$.

Using Bernoulli equation from section ② to ③ , velocity at the top of water column is zero, then

$$\frac{V_2^2}{2g} = H + \frac{V_3^2}{2g}$$

The height the jet can achieve is $H = 8.42\text{m}$.

Example 6.3.9 A small-diameter horizontal tube is connected to a supply reservoir as shown in Fig.X6.3.9. If 50L is captured at the outlet in 6s, estimate the viscosity of the gasoline ($\rho = 780\text{kg/m}^3$).

Solution The pressure at entrance of the tube is the static pressure at this elevation in the tank,

$$p_1 = \rho g H$$

Fig.X6.3.9

Using Bernoulli equation between the entrance and exit of the tube, the velocity in the tube is constant, and we must have

$\dfrac{p_1}{\rho g} = h_f$, using the Darcy's formula and rewriting, $H = \dfrac{64\mu}{\rho d V} \dfrac{L}{d} \dfrac{V^2}{2g}$

From $V = \dfrac{4q}{\pi d^2}$, it is, $H = \dfrac{128\mu q}{\pi \rho g} \dfrac{L}{d^4}$

Calculating results $\mu = \dfrac{H \pi \rho g d^4}{128 q L} = 0.72 \times 10^{-3}\,\text{Pa} \cdot \text{s}$

例 6.3.8　喷泉出水管长 h=0.5m，入口直径 d_1=40mm，出口直径 d_2=20mm，如图 X6.3.8 所示。测得入口压强 98kPa，喷嘴处损失水头 H=1.6m。求喷泉可能喷射的最大高度。

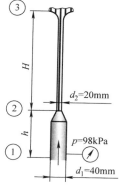

解　在截面①—②间建立伯努利方程

$$\frac{p_1}{\rho g}+\frac{V_1^2}{2g}=z_2+\frac{V_2^2}{2g}+h_L，根据连续性方程可知 V_2=4V_1$$

所以　$\dfrac{p_1}{\rho g}+\dfrac{V_1^2}{2g}=h+\dfrac{16V_1^2}{2g}+h_L$

解得 $V_1=3.21\text{m/s}$，$V_2=12.85\text{m/s}$。

在截面②—③间建立伯努利方程，在水柱最高点速度为 0，所以

图 X6.3.8

$$\frac{V_2^2}{2g}=H+\frac{\cancel{V_3^2}}{2g}$$

解得最大射流高度 H=8.42m。

例 6.3.9　一段水平管道与一个储油罐相连，如图 X6.3.9 所示。若在水平管内测得 6s 内流量为 50L，估算汽油的黏度（$\rho=780\text{kg/m}^3$）。

解　管道入口处静压强由罐内油液液面高度决定

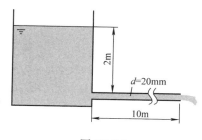

$$p_1=\rho gH$$

图 X6.3.9

对管道进、出口建立伯努利方程，两截面油液流速相等，所以

$\dfrac{p_1}{\rho g}=h_f$，根据达西公式，$H=\dfrac{64\mu}{\rho dV}\dfrac{L}{d}\dfrac{V^2}{2g}$

由 $V=\dfrac{4q}{\pi d^2}$，可得，$H=\dfrac{128\mu q}{\pi\rho g}\dfrac{L}{d^4}$

计算结果 $\mu=\dfrac{H\pi\rho g d^4}{128qL}=0.72\times10^{-3}\text{Pa}\cdot\text{s}$

6.4　Turbulent Flow in Circular Pipe

Most flows encountered in engineering practice are turbulent, it is characterized by disorderly, energy dissipation and rapid fluctuations throughout the flow. The best approach in the turbulent case turns out to be to identify the key variables and functional forms using dimensional analysis, and then to use experimental data to determine the numerical values of any constants.

In all that follows we shall consider in this chapter are only fully developed turbulent flow.

6.4.1　Velocity profile

Turbulent flow is a complex mechanism, therefore, we must rely on experiments and the empirical or semi-empirical correlations developed for various situations. All three velocity components in a turbulent flow are nonzero, and we introduce the notion of a time-average quantity. In a fully developed turbulent pipe flow, the average velocity component in the axial direction would be nonzero while the other two are zero. The expressions for the velocity profile in a turbulent flow are based on both analysis and measurements, and determined from experimental data.

Consider fully developed turbulent flow in a pipe. Typical velocity profiles for fully developed turbulent flow is given in Fig.6.4.1, note that u in the turbulent case is the time-averaged velocity component in the axial direction (and thus drop the overbar from $\bar{u}$ for simplicity).

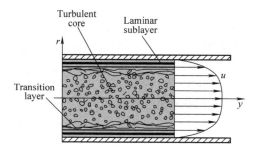

Fig.6.4.1　Velocity profile in developed turbulent flow

Comparing the turbulent-flow velocity profiles in Fig. 6.4.1 with the laminar-flow parabolic velocity profile we see that, the turbulent-flow profiles are much fuller, flatter near the central portion of the pipe and steeper near the wall. Turbulent flow along a wall can be considered to consist of three regions, characterized by the distance from the wall.

As velocity must be zero at a wall, turbulence there is inhibited so that a laminar-like sublayer occurs immediately next to the wall, where viscous effects are dominant is the *viscous sublayer*. This viscous sublayer is extremely thin, typically much less than 1 percent of the pipe diameter. Considering that velocity changes from zero to nearly the core region value across a layer that is sometimes only a few hundredths of a millimeter, the velocity profile in this layer must be very nearly linear. The turbulent shear reaches a maximum near the wall in the viscous sublayer. The viscous sublayer plays a dominant role on flow characteristics, dampens any eddy motion.

The thickness of the viscous sublayer can be determined from

6.4　圆管紊流

工程实践中遇到的大部分流动都是紊流，流动过程充满无序性、耗能性和快速扩散性。研究紊流的最适合方法就是识别关键变量并用量纲分析法建立其函数关系，再结合实验方法确定常系数的数值。

本章内容分析的都是充分发展的紊流。

6.4.1　速度分布

紊流的机理极其复杂，所以我们必须依赖实验方法或经验、半经验公式。紊流中三个方向的速度分量都不为零，必须使用时间平均值描述。充分发展的圆管紊流，沿管道中心线方向的速度分量不为零，其他两个方向的速度平均值为零。紊流速度分布规律表达式的建立结合了分析法加测量法以及实验数据。

图 6.4.1 绘出了圆管充分发展紊流速度分布情况，图中标注的速度 u 为沿管道中心方向瞬时速度的时间平均值（为方便省略了平均值$\bar{u}$上方的横线）。

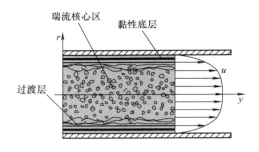

图 6.4.1　圆管充分发展瑞流的速度分布

把图 6.4.1 中瑞流速度曲线与层流速度曲线进行比较，我们可以发现瑞流速度曲线包围的面积更宽，中心附近曲线更平，而靠近管壁时则斜率更大。沿圆管的紊流从管壁向中心可分为三种不同的流态。

管壁处的流速一定为零，所以紧靠管壁的一层区域内流速较低，呈现层流特征，这一层受黏性力主导，称为**黏性底层**。黏性底层厚度非常小，通常不超过管道直径的 1%。在这样厚度可能仅有几百微米的流体层内，流速会迅速从零增加到接近核心区的紊流速度值，所以黏性底层内速度接近线性变化。黏性底层内剪切应力非常大，阻挡了紊流区的全部旋涡，从而将流动分为不同特征区域。

黏性底层的厚度可由下式计算

$$\delta = 32.8 \frac{d}{Re\sqrt{\lambda}} \tag{6.4.1}$$

or

$$\delta = 34.2 \frac{d}{Re^{0.875}} \tag{6.4.2}$$

Thus we conclude that the viscous sublayer is suppressed and it gets thinner as the Reynolds number increases.

At a greater distance from the wall the viscous effect becomes negligible, and the turbulent effects dominate over viscous effects. The characteristics of the flow in this layer are very important since they set the stage for flow in the most of the pipe. This layer is called the *turbulent core*. As shown in Fig.6.4.2, the turbulent velocity profile is fuller than the laminar one, and it becomes more flat and more uniform so that the kinetic energy and momentum correction factors α and β there are almost unity. Moreover, the velocity profile is very sensitive to the magnitude of average wall roughness height ε. Any irregularity or roughness on the surface of pipe wall disturbs the turbulent core and affects the flow strongly.

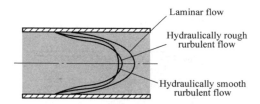

Fig.6.4.2　Velocity profile for laminar flow and turbulent flows

As noted in the preceding, the laminar shear is significant only near the wall in the viscous sublayer with thickness δ. If the thickness δ is sufficiently large, it submerges the wall roughness elements, so that they have negligible effect on the flow, as in Fig.6.4.3a, where ε is the average roughness height. It is as if the wall were smooth. Such a condition is often referred to as being *hydraulically smooth*. The flow and the pipe here are called hydraulically smooth flow and hydraulically smooth pipe respectively. Glass and copper surfaces are generally considered to be hydraulically smooth.

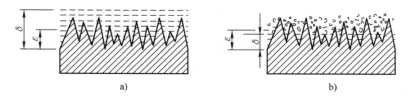

Fig.6.4.3　Hydraulically smooth and hydraulically rough flow

a) hydraulically smooth　b) hydraulically rough

$$\delta = 32.8 \frac{d}{Re\sqrt{\lambda}} \qquad (6.4.1)$$

或

$$\delta = 34.2 \frac{d}{Re^{0.875}} \qquad (6.4.2)$$

可见，雷诺数越大的流动，其黏性底层的厚度越小。

远离管壁，紊流的效应逐渐凸显，并将完全抑制黏性力使其影响几乎被消除。这一层的流动特征代表了几乎整个管路截面上的流动特点，称为**紊流核心区**。由图 6.4.2 可以看到，紊流的速度曲线较层流更加宽而平坦，速度分布更加均匀，所以在这一区域动能修正系数 α 和动量修正系数 β 都几乎等于 1。此外，紊流的速度分布对管壁粗糙度 ε 非常敏感，管壁表面的不规则或粗糙度会对紊流核心区的流速产生很大的干扰。

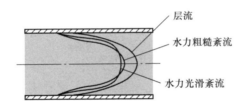

图 6.4.2　紊流和层流的速度曲线

如前所述，厚度为 δ 的黏性底层覆盖在管道内壁表面。那么，如果黏性底层的厚度足够大，就可能淹没管壁的粗糙表面。如图 6.4.3a 所示，图中 ε 表示平均粗糙高度，此时管壁粗糙度几乎不会对流动产生影响，流体好像在光滑的液体壁面内流动，这种情况称为**水力光滑**，这时的流动和管道分别称为水力光滑流动和水力光滑管。玻璃管和铜管的表面通常可以看作是水力光滑的。

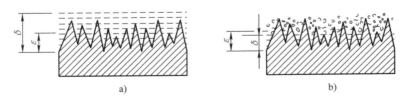

图 6.4.3　水力光滑流动和水力粗糙流动

a) 水力光滑　b) 水力粗糙

In reality, roughness is a relative concept, and it has significance when its average height ε is comparable to the thickness of the viscous sublayer. If the viscous sublayer is relatively thin, the roughness elements protrude out of this layer, as Fig.6.4.3b, the laminar layer is broken up and the surface is no longer hydraulically smooth, also the friction is independent of Reynolds number, so we call the wall *hydraulically rough*. We can see from Fig.6.4.2 that the velocity profile for the hydraulically smooth flow is flatter near the center than for the hydraulically rough flow. Obviously, the velocity profile in laminar flow is independent of pipe roughness.

The relative roughness ε/d which is the ratio of the average height of roughness of the pipe to the pipe diameter, and the Reynolds number can be used to determine if a pipe is hydraulically smooth or rough. This will be observed from the friction factor presented in the next section.

Between the viscous sublayer and the turbulent core, there must be a *transition layer*. The flow in this region may be laminar or turbulent, depending on flow disturbances, or it may alternate between laminar and turbulent. The transition layer can always simply go into the turbulent core.

6.4.2 Losses in turbulent pipe flow

When we select a pump, the pressure change must be calculated if the loss heads including frictional loss and minor loss are known in a developed flow, by energy equation

$$h_L = h_f + h_j = \frac{\Delta p}{\rho g} \tag{6.4.3}$$

The friction loss in a developed turbulent pipe flow is related to the friction factor λ, which depends on the Reynolds number and the relative roughness ε/d. The functional form of this dependence are all obtained from experiments. The friction factor can be calculated based on experimental data.

For hydraulically smooth pipe flow

$$\frac{1}{\sqrt{\lambda}} = 2\lg\left(Re\sqrt{\lambda}\right) - 0.8 \tag{6.4.4}$$

For hydraulically rough pipe flow

$$\frac{1}{\sqrt{\lambda}} = 2\lg\left(\frac{1}{2} \Big/ \frac{\varepsilon}{d}\right) + 1.74 \tag{6.4.5}$$

Equivalent roughness values for some commercial pipes are given in Tab.6.4.1, note that these values are for new pipes, and the relative roughness of pipes may increase with use as a result of corrosion, scale buildup, and precipitation.

实际上，粗糙度是一个相对的概念，平均粗糙高度 ε 和黏性底层的厚度的关系更能说明问题。如果黏性底层厚度相对较小，管壁粗糙凸起就会突出于黏性层，如图 6.4.3b 所示，此时黏性底层不再连续，流体也不能被视为在光滑的管道内流动，这种状态被称为**水力粗糙**，水力粗糙流动的摩擦损失与雷诺数无关。由图 6.4.2 可见，水力光滑流动的速度曲线中心部分较水力粗糙流动更加平、宽。同样可见，层流的速度曲线形状也与管壁粗糙度无关。

管壁粗糙度和管道直径之比 ε/d 称为相对粗糙度，通常可以用相对粗糙度和雷诺数判断水力光滑或水力粗糙流动。下一小节将介绍相对粗糙度与沿程损失系数之间的关系。

黏性底层和紊流核心区之间还存在一个**过渡层**。这一层的流态受扰动影响，可能为层流也可能为紊流，或者在二者之间切换。过渡层通常被划入紊流核心区。

6.4.2　圆管紊流的损失

选用泵时，必须考虑沿程损失和局部损失导致的总水头损失（压降），对充分发展流动，可用下式计算总水头损失

$$h_L = h_f + h_j = \frac{\Delta p}{\rho g} \qquad (6.4.3)$$

圆管紊流的沿程损失取决于沿程损失系数 λ，实验证明沿程损失系数由雷诺数和相对粗糙度 ε/d 决定，其经验关系式为

水力光滑流动：

$$\frac{1}{\sqrt{\lambda}} = 2\lg\left(Re\sqrt{\lambda}\right) - 0.8 \qquad (6.4.4)$$

水力粗糙流动：

$$\frac{1}{\sqrt{\lambda}} = 2\lg\left(\frac{1}{2}\bigg/\frac{\varepsilon}{d}\right) + 1.74 \qquad (6.4.5)$$

表 6.4.1 给出了一些常用材料管道的表面粗糙度，表内数值均为新管时的数值，如果管道已经使用发生了锈蚀、结垢和沉积等现象，粗糙度值会有所增大。

Table 6.4.1   Equivalent roughness values ε for new commercial pipes

Material	Roughness/mm
Glass, plastic	0.0015
Smoothed copper or brass tubing	0.0015
Stainless steel	0.002
Rubber	0.01
Commercial steel, welded-steel pipe	0.046
Galvanized iron	0.15
Cast iron	0.26
Concrete	$0.5 \sim 3$
Asphalt-dipped cast iron	0.12

The friction factor for pipe flow can also be determined by using the Mood chart, which is one of the most widely used in engineering. The Mood chart is given in the Appendix B.

We now know how to calculate the frictional losses due to wall shear in a pipe. In general, the losses in gradually various flow are very small, relatively large losses are associated with sudden enlargements or contractions. Pipe systems, however, include enlargements, contractions, as well as valves, elbows, bends, and other fittings that cause additional losses, referred to as *minor losses*, such losses can even exceed the frictional losses.

A minor loss is expressed in terms of a loss coefficient ζ, defined by

$$h_j = \zeta \frac{V^2}{2g} \tag{6.4.6}$$

Values of ζ are determined experimentally for the various fittings and geometry changes of interest in piping systems. The loss coefficients for various geometries are presented in Tab.6.4.2.

Tab.6.4.2   Nominal Loss Coefficients ζ (Turbulent Flow)

Type of fitting	Sketch	Loss coefficient ζ	Formula
Sudden enlargement		$\zeta = \left(\dfrac{A_2}{A_1} - 1 \right)^2$	$h_j = \zeta \dfrac{V_2^2}{2g}$
		$\zeta' = \left(1 - \dfrac{A_1}{A_2} \right)^2$	$h_j = \zeta' \dfrac{V_1^2}{2g}$
Gradually enlargement		$\zeta = k \left(\dfrac{A_2}{A_1} - 1 \right)^2 + \dfrac{\lambda}{8 \tan \frac{\theta}{2}} \left[\left(\dfrac{A_2}{A_1} \right)^2 - 1 \right]$	$h_j = \zeta \dfrac{V_1^2}{2g}$

θ	8°	10°	12°	15°	20°	25°
k	0.14	0.16	0.22	0.30	0.42	0.62

表 6.4.1　常用管道（新）的平均表面粗糙度 ε

材料	粗糙度 /mm
玻璃，塑料	0.0015
光滑的铜管	0.0015
不锈钢	0.002
橡胶	0.01
焊接钢管	0.046
镀锌管	0.15
铸铁管	0.26
水泥管	0.5 ~ 3
沥青漆铸铁管	0.12

工程中也常用莫迪图来确定管流的沿程损失系数。莫迪图见附录 B。

我们可以通过以上公式和参数计算沿程损失。一般来说，缓变流的沿程损失很小，相对而言管径突变，包括突扩和突缩，导致的损失更加明显。管流中的管径的扩大或缩小，还有阀门、弯头、接头等产生的能量损失，可能会大于沿程损失，这些损失称为**局部损失**。

局部损失可以由局部损失系数 ζ 计算，计算式为

$$h_j = \zeta \frac{V^2}{2g} \tag{6.4.6}$$

局部损失系数的取值与流场几何特征的变化情况有关，可由实验获得，表 6.4.2 列出了不同管流几何参数变化对应的局部损失系数。

表 6.4.2　紊流的局部损失系数 ζ

管件名称	简图	局部损失系数 ζ						公式
突扩		$\zeta = \left(\dfrac{A_2}{A_1} - 1 \right)^2$						$h_j = \zeta \dfrac{V_2^2}{2g}$
		$\zeta' = \left(1 - \dfrac{A_1}{A_2} \right)^2$						$h_j = \zeta' \dfrac{V_1^2}{2g}$
渐扩		$\zeta = k\left(\dfrac{A_2}{A_1} - 1 \right)^2 + \dfrac{\lambda}{8\tan\dfrac{\theta}{2}}\left[\left(\dfrac{A_2}{A_1} \right)^2 - 1 \right]$						$h_j = \zeta \dfrac{V_1^2}{2g}$
		θ	8°	10°	12°	15°	20°	25°
		k	0.14	0.16	0.22	0.30	0.42	0.62

（续）

Type of fitting	Sketch	Loss coefficient ζ	Formula
Sudden contraction		$\zeta = 0.5\left(1 - \dfrac{A_2}{A_1}\right)$	$h_j = \zeta \dfrac{V_2^2}{2g}$

| Gradually contraction | | $\zeta = k\left(\dfrac{1}{\varepsilon}-1\right)^2 + \dfrac{\lambda}{8\tan\dfrac{\theta}{2}}\left(1-\dfrac{A_2^2}{A_1^2}\right)$

$\varepsilon = 0.57 + \dfrac{0.043}{1.1-\left(A_2/A_1\right)}$ | $h_j = \zeta \dfrac{V_1^2}{2g}$ |

θ	10°	20°	40°	80°	100°	140°
k	0.4	0.25	0.2	0.3	0.4	0.6

Type of fitting	Sketch	Loss coefficient ζ	Formula
Square-edged entrance		$\zeta = 0.5$	
Well-rounded entrance		$\zeta = 0.2$	$h_j = \zeta \dfrac{V_2^2}{2g}$

Reentrant entrance

δ/D	b/D				
	0	0.002	0.01	0.05	0.5
0	0.5	0.57	0.63	0.80	1.00
0.008	0.5	0.53	0.58	0.74	0.88
0.016	0.5	0.51	0.53	0.58	0.77
0.024	0.5	0.50	0.51	0.53	0.68
0.030	0.5	0.50	0.51	0.52	0.61
0.050	0.5	0.50	0.50	0.50	0.53

Orifice outlet

A_2/A_1	0.1	0.2	0.3	0.4	0.5
ζ	268	66.5	28.9	15.5	9.81
A_2/A_1	0.6	0.7	0.8	0.9	
ζ	5.8	3.7	2.38	1.56	

$h_j = \zeta \dfrac{V_1^2}{2g}$

（续）

管件名称	简图	局部损失系数 ζ	公式
突缩		$\zeta = 0.5\left(1 - \dfrac{A_2}{A_1}\right)$	$h_j = \zeta \dfrac{V_2^2}{2g}$
渐缩		$\zeta = k\left(\dfrac{1}{\varepsilon} - 1\right)^2 + \dfrac{\lambda}{8\tan\dfrac{\theta}{2}}\left(1 - \dfrac{A_2^2}{A_1^2}\right)$ $\varepsilon = 0.57 + \dfrac{0.043}{1.1 - (A_2/A_1)}$ <table><tr><td>θ</td><td>10°</td><td>20°</td><td>40°</td><td>80°</td><td>100°</td><td>140°</td></tr><tr><td>k</td><td>0.4</td><td>0.25</td><td>0.2</td><td>0.3</td><td>0.4</td><td>0.6</td></tr></table>	$h_j = \zeta \dfrac{V_1^2}{2g}$
锐边入口		$\zeta = 0.5$	
圆角入口		$\zeta = 0.2$	$h_j = \zeta \dfrac{V_2^2}{2g}$

突出入口

δ/D	b/D				
	0	0.002	0.01	0.05	0.5
0	0.5	0.57	0.63	0.80	1.00
0.008	0.5	0.53	0.58	0.74	0.88
0.016	0.5	0.51	0.53	0.58	0.77
0.024	0.5	0.50	0.51	0.53	0.68
0.030	0.5	0.50	0.51	0.52	0.61
0.050	0.5	0.50	0.50	0.50	0.53

孔口出流

A_2/A_1	0.1	0.2	0.3	0.4	0.5
ζ	268	66.5	28.9	15.5	9.81
A_2/A_1	0.6	0.7	0.8	0.9	
ζ	5.8	3.7	2.38	1.56	

$h_j = \zeta \dfrac{V_1^2}{2g}$

（续）

Type of fitting	Sketch	Loss coefficient ζ						Formula
Orifice plate		$\zeta=\left(\dfrac{A_2}{\varepsilon A_1}-1\right)^2$, $\varepsilon=0.57+\dfrac{0.043}{1.1-A_2/A_1}$						$h_j=\zeta\dfrac{V_2^2}{2g}$

Type of fitting	Sketch	Loss coefficient ζ						Formula
Gate valve		h/d	Fully open	6/8	4/8	3/8	2/8	1/8	$h_j=\zeta\dfrac{V^2}{2g}$
		ζ	0.11	0.26	2.06	5.52	17.0	97.8	

Type of fitting	Sketch	Loss coefficient ζ					Formula
Bend		$\zeta_\theta=\zeta_{90°}(1-\cos\theta)$					$h_j=\zeta\dfrac{V^2}{2g}$
		D	20	25	34	39	49
		$\zeta_{90°}$	1.7	1.3	1.1	1.0	0.83

Type of fitting	Sketch	Loss coefficient ζ	Formula
Elbow		$\zeta_\theta=\zeta_{90°}\alpha$ $\zeta_{90°}=\left[0.20+0.001(100\lambda)^3\right]\sqrt{d/R}$ $\theta<90°,\alpha=\sin\theta$ $\theta>90°,\alpha=0.7+0.35\dfrac{\theta}{90°}$	$h_j=\zeta\dfrac{V^2}{2g}$

Type of fitting	Sketch	Loss coefficient ζ	Formula
Tees		$h_{1-3}=2\dfrac{V_3^2}{2g}$, $h_{1-2}=\dfrac{V_1^2-V_2^2}{2g}$	

In piping systems involving intermediate lengths ($L\approx100d$) of pipe between minor losses, the minor losses may be of the same order of magnitude as the frictional losses; for relatively short lengths the minor losses may be substantially greater than the frictional losses; and for long lengths ($L>1000d$) of pipe, the minor losses are usually neglected.

Example 6.4.1 Water at 20℃ is transported for 500m in a 40-mm-diameter welded-steel horizontal pipe with a flow rate of 3L/s. Calculate the pressure drop over the 500m length of pipe, and the required pumping power input for flow of the pipe.

Solution The average velocity in the pipe is

$$V=\frac{4q}{\pi d^2}=2.39\text{m/s}$$

The Reynolds number is

$$Re=\frac{dV}{v}=\frac{0.04\times2.39}{10^{-6}}=95.6\times10^3$$

The roughness of new welded-steel pipe is 0.046mm.

（续）

管件名称	简图	局部损失系数 ζ						公式
孔板	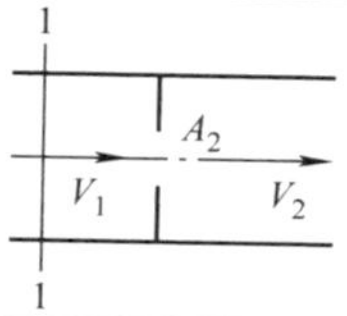	$\zeta = \left(\dfrac{A_2}{\varepsilon A_1} - 1\right)^2$, $\varepsilon = 0.57 + \dfrac{0.043}{1.1 - A_2/A_1}$						$h_j = \zeta \dfrac{V_2^2}{2g}$
闸阀	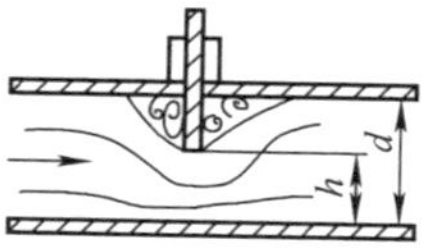	h/d	全开	6/8	4/8	3/8	2/8	1/8	$h_j = \zeta \dfrac{V^2}{2g}$
		ζ	0.11	0.26	2.06	5.52	17.0	97.8	
弯管		$\zeta_\theta = \zeta_{90°}(1 - \cos\theta)$						$h_j = \zeta \dfrac{V^2}{2g}$
		D	20	25	34	39	49	
		$\zeta_{90°}$	1.7	1.3	1.1	1.0	0.83	
弯头	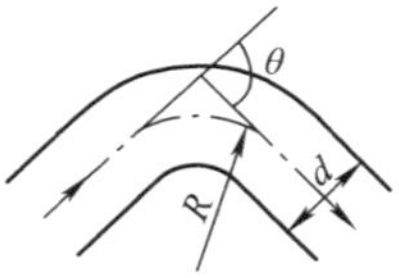	$\zeta_\theta = \zeta_{90°}\alpha$ $\zeta_{90°} = \left[0.20 + 0.001(100\lambda)^3\right]\sqrt{d/R}$ $\theta < 90°, \alpha = \sin\theta$ $\theta > 90°, \alpha = 0.7 + 0.35\dfrac{\theta}{90°}$						$h_j = \zeta \dfrac{V^2}{2g}$
三通	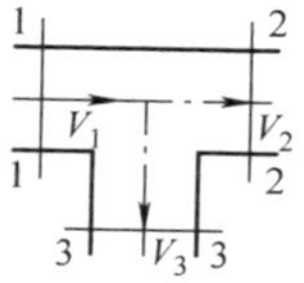	$h_{1\text{-}3} = 2\dfrac{V_3^2}{2g}$, $h_{1\text{-}2} = \dfrac{V_1^2 - V_2^2}{2g}$						

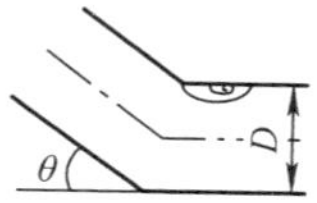

中等长度的管道（$L \approx 100d$）中局部损失可能与沿程损失大小相当；相对短些的管道中局部损失的能量可能高于沿程损失；而长管（$L > 1000d$）的局部损失通常可以忽略。

例 6.4.1　用水平的长度 500m 直径 40mm 焊接钢管输水，水温 20℃，流量 3L/s。计算（1）500m 长度管道的压降；（2）应选用泵的功率。

解　（1）管内平均流速

$$V = \frac{4q}{\pi d^2} = 2.39\,\text{m/s}$$

雷诺数

$$Re = \frac{dV}{\nu} = \frac{0.04 \times 2.39}{10^{-6}} = 95.6 \times 10^3$$

焊接钢管（有缝钢管）的粗糙度为 0.046mm

By Eq.6.4.2, the thickness of laminar sublayer is

$$\delta = 34.2 \frac{d}{Re^{0.875}} = 6 \times 10^{-5}\,\text{m} < \varepsilon$$

It is hydraulically rough flow in the pipe.
The friction factor is calculated by Eq.6.4.5 gives

$$\frac{1}{\sqrt{\lambda}} = 2\lg\left(\frac{1}{2}\Big/\frac{\varepsilon}{d}\right) + 1.74 \quad \rightarrow \quad \lambda = 0.020$$

The head loss is calculated from Darcy's formula as

$$h_f = \lambda \frac{L}{d}\frac{V^2}{2g} = 72.78\text{m}$$

The pressure drop is found by Eq.6.4.3 to be

$$\Delta p = \rho g h_f = 714\text{kPa}$$

The power needed to overcome the frictional losses is

$$W = q\Delta p = 3 \times 10^{-3} \times 714\text{kW} = 2.142\text{kW}$$

Example 6.4.2   Water at 20 ℃ is transported in a 50mm-diameter plastic pipe with a flow rate of 0.005m³/s. Minor loss at a square bend is measured and a 90mm height difference is read (Fig.X6.4.2). The measuring liquid in the manometer is carbon tetrachloride with a density of 1600kg/m³. Find the minor loss factor at this bend.
Solution   The average velocity in the pipe is

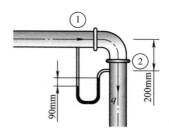

$$V = \frac{4q}{\pi d^2} = 2.55\text{m/s}$$

Fig.X6.4.2

Using energy equation between section ① and ②

$$z_1 + \frac{V_1^2}{2g} + \frac{p_1}{\rho g} = z_2 + \frac{V_2^2}{2g} + \frac{p_2}{\rho g} + h_j \quad \rightarrow \quad 0.2 + \frac{\Delta p}{\rho g} = h_j$$

The pressure drop between these two sections is measured

$$\Delta p = (\rho' - \rho)g\Delta h$$

The minor loss factor can be calculated from

根据式 6.4.2，黏性底层厚度为

$$\delta = 34.2 \frac{d}{Re^{0.875}} = 6 \times 10^{-5} \, \text{m} < \varepsilon$$

为水力粗糙流。

由式 6.4.5 计算沿程损失系数，得

$$\frac{1}{\sqrt{\lambda}} = 2 \lg \left(\frac{1}{2} \Big/ \frac{\varepsilon}{d} \right) + 1.74 \quad \rightarrow \quad \lambda = 0.020$$

由达西公式计算的沿程损失水头为

$$h_f = \lambda \frac{L}{d} \frac{V^2}{2g} = 72.78 \, \text{m}$$

由式 6.4.3 计算管内压强差

$$\Delta p = \rho g h_f = 714 \text{kPa}$$

（2）克服该压力损失所需的功率为

$$W = q \Delta p = 3 \times 10^{-3} \times 714 \text{kW} = 2.142 \text{kW}$$

例 6.4.2　用 50mm 直径的塑料管输送 20℃的水，流量为 0.005m³/s。在直角转弯处测量局部损失水头，测压管读数为 90mm（见图 X6.4.2），测压管内工作液体为四氯化碳，密度为 1600kg/m³。计算弯管处局部损失系数。

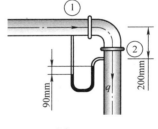

图 X6.4.2

解　管内平均流速为

$$V = \frac{4q}{\pi d^2} = 2.55 \text{m/s}$$

在截面①—②间建立能量方程

$$z_1 + \frac{V_1^2}{2g} + \frac{p_1}{\rho g} = z_2 + \frac{V_2^2}{2g} + \frac{p_2}{\rho g} + h_j \quad \rightarrow \quad 0.2 + \frac{\Delta p}{\rho g} = h_j$$

两截面间压强差为

$$\Delta p = (\rho' - \rho) g \Delta h$$

局部损失水头

$$h_{\mathrm{m}} = 0.2 + \frac{p'-p}{\rho}\ \Delta h = \zeta \frac{V^2}{2g}$$

The result is

$$\zeta = 0.766$$

Example 6.4.3 Estimate the flow rate in the piping system of Fig.X6.4.3a if the pump q-p curve and efficiency curve are as shown in Fig.X6.4.3b. Also, find the pump power requirement.

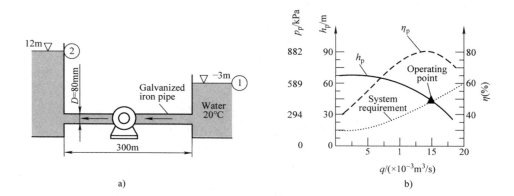

Fig.X6.4.3

Solution Assume that the Reynolds number is sufficiently large and the flow is hydraulically rough turbulent. So that the friction factor can be calculated

$$\frac{1}{\sqrt{\lambda}} = 2\lg\left(\frac{1}{2}\Big/\frac{\varepsilon}{d}\right) + 1.74 \quad \rightarrow \quad \frac{1}{\sqrt{\lambda}} = 2\lg\left(\frac{1}{2}\Big/\frac{0.15}{80}\right) + 1.74 \quad \rightarrow \quad \lambda = 0.023$$

The minor losses occur at the outlet (Sudden contraction) and the entrance (Square-edged entrance) from Tab.6.4.2, are

$$\zeta_{\mathrm{out}} = 0.5, \quad \zeta_{\mathrm{in}} = 0.5$$

The energy equation is applied between the two free surfaces, we have

$$z_1 + \frac{\cancel{p_1}}{\rho g} + \frac{\cancel{V_1^2}}{2g} + h_{\mathrm{p}} = z_2 + \frac{\cancel{p_2}}{\rho g} + \frac{\cancel{V_2^2}}{2g} + \zeta_1 \frac{V_1^2}{2g} + \zeta_2 \frac{V_2^2}{2g} + \lambda \frac{L}{d}\frac{V^2}{2g}$$

$$h_{\mathrm{p}} = z_2 - z_1 + \left(\zeta_1 + \zeta_2 + \lambda \frac{L}{d}\right)\frac{V^2}{2g} \quad \rightarrow \quad h_{\mathrm{p}} = 15 + \left(1 + \lambda \frac{L}{d}\right)\frac{V^2}{2g} = 15\mathrm{m} + 176364q^2$$

The curves of pump characteristic $h_{\mathrm{p}\text{-}q}$ and system requirement $h_{\mathrm{p}\text{-}q}$ intersect at operating point. Then it yields the pressure head and discharge of the pump

$$h_{\mathrm{m}} = 0.2 + \frac{\rho' - \rho}{\rho} \Delta h = \zeta \frac{V^2}{2g}$$

局部损失系数

$$\zeta = 0.766$$

例 6.4.3　图 X6.4.3a 所示的输水管，水泵流量 - 压力特性曲线和功率曲线如图 X6.4.3b 所示。求水泵驱动功率。

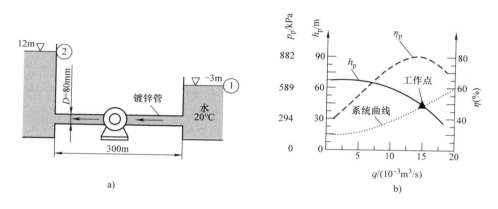

图 X6.4.3

解　假设雷诺数很大，管内为水力粗糙紊流。则沿程损失系数为

$$\frac{1}{\sqrt{\lambda}} = 2\lg\left(\frac{1}{2}\bigg/\frac{\varepsilon}{d}\right) + 1.74 \;\rightarrow\; \frac{1}{\sqrt{\lambda}} = 2\lg\left(\frac{1}{2}\bigg/\frac{0.15}{80}\right) + 1.74 \;\rightarrow\; \lambda = 0.023$$

局部损失主要发生在出口（突缩）和入口（锐边突扩），查表 6.4.2 得

$$\zeta_{\mathrm{out}} = 0.5, \quad \zeta_{\mathrm{in}} = 0.5$$

在两自由液面间建立能量方程，有

$$z_1 + \frac{p_1}{\rho g} + \frac{V_1^2}{2g} + h_{\mathrm{p}} = z_2 + \frac{p_2}{\rho g} + \frac{V_2^2}{2g} + \zeta_1 \frac{V_1^2}{2g} + \zeta_2 \frac{V_2^2}{2g} + \lambda \frac{L}{d} \frac{V^2}{2g}$$

$$h_{\mathrm{p}} = z_2 - z_1 + \left(\zeta_1 + \zeta_2 + \lambda \frac{L}{d}\right)\frac{V^2}{2g} \;\rightarrow\; h_{\mathrm{p}} = 15 + \left(1 + \lambda \frac{L}{d}\right)\frac{V^2}{2g} = 15 + 176364 q^2$$

系统工作曲线 $h_{\mathrm{p}\text{-}q}$ 和泵性能曲线 $h_{\mathrm{p}\text{-}q}$ 的交点为工作点，则泵输出压力水头和流量

339

$$H_p = 45\text{m}, \quad q = 0.013\text{m}^3/\text{s}$$

Check the Reynolds number, $Re = 2.39 \times 10^6$, this is sufficiently large, so that our assumption of turbulent is reasonable.

The power requirement of the pump is given by

$$W = \frac{\rho g H_p q}{\eta_p} = 7.2\text{kW}$$

where the efficiency of pump $\eta_p = 0.8$ is found from the characteristic curve at $q = 0.013\text{m}^3/\text{s}$.
Note that, since $L/d \gg 1000$, minor losses due to the entrance and exit are negligible.

6.5   Flow at Orifices, Tubes and Slender Holes

An orifice is an opening (usually circular) in the wall of a tank or in a plate normal to the axis of a pipe, the plate being either at the end of the pipe or in some intermediate location.

6.5.1   Velocity and discharge at orifices

A thin-wall orifice is characterized by the fact that the thickness of the wall to size of the opening ratio is very small, usually, $L/d < 0.5$. A standard thin-wall orifice is one with a sharp edge as in Fig.6.5.1a or an absolutely square shoulder as in Fig.6.5.1b, so that there is only a line contact with the fluid.

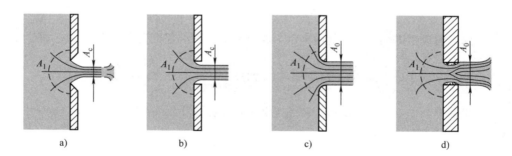

a)   b)   c)   d)

Fig.6.5.1   Orifices

a) sharp-sdged thin-wall orifice   b) square shoulder thin-wall orifice   c) round-edged orifice
d) thick-wall orifice

The jet contracts where the streamlines converge in approaching a thin-wall orifice, as shown in Fig.6.5.2. The streamlines continue to converge beyond the upstream section of the orifice until they reach the section A_c, where they become parallel. The section A_c is then a section of minimum area. Beyond A_c the streamlines commonly diverge. Therefore the only minor loss at the thin-wall orifice due to the contraction of streamlines, and the friction can be considered does not exist.

$$H_\mathrm{p} = 45\mathrm{m}, \quad q = 0.013\mathrm{m^3/s}$$

复核雷诺数，$Re = 2.39 \times 10^6$，紊流假设成立。

由水泵性能曲线查得，流量 $q = 0.013\mathrm{m^3/s}$ 时水泵效率 $\eta_\mathrm{p} = 0.8$

水泵所需驱动功率为

$$W = \frac{\rho g H_\mathrm{p} q}{\eta_\mathrm{p}} = 7.2\mathrm{kW}$$

由于水管的长径比 $L/d \gg 1000$，为长管，所以进、出口处的局部损失可以忽略。

6.5　孔口、管口和细长管出流

孔口常于罐体壁面或与管路垂直的孔板处，通常为圆形，孔板可位于管道端点或中间位置。

6.5.1　孔口流速和流量

薄壁小孔是指孔板厚度与孔口直径之比很小的薄壁孔，通常孔的长径比 $L/d < 0.5$。标准的薄壁小孔如图 6.5.1a 所示有锐利的边缘，或如图 6.5.1b 所示为方肩边缘。总之，流体与孔口仅为线接触。

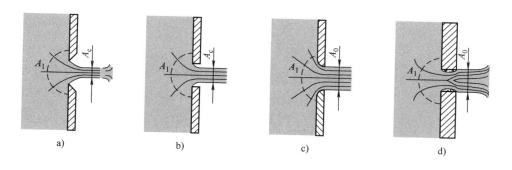

图 6.5.1　孔口

a）锐边薄壁孔　b）方肩薄壁孔　c）圆边薄壁孔　d）厚壁孔

如图 6.5.2 所示，液流经过薄壁孔时流线会发生收缩，而且由于惯性，流线在经过孔口后会继续收缩一段距离，直至截面面积收缩至 A_c 之后，流线变为平行。所以面积 A_c 为流线的最小截面积，离开 A_c 截面较远距离后，流线逐渐发散。所以，对于薄壁小孔，仅有由于流线收缩而产生的局部损失，没有沿程损失。

The area of orifice is denoted by A_0, and at sections A_1 in Fig.6.5.2 where the streamlines are curved, the effective cross-sectional area is greater than at the orifice section A_0 and the minimum section A_c, and hence the average velocity at section A_1 is considerably less than the jet velocity. Jet velocity is defined as the average velocity at the minimum section A_c in Fig.6.5.1a and b, and at the downstream edge of the orifices in Fig.6.5.1c and d.

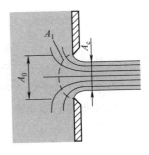

Fig.6.5.2   Jet contraction

Now let us derive the jet velocity and discharge at a thin-wall orifice. Bernoulli equation is applied between section A_1 and A_c, that is

$$\frac{p_1}{\rho g} + \frac{V_1^2}{2g} = \frac{p_c}{\rho g} + \frac{V_c^2}{2g} + \zeta \frac{V_c^2}{2g} \tag{6.5.1}$$

From continuum equation

$$V_1 A_1 = V_c A_c = V_c C_c A_0 \tag{6.5.2}$$

where the ratio of the area A_c of a jet to the area A_0 of the orifice is called the coefficient of contraction, then $A_c = C_c A_0$.

Solving for V_c yields

$$V_c = \frac{1}{\sqrt{1 - \left(\frac{C_c A_0}{A_1}\right)^2 + \zeta}} \sqrt{\frac{2(p_1 - p_c)}{\rho}} \tag{6.5.3}$$

The quadratic term under the root sign is negligible since $A_c \ll A_1$, and the Eq.6.5.3 can be written as

$$V_c = \frac{1}{\sqrt{1 + \zeta}} \sqrt{\frac{2(p_1 - p_c)}{\rho}} = C_v \sqrt{\frac{2\Delta p}{\rho}} \tag{6.5.4}$$

where $C_v = (1 + \zeta)^{-1/2}$ is called the coefficient of velocity.

The discharge at a thin-wall orifice is given from

$$q = V_c A_c = C_v C_c A_0 \sqrt{\frac{2\Delta p}{\rho}} = C_d A_0 \sqrt{\frac{2\Delta p}{\rho}} \tag{6.5.5}$$

where $C_d = C_c C_v$ is defined as the coefficient of discharge.

假设孔口的截面积为 A_0，图 6.5.2 中流线开始弯曲处的截面积为 A_1。截面 A_1 处的有效截面积远远大于孔口截面积 A_0 和最小收缩截面积 A_c，因此可以假设截面 A_1 处的流速远远小于孔口出流速度。对于图 6.5.1a、b 中的薄壁孔，定义最小截面 A_c 处的流速为出流速度，而图 6.5.1c、d 中，定义小孔出口边缘处的流速为出流速度。

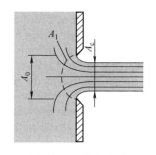

图 6.5.2　射流收缩

首先推导薄壁小孔的出流速度和流量。在截面 A_1 和 A_c 间建立伯努利方程，有

$$\frac{p_1}{\rho g}+\frac{V_1^2}{2g}=\frac{p_c}{\rho g}+\frac{V_c^2}{2g}+\zeta\frac{V_c^2}{2g} \tag{6.5.1}$$

根据连续性方程，可知

$$V_1 A_1 = V_c A_c = V_c C_c A_0 \tag{6.5.2}$$

式中，取收缩截面面积 A_c 和孔口截面面积 A_0 之比为收缩系数 C_c，则 $A_c=C_c A_0$。求解 V_c 得

$$V_c=\frac{1}{\sqrt{1-\left(\dfrac{C_c A_0}{A_1}\right)^2+\zeta}}\sqrt{\frac{2\left(p_1-p_c\right)}{\rho}} \tag{6.5.3}$$

因为 $A_c \ll A_1$，所以根号下的二次方项非常小，可以忽略，则式 6.5.3 可化简为

$$V_c=\frac{1}{\sqrt{1+\zeta}}\sqrt{\frac{2\left(p_1-p_c\right)}{\rho}}=C_v\sqrt{\frac{2\Delta p}{\rho}} \tag{6.5.4}$$

式中，$C_v=(1+\zeta)^{-1/2}$ 称为流速系数。

薄壁小孔的出口流量为

$$q=V_c A_c = C_v C_c A_0\sqrt{\frac{2\Delta p}{\rho}}=C_d A_0\sqrt{\frac{2\Delta p}{\rho}} \tag{6.5.5}$$

式中，$C_d=C_c C_v$，称为出流系数。

Typical values of C_c, C_v and C_d are listed in Tab.6.5.1 for different type of orifices.

Those shown in Fig.6.5.1c and Fig.6.5.1d are not standard thin-wall orifice, because the flow through them is affected by the thickness of the plate, the roughness of the surface, and, for Fig.6.5.1c, the radius of curvature. The orifice as shown in Fig.6.5.1c, there is no contraction as the rounded entry to the opening permits the streamlines to gradually converge to the cross-sectional area of the orifice, will always be considered as a thick-wall orifice, even the wall is not really thick.

Hence such thick-wall orifices should not be applied if high accuracy is desired.

Tab.6.5.1 Value of coefficients for different type of orifices

	C_c	C_v	C_d
Sharp-edged thin-wall orifice	0.62	0.98	0.61
Square shoulder orifice	0.62	0.98	0.61
Round-edged orifice	1.0	0.86	0.86
Thick-wall orifice	1.0	0.98	0.98

6.5.2 Velocity and discharge at tubes and slender holes

A tube is a short pipe whose length is about two or three diameters, usually $0.5 < L/d < 4$. There is no sharp distinction between a tube and a thick-walled orifice, especially that frictions and minor losses exist and affect both types of flow. We see from Fig.6.5.3, streamlines contract at the entrance of the tube to the minimum area at A_c, and then diverge and will be influenced by the wall shear.

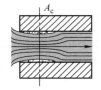

Fig.6.5.3 Jet flow contracts and diverges in a tube

Average velocity at cross section of a tube can be expressed as

$$V = \sqrt{\frac{2\Delta p}{\rho}} \bigg/ \sqrt{(1+\zeta) + \lambda \frac{l}{d} - \left(\frac{C_c A_0}{A_1}\right)^2} = C_v \sqrt{\frac{2\Delta p}{\rho}} \tag{6.5.6}$$

Here the coefficient of velocity C_v is composed of frictional loss
While the discharge has the same expression with orifices, that is

$$q = V_c A_c = C_c C_v A_0 \sqrt{\frac{2\Delta p}{\rho}} = C_d A_0 \sqrt{\frac{2\Delta p}{\rho}} \tag{6.5.7}$$

The value of coefficient of discharge for a turbulent flow in a tube where $Re > 4000$ is about $C_d \approx 0.8$.

Flow in a slender hole (deep hole) can be considered as a part of circular pipe flow. Discharge of viscous laminar flow in a circular pipe (see 6.3) is represented by

$$q = \frac{\pi d^4}{128 \mu l} \Delta p \tag{6.5.8}$$

We can find that the discharge in a circular pipe as well as a slender hole is inversely proportional to the viscosity of fluid, which is sensitive to temperature. This is why a hydraulic element would prefer a thin-wall orifice than a slender hole.

各种典型孔口的 C_c、C_v 和 C_d 值见表 6.5.1。

图 6.5.1c 和图 6.5.1d 中的孔口不是标准的薄壁小孔，因为经这种孔口的出流会受孔板厚度、孔壁粗糙度甚至图 6.5.1c 中边缘半径的影响。图 6.5.1c 中由于孔板前至孔口处圆滑过渡，流线在圆弧面上逐渐收缩至与孔口截面积相同，与典型的薄壁孔不符，所以尽管其孔壁厚度并不大，依然被视为厚壁孔。厚壁孔不适合用于有高精度要求的场合。

表 6.5.1　各种孔口的系数值

	C_c	C_v	C_d
锐边薄壁孔	0.62	0.98	0.61
方肩薄壁孔	0.62	0.98	0.61
圆边孔	1.0	0.86	0.86
厚壁孔	1.0	0.98	0.98

6.5.2　管口和细长管出流

管口指的是长度约为直径的 2 倍或 3 倍的短管，通常 $0.5 < L/d < 4$，其与厚壁孔的区别并不明显，共同点在于二者都存在沿程损失和局部损失。由图 6.5.3 可以看出，流线在进入孔口后收缩至最小截面 A_c，然后开始发散并与孔壁发生剪切摩擦。

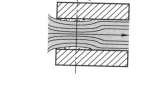

图 6.5.3　短孔内射流的
收缩和发散

管口截面的平均流速计算式为

$$V = \sqrt{\frac{2\Delta p}{\rho}} \bigg/ \sqrt{(1+\zeta) + \lambda\frac{l}{d} - \left(\frac{C_c A_0}{A_1}\right)^2} = C_v\sqrt{\frac{2\Delta p}{\rho}} \qquad (6.5.6)$$

式中的流速系数包含了局部损失和沿程的影响，$C_v = (1+\zeta+\lambda L/d)^{-1/2}$。

管口流量表达式与孔口出流的流量式具有相同形式为

$$q = V_c A_c = C_c C_v A_0\sqrt{\frac{2\Delta p}{\rho}} = C_d A_0\sqrt{\frac{2\Delta p}{\rho}} \qquad (6.5.7)$$

当雷诺数大于 4000 时，管口出流的出流系数 $C_d \approx 0.8$。

细长孔可以视为一段圆管，适用圆管黏性层流的流量表达式（见 6.3 节）：

$$q = \frac{\pi d^4}{128\mu l}\Delta p \qquad (6.5.8)$$

由上式可见，细长孔和圆管的通流量与流体黏度成反比，而黏度对温度非常敏感。这也是液压元件中应该选用薄壁小孔通流而非细长孔的主要原因。

Applications: Discharge at opening of hydraulic valves.

a. Discharge of sloop valves

A sloop valve, shown in Fig.A6.5.1a, with a d-diameter core makes an x_T-wide opening when the core slides leftward. The gap between core and seat is C_T. Discharge of this sloop valve can be determined by a thin-wall orifice form.

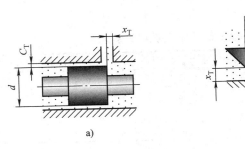

Fig.A6.5.1   Discharge of hydraulic valves at an orifice

a) sloop valve   b) poppet valve

From Eq.6.5.5, the discharge

$$q = C_d A_0 \sqrt{\frac{2\Delta p_c}{\rho}}$$

Where the area of flow is given by

$$A_0 = \pi d \sqrt{x_T^2 + C_T^2} \approx \pi d k x_T$$

where the factor $k \geqslant 1$.

Then the discharge at sloop valve is

$$q = C_d w x_T \sqrt{\frac{2\Delta p}{\rho}}, \text{ where } w = k\pi d$$

Note that the discharge at a sloop valve is directly proportional to the width of opening x_T.

b. Discharge of poppet valves

A poppet valve, shown in Fig.A6.5.1b, with a 2ϕ-cone angle core makes an x_T-wide opening when the core is pushed upward. Discharge of this poppet valve can be determined also by an orifice form.

From Eq.6.5.5, the discharge

$$q = C_d A_0 \sqrt{\frac{2\Delta p_c}{\rho}}$$

Where the area of flow is given by

$$A_0 = \pi \frac{d_1 + d_2}{2} h = \pi d_m x_T \sin\phi$$

应用案例：液压阀阀口通流量

a. 滑阀出流量

分析图 A6.5.1a 中的滑阀。当直径为 d 的阀芯向左滑动时，阀口开启形成宽度为 x_T 的开口，阀芯和阀座间的间隙为 C_T。滑阀开口的通流量适用薄壁孔口流量计算式。

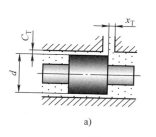
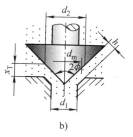

图 A6.5.1　液压阀阀口通流量

a）滑阀　b）锥阀

根据式 6.5.5，通流量为

$$q = C_d A_0 \sqrt{\frac{2\Delta p_c}{\rho}}$$

阀口通流截面积为

$$A_0 = \pi d \sqrt{x_T^2 + C_T^2} \approx \pi d k x_T$$

式中，系数 $k \geqslant 1$。

得到滑阀通流量计算式

$$q = C_d w x_T \sqrt{\frac{2\Delta p}{\rho}}$$

式中，$w = k\pi d$。

可知，滑阀通流量与阀口开启宽度 x_T 成正比。

b. 锥阀通流量

如图 A6.5.1b 所示的锥阀，阀芯锥角为 2ϕ，当油压将锥阀芯向上推动时可形成宽度为 x_T 的开口。锥阀的通流口同样可视为薄壁孔。

由式 6.5.5 可知，通流量为

$$q = C_d A_0 \sqrt{\frac{2\Delta p_c}{\rho}}$$

阀口通流截面积为

$$A_0 = \pi \frac{d_1 + d_2}{2} h = \pi d_m x_T \sin\phi$$

where

$$d_{\mathrm{m}} = \frac{d_1 + d_2}{2}$$

Then the discharge at poppet valve is

$$q = C_{\mathrm{d}} \pi d_{\mathrm{m}} x_{\mathrm{T}} \sin\phi \sqrt{\frac{2\Delta p}{\rho}} = C_{\mathrm{d}} w x_{\mathrm{T}} \sqrt{\frac{2\Delta p}{\rho}}$$

where $w = \pi d_{\mathrm{m}} \sin\phi$.

Also is directly proportional to the width of opening x_{T}.

6.6 Developed Flow in Gaps

In this section, we will discuss developed steady laminar flow of incompressible Newtonian fluid in narrow gaps.

6.6.1 Developed flow between parallel plates

Consider the incompressible, steady, developed flow of a fluid between parallel plates, with the upper plate moving at velocity u_0 parallel to the lower one, and the bottom plate is stationary, as shown in Fig.6.6.1. The distance between these two plates is h. We will calculate the velocity distribution and flowrate.

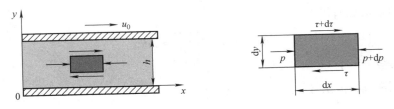

Fig.6.6.1   Flow between parallel plates

We assume that the plates are infinite in x and y, so that for a given geometry as sketched in Fig.6.6.1 must be in a developed flow. Set up an elemental volume of unit depth in the z-direction (into the paper), if we sum forces in the x-direction, we can write

$$p\mathrm{d}y + (\tau + \mathrm{d}\tau)\mathrm{d}x = (p + \mathrm{d}p)\mathrm{d}y + \tau\mathrm{d}x \tag{6.6.1}$$

Eq.6.6.1 can be reduced as

$$\frac{\mathrm{d}p}{\mathrm{d}x} = \frac{\mathrm{d}\tau}{\mathrm{d}y} \tag{6.6.2}$$

Using the shear stress of an one-dimensional flow, $\tau = \mu \dfrac{\mathrm{d}u}{\mathrm{d}y}$, there results

$$\frac{\mathrm{d}p}{\mathrm{d}x} = \mu \frac{\mathrm{d}^2 u}{\mathrm{d}y^2} \tag{6.6.3}$$

式中，阀口平均直径$d_m = \dfrac{d_1 + d_2}{2}$。

所以，得出锥阀阀口的通流量为

$$q = C_d \pi d_m x_T \sin\phi \sqrt{\frac{2\Delta p}{\rho}} = C_d w x_T \sqrt{\frac{2\Delta p}{\rho}}$$

式中，$w = \pi d_m \sin\phi$。

通流量同样与阀口开启宽度 x_T 成正比。

6.6　缝隙出流

本节我们讨论不可压缩牛顿流体在狭缝中的稳定发展层流。

6.6.1　平行平板缝隙流动

如图 6.6.1 所示，距离为 h 的两平行平板，下平板固定不动，上平板以速度 u_0 平行于下板移动。我们将对不可压缩流体在平行平板间缝隙内做稳定发展流动时的流速分布和流量进行分析。

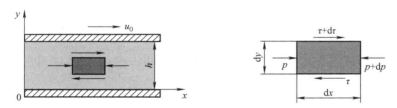

图 6.6.1　平行平板间缝隙流动

假设平板在 x 和 y 方向的尺寸无限大，这样可将图 6.6.1 中的流动视为充分发展流动。选取一个 z 方向（垂直页面方向）尺寸为单位值的流体微元，微元在 x 方向受力平衡，力的平衡式为

$$p\,\mathrm{d}y + (\tau + \mathrm{d}\tau)\,\mathrm{d}x = (p + \mathrm{d}p)\,\mathrm{d}y + \tau\,\mathrm{d}x \tag{6.6.1}$$

式 6.6.1 可化简为

$$\frac{\mathrm{d}p}{\mathrm{d}x} = \frac{\mathrm{d}\tau}{\mathrm{d}y} \tag{6.6.2}$$

代入一元流动的剪切应力定义式 $\tau = \mu \dfrac{\mathrm{d}u}{\mathrm{d}y}$，得

$$\frac{\mathrm{d}p}{\mathrm{d}x} = \mu \frac{\mathrm{d}^2 u}{\mathrm{d}y^2} \tag{6.6.3}$$

Rewrite pressure gradient with the form of $\dfrac{dp}{dx} = -\dfrac{\Delta p}{L}$ (see 6.3, obviously the pressure gradient is negative), integrate twice and yields

$$u = -\frac{1}{2\mu}\frac{\Delta p}{L}y^2 + C_1 y + C_2 \tag{6.6.4}$$

where C_1 and C_2 are integration constants, set $u = 0$ at $y = 0$ and $u = u_0$ at $y = h$, we have

$$C_1 = \frac{u_0}{h} + \frac{1}{2\mu}\frac{\Delta p}{L}h, \quad C_2 = 0 \tag{6.6.5}$$

Thus the velocity distribution is

$$u = \frac{1}{2\mu}\frac{\Delta p}{L}(h-y)y + \frac{y}{h}u_0 \tag{6.6.6}$$

Note that the velocity of fluid particles is composed of two parts, respectively due to the pressure gradient and the shear motion of plate. If there is no applied pressure gradient pushing the flow in the x-direction, i.e., $\Delta p/L = 0$, the fluid motion is due to the motion of the upper plate only, and be of linear profile, is called a *Couette flow*. If the motion is due to the pressure gradient only, i.e., $u_0 = 0$, it is a *Poiseuille flow* of parabolic profile.

The velocity profile of Couette flow between parallel plates with an applied negative pressure gradient in shown in Fig.6.6.2.

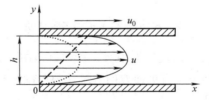

Fig.6.6.2   Velocity profile in the gap of two parallel plates

The dashed line indicates the profile for a zero pressure gradient, i.e., a Couette flow; and the dotted line indicates the profile for a negative pressure gradient with both plates are stationary, i.e., a Poiseuille flow.

Obviously, the flow rate is also consisted of this two parts.

$$q = \int_0^h ub\,dy = \frac{bh^3}{12\mu}\frac{\Delta p}{L} + \frac{u_0}{2}bh \tag{6.6.7}$$

where b is the width of the plate (into the paper).

用压强变化率代替压力梯度，$\dfrac{\mathrm{d}p}{\mathrm{d}x} = -\dfrac{\Delta p}{L}$（见 6.3 节，显然压力梯度为负值），积分得

$$u = -\frac{1}{2\mu}\frac{\Delta p}{L}y^2 + C_1 y + C_2 \qquad (6.6.4)$$

式中，C_1 和 C_2 为积分常数，代入边界条件：$y = 0$ 处 $u = 0$ 和 $y = h$ 处 $u = u_0$，解得

$$C_1 = \frac{u_0}{h} + \frac{1}{2\mu}\frac{\Delta p}{L}h, \qquad C_2 = 0 \qquad (6.6.5)$$

因此，平行平板缝隙内速度分布规律为

$$u = \frac{1}{2\mu}\frac{\Delta p}{L}(h - y)y + \frac{y}{h}u_0 \qquad (6.6.6)$$

由上式可见，流体质点的速度由两部分构成，一部分由压力梯度引起，另一部分由平板的剪切运动引起。如果 x 方向无压力驱动流体运动，即 $\Delta p/L = 0$，则流体流动仅由平板的剪切运动导致，流速为线性变化，称为 *Couette* **流动**。如果流动仅由压强差驱动，即 $u_0 = 0$，则流速呈抛物面分布，称为 *Poiseuille* **流动**。

图 6.6.2 所示为压力梯度为负的压差流动和剪切流动共同作用的速度分布曲线。

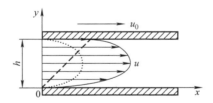

图 6.6.2　两平行平板缝隙间流速分布

图中虚线为剪切流动（压强差为 0）的速度曲线；点线为两平板都固定不动时负压力梯度的压差流动速度曲线。

显然，平板间的流量也由两部分构成。

$$q = \int_0^h ub\,\mathrm{d}y = \frac{bh^3}{12\mu}\frac{\Delta p}{L} + \frac{u_0}{2}bh \qquad (6.6.7)$$

式中，b 为平板的宽度（垂直页面方向）。

6.6.2 Developed flow between concentric cylinders

Now we derive the velocity distribution and flow rate of a developed flow between two concentric cylinders with a small gap fulfilled with viscous fluid. Such flow may occurs inside the gap between a core and the seat of a sloop valve shown in FigA6.5.1.

An element in the form of a thin cylindrical shell with r-radius, dr-thick and dx-long will be used, as shown in Fig.6.6.3.

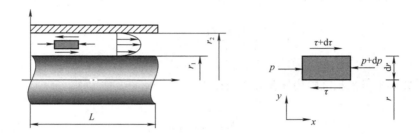

Fig.6.6.3　Developed flow between concentric cylinders

We sum the forces and obtain

$$p2\pi r dr + (\tau + d\tau)2\pi r dx = (p + dp)2\pi r dr + 2\pi r \tau dx \tag{6.6.8}$$

Eq.6.6.8 is reduced to

$$\frac{dp}{dx} = \frac{d\tau}{dr} \tag{6.6.9}$$

Substituting into $\tau = \mu \dfrac{du}{dr}$, and integrate twice and yields

$$u = \frac{1}{2\mu}\frac{dp}{dx}r^2 + C_1 r + C_2 \tag{6.6.10}$$

Set $u = 0$ at $r = r_1$ and $u = 0$ at $r = r_2$ by no slip condition, solving the integration constants C_1 and C_2, we have

$$C_1 = -\frac{1}{2\mu}\frac{dp}{dx}(r_1 + r_2), \quad C_2 = \frac{1}{2\mu}\frac{dp}{dx}r_1 r_2 \tag{6.6.11}$$

Then the velocity distribution is

$$u(r) = \frac{1}{2\mu}\frac{dp}{dx}\left[r^2 - (r_1 + r_2)r + r_1 r_2\right] \tag{6.6.12}$$

This is integrated to give the flow rate

6.6.2　同心圆环缝隙流动

接下来我们讨论充满黏性流体的同心圆环形缝隙内层流的速度和流量。图 A6.5.1 所示的滑阀阀芯 - 阀座缝隙内就可能发生这样的流动。

取流体微元的半径为 r，厚度为 dr，长度为 dx 的薄壁圆环，如图 6.6.3 所示。

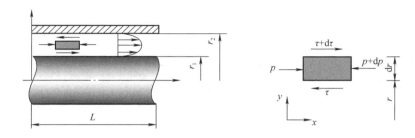

图 6.6.3　同心圆环缝隙内层流

列出 x 方向力的平衡方程，为

$$p2\pi rdr + (\tau + d\tau)2\pi rdx = (p + dp)2\pi r\tau dr + 2\pi rdx \quad\quad (6.6.8)$$

式 6.6.8 化简为

$$\frac{dp}{dx} = \frac{d\tau}{dr} \quad\quad (6.6.9)$$

代入 $\tau = \mu\dfrac{du}{dr}$，并积分得

$$u = \frac{1}{2\mu}\frac{dp}{dx}r^2 + C_1 r + C_2 \qu\quad (6.6.10)$$

代入边界条件：根据壁面不滑移定理，$r = r_1$ 和 $r = r_2$ 处，$u = 0$，解得积分常数 C_1 和 C_2

$$C_1 = -\frac{1}{2\mu}\frac{dp}{dx}(r_1 + r_2) \quad\quad C_2 = \frac{1}{2\mu}\frac{dp}{dx}r_1 r_2 \quad\quad (6.6.11)$$

解得速度分布函数

$$u(r) = \frac{1}{2\mu}\frac{dp}{dx}\left[r^2 - (r_1 + r_2)r + r_1 r_2\right] \ququad (6.6.12)$$

对流速积分得流量

$$q = \int_{r_1}^{r_2} 2\pi r u r dr = \frac{\pi}{\mu}\frac{dp}{dx}\int_{r_1}^{r_2}\left[r^2 - (r_1 + r_2)r + r_1 r_2\right]r dr$$

$$= -\frac{\pi}{12\mu}\frac{dp}{dx}(r_2 + r_1)(r_2 - r_1)^3 \tag{6.6.13}$$

or

$$q = \frac{\pi}{12\mu}\frac{\Delta p}{L}(r_2 + r_1)(r_2 - r_1)^3 \tag{6.6.14}$$

For a number of situations, such as a shaft rotating in a sliding bearing, the outer cylinder is fixed, while the inner rotates with an angular velocity of ω. This flow can be considered as it between two parallel plates if the gap is very small relative to the radius of the cylinder, i.e., $r_2 - r_1 \ll R$, and, $r_1 \approx r_2 = R$, then the flow rate is given as

$$q = \frac{\pi d \delta^3}{12\mu}\frac{\Delta p}{L} + \frac{\pi d \delta}{2}\omega R \tag{6.6.15}$$

where δ is the width of gap, L is the length and d is diameter of a cylinder.

Obviously, the two terms at right hand side represent the flow rate of Poiseuille and Couette flow respectively.

6.6.3 Developed flow between eccentric cylinders

Consider the flow between the journal and bearing, but instead of gap being constant, let there be a coaxial error around the circumference (Fig.6.6.4). Let us think the circular gap as a series of parallel plates of db length (obviously db = rdθ) and have h gaps.

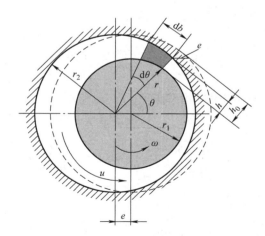

Fig.6.6.4   Flow between eccentric cylinders.

Then h must change with θ, that is

$$q = \int_{r_1}^{r_2} 2\pi r u \mathrm{d}r = \frac{\pi}{\mu}\frac{\mathrm{d}p}{\mathrm{d}x}\int_{r_1}^{r_2}\left[r^2 - (r_1 + r_2)r + r_1 r_2\right]r\mathrm{d}r$$

$$= -\frac{\pi}{12\mu}\frac{\mathrm{d}p}{\mathrm{d}x}(r_2 + r_1)(r_2 - r_1)^3 \qquad (6.6.13)$$

或

$$q = \frac{\pi}{12\mu}\frac{\Delta p}{L}(r_2 + r_1)(r_2 - r_1)^3 \qquad (6.6.14)$$

通常，如轴和滑动轴承的配合，轴承座孔固定不动，而轴颈以角速度 ω 旋转。这样的流场如果轴颈和轴承间的间隙非常小，可以把同心圆环缝隙近似展开为平行平板间的流动，即，$r_2 - r_1 << r_1$，且，$r_1 \approx r_2 = R$，这时，流量表达式可写为

$$q = \frac{\pi d \delta^3}{12\mu}\frac{\Delta p}{L} + \frac{\pi d \delta}{2}\omega R \qquad (6.6.15)$$

式中，δ 是间隙宽度；L 是圆柱面长度；d 是圆柱直径。

显然，等式右边的两项同样分别代表压差流动和剪切流动的流量。

6.6.3　偏心圆环缝隙流动

继续讨论轴颈和轴承的配合，如果二者间存在同轴度误差，如图 6.6.4 所示，则环形缝隙变为不均匀缝隙。这时，我们可以把偏心环形缝隙看作一系列长度为 $\mathrm{d}b$、缝隙宽度为 h 的微小平行平板连接而成，显然，在柱坐标系，$\mathrm{d}b = r\mathrm{d}\theta$。

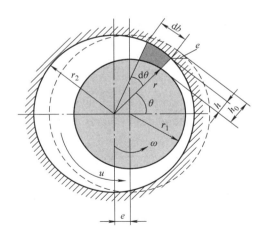

图 6.6.4　偏心圆环缝隙流动

此时，间隙 h 随角度 θ 变化

$$h \approx h_0 - e\cos\theta = h_0\left(1 - \varepsilon\cos\theta\right) \tag{6.6.16}$$

where, $h_0 = r_2 - r_1$, equals the gap width between concentric cylinders: and e is eccentric distance between inner and outer cylinders, and ε is called eccentricity ratio, defined as $\varepsilon = e/h_0$.

Take the width of the plates (into the paper) as unity, then the flow rate between each elemental parallel plates from Eq.6.6.7 is

$$\mathrm{d}q = \frac{rh^3\mathrm{d}\theta}{12\mu}\frac{\Delta p}{l} + h\frac{r\mathrm{d}\theta}{2}u_0 \tag{6.6.17}$$

This is integrated from 0 to 2π about $\mathrm{d}\theta$ to give the flow rate

$$q_\varepsilon = \left(1 + 1.5\varepsilon^2\right)\frac{\pi dh_0^3}{12\mu}\frac{\Delta p}{l} + \frac{\pi dh_0}{2}u_0 \tag{6.6.18}$$

Now we discuss the pressure distribute inside the eccentric gap. Expand half of these eccentric cylinders as un-parallel plates as shown in Fig.6.6.5.

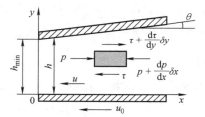

Fig.6.6.5   Flow between un-parallel plates

The velocity profile between each elemental un-parallel plates from Eq.6.6.6 is

$$u = -\frac{1}{2\mu}\frac{\mathrm{d}p}{\mathrm{d}x}(h-y)y + \frac{y}{h}u_0 \tag{6.6.19}$$

While the gap width h is the function of x-coordinate, $h = h_{min} + x\tan\theta$.

Then the flow rate with unity width plates is expressed as

$$q = -\frac{h^3}{12\mu}\frac{\mathrm{d}p}{\mathrm{d}x} + \frac{u_0}{2}h \tag{6.6.20}$$

The pressure must reaches its maximum at $h = h_{min}$, and the pressure gradient where is set to be zero, it is $\mathrm{d}p/\mathrm{d}x = 0$, then

$$q_{h_{min}} = \frac{1}{2}u_0 h_{min} \tag{6.6.21}$$

From continuity, $q_{h_{min}} = q_h$, we have

$$h \approx h_0 - e\cos\theta = h_0\left(1 - \varepsilon\cos\theta\right) \qquad (6.6.16)$$

式中，$h_0 = r_2 - r_1$，为不存在偏心误差时的间隙值；e 是内、外圆柱面间的偏心距；ε 是偏心率，定义为 $\varepsilon = e/h_0$。

假设圆柱面长度（垂直页面方向）为单位值，根据式 6.6.7，每一对平行平板微元间的流量为

$$\mathrm{d}q = \frac{rh^3\mathrm{d}\theta}{12\mu}\frac{\Delta p}{l} + h\frac{r\mathrm{d}\theta}{2}u_0 \qquad (6.6.17)$$

对此流量微元在 $0 \sim 2\pi$ 间积分，得偏心缝隙间的流量为

$$q_\varepsilon = \left(1 + 1.5\varepsilon^2\right)\frac{\pi dh_0^3}{12\mu}\frac{\Delta p}{l} + \frac{\pi dh_0}{2}u_0 \qquad (6.6.18)$$

进一步讨论偏心缝隙内的压强分布。将偏心圆环间隙的一半展开成一对不平行平板，如图 6.6.5 所示。

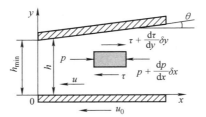

图 6.6.5　不平行平板缝隙流动

由推导式 6.6.6 的方法同样可推导出两平板间的速度分布

$$u = -\frac{1}{2\mu}\frac{\mathrm{d}p}{\mathrm{d}x}\left(h - y\right)y + \frac{y}{h}u_0 \qquad (6.6.19)$$

但是需要注意，此时平板间隙 h 随 x 坐标值变化，即 $h = h_{\min} + x\tan\theta$。则单位长度的平板间流量为

$$q = -\frac{h^3}{12\mu}\frac{\mathrm{d}p}{\mathrm{d}x} + \frac{u_0}{2}h \qquad (6.6.20)$$

显然，在间隙最小 $h = h_{\min}$ 处，压强达到最大值，压力梯度为零 $\mathrm{d}p/\mathrm{d}x = 0$，即

$$q_{h_{\min}} = \frac{1}{2}u_0 h_{\min} \qquad (6.6.21)$$

根据连续性方程，有 $q_{h_{\min}} = q_h$，所以

$$\frac{1}{2}u_0 h_{min} = -\frac{h^3}{12\mu}\frac{dp}{dx} + \frac{u_0}{2}h \tag{6.6.22}$$

Therefore, the pressure gradient takes the form of

$$\frac{dp}{dx} = 6\mu u_0 \frac{h - h_{min}}{h^3} \tag{6.6.23}$$

Eq.6.6.23 is called one-dimensional Reynolds formula.

From Eq.6.6.23 we find that, rotation of journal can take pressure oil into the wedge gap between the journal and the bearing when some conditions are met. This pressure can separates the two metal parts and reduces the friction even under certain loads, is the fundamentals of hydrodynamic lubrication sliding bearings.

Exercises 6

6.1　A liquid flows in a pipe at a Reynolds number of 9000, it states that the flow is_____.

A. laminar　　　　B. transition　　　C. turbulent　　　D. any of above

6.2　In a laminar flow in a pipe, the frictional head loss is_____.

A. proportional to the average veloctity　　B. proportional to the square of the average velocity
C. proportional to the diameter of the pipe　D.proportional to the Re number

6.3　A liquid with a density of 860 kg/m^3 and a kinematic viscosity of 60×10^{-6} m^2/s flows in a 50mm-diameter pipe at an average velocity is 2.4m/s. The pressure drop over 20m of pipe is_____.

A. 0.857kPa　　　B. 3.76kPa　　　　C. 31.7kPa　　　　D. 234.8kPa

6.4　A pressure drop of 0.5kPa occurs over a section of 20mm-diameter pipe transporting water at 20℃ , if the Reynolds number is 1600, the length of the horizontal pipe is about_____.

A. 3.2m　　　　　B. 78m　　　　　C. 125m　　　　　D. 400m

6.5　Consider fully developed laminar flow in a circular pipe with negligible entrance effects. If the length of the pipe is doubled, the frictional head loss will_____.

A. reduced by half　B. remain constant　C. double　　　　D. more than double

6.6　Consider fully developed laminar flow in a circular pipe. If the diameter of the pipe is reduced by half while the flow rate and the pipe length are held constant, the frictional head loss will_____.

A. double　　　　　　　　　　　B. quadruple
C. increase by a factor of 8　　　　D. increase by a factor of 16

6.7　Consider fully developed laminar flow in a circular pipe. If the viscosity of the fluid is reduced by half by heating while the flow rate is held constant, the frictional head loss will_____.

A. reduced by half　B. remain constant　C. double　　　　D. quadruple

6.8　In fully developed laminar flow in a circular pipe, the velocity at $R/2$ (midway between the wall surface and the centerline) is measured to be 6m/s. The velocity at the center of the pipe is_____.

A. 3m/s　　　　　B. 8m/s　　　　　C. 12m/s　　　　　D. 36m/s

$$\frac{1}{2}u_0 h_{\min} = -\frac{h^3}{12\mu}\frac{\mathrm{d}p}{\mathrm{d}x} + \frac{u_0}{2}h \qquad (6.6.22)$$

由此得出压力梯度变化规律

$$\frac{\mathrm{d}p}{\mathrm{d}x} = 6\mu u_0 \frac{h - h_{\min}}{h^3} \qquad (6.6.23)$$

式 6.6.23 称为一维雷诺方程。

由式 6.6.23 可知，满足一定条件下，轴颈的旋转运动可以将油液以一定压强代入其与轴承座孔间的楔形间隙中，达到一定压强的油液可以在承受载荷的同时将轴颈和轴承的金属表面分隔开，这就是液体动压滑动轴承的基本原理。

习题六

6.1　管道内液体流动的雷诺数为 9000，液体的流动状态为_____。

A. 层流　　　　　　　B. 过渡流　　　　　C. 紊流　　　　　　　D. 以上全部

6.2　圆管层流的损失水头_____。

A. 与速度成正比　　　　　　　　　　B. 与速度的平方成正比

C. 与管道直径成正比　　　　　　　　D. 与雷诺数成正比

6.3　密度为 $860\mathrm{kg/m^3}$、运动黏度为 $60\times10^{-6}\mathrm{m^2/s}$ 的液体在直径 50mm 的圆管内流动，平均流速为 2.4m/s。则距离 20m 的截面间压强下降_____。

A. 0.857kPa　　　　B. 3.76kPa　　　　　C. 31.7kPa　　　　　D. 234.8kPa

6.4　输水管内水流温度为 20℃，管道直径为 20mm、水流雷诺数为 1600，测得一段管道上压强差为 0.5kPa，则这一段管道的长度是 _____。

A. 3.2m　　　　　　B. 78m　　　　　　　C. 125m　　　　　　D. 400m

6.5　圆管内充分发展层流，忽略入口效应。若管道长度加倍，则沿程损失将_____。

A. 减半　　　　　　B. 保持不变　　　　　C. 加倍　　　　　　D. 比加倍还多

6.6　圆管内充分发展层流，若保持流量和管道长度不变，管道直径减半，则沿程损失与管道直径没有变化前相比将 _____。

A. 乘以 2　　　　　B. 乘以 4　　　　　　C. 乘以 8　　　　　　D. 乘以 16

6.7　圆管充分发展层流，若流量保持不变，而因温度升高导致流体黏度降低至原来的一半，则沿程损失将 _____。

A. 减半　　　　　　B. 保持不变　　　　　C. 增至 2 倍　　　　　D. 增至 4 倍

6.8　圆管充分发展层流，从管道中心向管壁方向，至半径为管壁半径 R 的一半，即 $r = R/2$ 处测得流速为 6m/s，则管道中心处流速为 _____。

A. 3m/s　　　　　　B. 8m/s　　　　　　　C. 12m/s　　　　　　D. 36m/s

6.9　With laminar flow in a circular pipe, at _____ distance from the centerline (in terms of the pipe radius R) the average velocity occurs.

A. 0.25R　　　　B. 0.5R　　　　C. 0.707R　　　　D. R

6.10　The energy requirement of an 83% efficient pump that transports 600L/min of water have a pressure rise of 800kPa is at least _____.

A. 6.6kW　　　　B. 8kW　　　　C. 9.4kW　　　　D. 10kW

6.11　The energy requirement of an 83% efficient pump that transports 600L/min of water to a specific location that 30m above is at least _____.

A. 2.9kW　　　　B. 3.5kW　　　　C. 147kW　　　　D. 177kW

6.12　Consider a waterfall flowing downward at an average speed of 5m/s at a flow rate of 200m^3/s to a location 20m above the pool surface. The power generation potential of the waterfall is about _____.

A. 100kW　　　　B. 200kW　　　　C. 39MW　　　　D. 200MW

6.13　Consider the flow of air and water in pipes of the same diameter, at the same temperature, and at the same average velocity. Which flow is more likely to be turbulent? Why?

6.14　Find the radius in a developed laminar flow in a pipe where: (a) the velocity is equal to the average velocity. (b) the shear stress is equal to one-half the wall shear stress.

6.15　Find the ratio of the total flow rate through a pipe of radius r_0 to the flow rate through an annulus with inner and outer radii of $r_0/2$ and r_0. Assume developed laminar flow with the same pressure gradient.

6.16　Water at 20℃ is flowing in a 100mm diameter pipe with a rate of 0.005m^3/s. Is the flow laminar or turbulent?

6.17　Water at 15℃ ($\rho = 999.1$kg/m^3 and $\mu = 1.139\times10^{-3}$Pa · s) is flowing steadily in a 50mm-diameter, 20m-long stainless steel pipe at a flow rate of 0.01m^3/s. Determine (a) the pressure drop, (b) the head loss, and (c) the pumping power requirement to overcome this pressure drop.

6.18　A liquid with a density of 900kg/m^3 and viscosity of 36×10^{-3}Pa·s flows in a 60mm-diameter pipe, calculate (a) the Reynolds number if the average velocity is 2m/s, (b) the maximum allowable average velocity that a laminar flow exits.

6.19　Water at 20℃ is flowing in a pipe with a flow rate of 0.005m^3/s, what is the allowable minimum diameter with which water can flow in a laminar state.

6.20　Oil with a kinematic viscosity of 200cSt is flowing through a 100mm-diameter pipe. Below what velocity will the flow be laminar?

6.21　An oil with kinematic viscosity of 0.00032m^2/s weighs 930kg/m^3. What will be its flow rate and head loss in a 1000m length of an 80mm-diameter pipe when the Reynolds number is 1000?

6.22　Stream with a density of 680kg/m^3 is flowing with a velocity of 0.5m/s through a circular pipe with λ=0.015. What is the shear stress at the pipe wall?

6.23　Oil ($S_g = 0.92$) of viscosity 380×10^{-6}m^2/s flows in a 100mm-diameter pipe at a flow rate of 0.64L/s. Find the head loss per unit length.

6.24　Water at 20℃ flows in a wide river. Using a Reynolds number ($Re = Vh/v$), calculate the average velocity V that will result in a laminar flow if the depth h of the river is 5m.

6.9　半径为 R 的圆管层流，自管道中心起半径为 _____ 处，实际流速等于平均流速。

A. 0.25R　　　　B. 0.5R　　　　C. 0.707R　　　　D. R

6.10　用效率为 83% 的水泵输水，流量 600L/min，出口水压需升高 800kPa，则选取的水泵功率不应小于 _____。

A. 6.6kW　　　　B. 8kW　　　　C. 9.4kW　　　　D. 10kW

6.11　用效率为 83% 的水泵输水，流量 600L/min，出口水位高于进口 30m，则选取的水泵功率不应小于 _____。

A. 2.9kW　　　　B. 3.5kW　　　　C. 147kW　　　　D. 177kW

6.12　落差 20m 的瀑布，平均流速 5m/s，流量 $200m^3/s$，该瀑布可能产生的最大发电功率为 _____。

A. 100kW　　　　B. 200kW　　　　C. 39MW　　　　D. 200MW

6.13　在直径相同的管道内分别流动有平均速度相同、温度相等的空气和水，哪一种流体更容易发生紊流，为什么？

6.14　充分发展层流圆管内，找出满足以下条件的位置的半径（1）实际流速等于平均流速；（2）剪切应力等于管壁处剪切应力的一半。

6.15　一半径为 r_0 的圆管和一个内径为 $r_0/2$、外径为 r_0 的环形管，假设两个管道内均为充分发展层流且压力梯度相同，试求两个管道内流量之比。

6.16　水温 20℃、直径 100mm 的圆管内，流量为 $0.005m^3/s$，则该流动为层流还是紊流？

6.17　温度为 15℃ 的水（ρ=999.1kg/m^3 and μ=1.139×10^{-3}Pa·s）在直径 50mm、长度 20m 的不锈钢管内稳定流动，流量为 $0.01m^3/s$。求（1）压降；（2）损失水头；（3）为克服压降所需水泵的功率。

6.18　密度为 900 kg/m^3、黏度为 36×10^{-3}Pa·s 的液体在直径 60mm 的管道内流动，（1）当平均流速为 2m/s 时，是哪种流动状态？（2）能够维持层流状态的最大流速是多少？

6.19　管道内 20℃的水流量为 $0.005m^3/s$，要保证水流为层流，管道的最小直径是多少？

6.20　黏度为 200cSt 的油液在 100mm 直径的管道内流动，保证层流的最小流量是多少？

6.21　密度为 930kg/m^3、运动黏度为 $0.00032m^2/s$ 的油液，在直径为 80mm、长度为 1000m 的管道内流动，雷诺数为 1000。计算油液的流量和水头损失。

6.22　密度为 680kg/m^3 的流体以 0.5m/s 的流速在圆管内流动，且沿程损失系数 λ=0.015，求管道内壁处所受剪切应力。

6.23　比重 S_g = 0.92、黏度为 380×10^{-6}m^2/s 的油液、在直径 100mm 的圆管内流动，流量为 0.64L/s。求单位长度管道的损失水头。

6.24　宽阔河流水温为 20℃，使用雷诺数计算公式 $Re=Vh/\nu$。若河水深

6.25   Water at 20℃ flows in a 120mm-diameter Galvanized iron pipe with $V = 8$m/s. Head loss measurements indicate that $\lambda = 0.020$. (a) What is the thickness of the viscous sublayer? (b) Is the pipe behaving as a hydraulically smooth or hydraulically rough pipe?

6.26   When fluid of density 850kg/m³ flows in a 100mm-diameter pipe, the frictional stress at the pipe wall is 50N/m². Calculate the friction head loss per meter of pipe. If the flow rate is 50L/s, how much power is lost per meter of pipe?

6.27   Water flows at a flow rate of 30L/s through a horizontal pipe whose diameter is constant at 50mm. The pressure drop across a valve in the pipe is measured to be 2.5kPa. Determine the irreversible head loss of the valve, and the useful pumping power needed to overcome the resulting pressure drop.

6.28   Water flows at the flow rate of 0.6m³/min in a 40mm-diameter horizontal pipeline with a pressure of 690kPa. If the pressure after an enlargement to 60mm diameter is measured to be 700kPa, calculate the head loss across the enlargement.

6.29   Calculate the pressure p_1 shown in Fig.E6.29 needed to maintain a flow rate of 0.1m³/s of water in a 90mm-diameter horizontal pipe leading to a nozzle if a loss coefficient based on V_1 is 0.25 between the pressure gage and the exit.

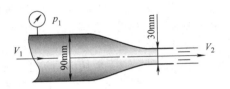

Fig.E6.29

6.30   Water is being pumped from a large lake to a reservoir 25m above at a flow rate of 25L/s by a 10kW pump. If the irreversible head loss of the piping system is 7m, determine the mechanical efficiency of the pump.

6.31   Water flows at a flow rate of 0.05m³/s in a horizontal pipe whose diameter is reduced from 200mm to 120mm by a reducer. If a pressure drop of 50kPa is measured after the reducer, determine the irreversible head loss in the reducer. Take the kinetic energy correction factors to be 1.06.

6.32   A large tank is initially filled with water 5m above the center of a sharp-edged 100mm-diameter orifice. The tank water surface is open to the atmosphere, and the orifice drains to the atmosphere. If the total irreversible head loss in the system is 0.3m, determine (a) the initial discharge velocity of water from the tank (b) the minor loss factor at the orifice. Take the kinetic energy correction factor at the orifice to be 1.2.

6.33   Water is pumped from a lower reservoir to a higher reservoir by a pump that provides 20kW of useful mechanical power to the water. The free surface of the upper reservoir is 50m higher than the surface of the lower reservoir. If the flow rate of water is measured to be 0.03m³/s, determine the irreversible head loss of the system and the lost mechanical power during this process.

6.34   A 250mm-diameter pipeline ($\lambda = 0.025$) is 3km long. When pumping 0.1m³/s of water through it, with a total actual lift of 12m, how much power is required? The pump efficiency is 75%.

6.35   Water in a partially filled large tank is to be supplied to the roof top, which is 8m above the water level in the tank, through a 25mm-internal-diameter pipe by maintaining a constant air pressure of 300kPa (gage) in the tank. If the head loss in the piping is 2m of water, determine the discharge rate of the supply of water to the roof top.

6.36   A 72-percent efficient 11kW pump is pumping water from a lake to a nearby pool at a flow rate of 0.025m³/s through a constant-diameter pipe. The free surface of

h=5m，计算河水为层流时的平均流速。

6.25　直径为 120mm 的镀锌管内水温 20℃、流速 V=8m/s，测得沿程损失系数 λ=0.020。求：（1）黏性底层的厚度。（2）管道为水力光滑管还是水力粗糙管？

6.26　密度为 850kg/m³ 的液体在直径为 100mm 的圆管内流动，管壁表面摩擦应力为 50N/m²，计算（1）单位长度管道上的沿程水头损失；（2）流量为 50L/s 时，单位长度管道上的能量损失。

6.27　直径为 50mm 的水平圆管内水流量为 30L/s，管道一阀门处测得压降为 2.5kPa。计算（1）该阀门产生的损失水头；（2）克服该损失所需的水泵功率。

6.28　直径为 40mm 的水平管道内测得压强为 690kPa，水流量为 0.6m³/min，之后管道直径扩大为 60mm 并测得压强 700kPa。计算该扩径产生的局部水头损失。

6.29　如图 E6.29 所示入口直径为 90mm 的水平喷嘴。若压力表至出口间相对 V_1 其损失系数为 0.25，计算维持流量 0.1m³/s 所需的压强 p_1。

6.30　用 10kW 水泵从河中抽水至 25m 高处的水罐，流量为 25L/s，测得管道的损失水头为 7m，计算水泵的效率。

6.31　水平管道某处有一个接头将管径由 200mm 缩小至 120mm，管内水流量为

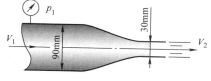

图 E6.29

0.05m³/s，接头后测得水压下降了 50kPa，取动能修正系数为 1.06，计算管接头产生的水头损失。

6.32　一个大型水罐底部有一个直径 100mm 的锐边孔口，罐内初始水位距出口中心高度为 5m。罐顶和出口处均为大气压，且全部水头损失为 0.3m。取孔口的动能修正系数为 1.2。求：（1）出口处的初始出流量；（2）孔口的局部损失系数。

6.33　用水泵从低处的水箱向高处水箱抽水，两个水箱高度差为 50m，水泵的有效功率为 20kW。已知抽水的流量为 0.03m³/s，求：（1）系统损失水头；（2）损失功率。

6.34　用直径 250mm、长度 3km 的管道（λ = 0.025）输水，要求流量 0.1m³/s，输水高度为 12m。所使用的水泵效率为 75%，计算水泵的驱动功率。

6.35　水罐通过一内径 25mm 的出水管向高于罐内水位 8m 的建筑物送水，罐内上部密闭加压，空气压力始终保持 300kPa。现已知总损失水头为 2m，求输水流量。

6.36　用功率 11kW、效率 72% 的水泵从湖中抽水距湖面至 20m 高处的水池，流量为 0.025m³/s，水管直径保持不变。求：（1）管路的损失水头；（2）克服该损失所需水泵功率。

the pool is 20m above that of the lake. Determine the irreversible head loss of the piping system, in m, and the mechanical power used to overcome it.

6.37   Water is pumped from a water pool to a storage tank 6m above at a flow rate of 50L/s while consuming 4.5kW of electric power. Neglect any losses and determine (a) the overall efficiency of the pump-motor unit and (b) the pressure difference between the inlet and the exit of the pump.

6.38   An oil pump is drawing 20kW of electric power while pumping oil with ρ = 860kg/m³ at a flow rate of 0.1m³/s. The diameter of the pipes is 100mm. If the pressure rise of oil in the pump is measured to be 250kPa and the motor efficiency is 90 percent, determine the mechanical efficiency of the pump.

6.39   Water enters a hydraulic turbine through a 50cm-diameter pipe at a flow rate of 1m³/s and exits through a 40cm-diameter pipe. The pressure drop in the turbine is measured by 180kPa. For a combined turbine-generator efficiency of 80 percent, determine the net electric power output. Disregard the effect of the kinetic energy correction factors.

6.40   The pump shown in Fig.E6.40 is 85% efficient. If d_1 = 60mm and d_2 = 120mm, calculate the required energy input in kW.

6.41   Water flows at a flow rate of 0.025m³/s in a horizontal pipe whose diameter increases from 60 to 120mm by an enlargement section. If the head loss across the enlargement section is 0.5m and the kinetic energy correction factor at both the inlet and the outlet is 1.05, determine the pressure change.

6.42   A 3m-high large tank shown in Fig.E6.42 is initially filled with water. The tank water surface is open to the atmosphere, and a sharp-edged 10mm-diameter orifice at the bottom drains to the atmosphere through a horizontal 60m-long pipe. If the total irreversible head loss of the system is determined to be 1.2m, determine the initial velocity of the water from the tank. Disregard the effect of the kinetic energy correction factors.

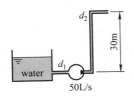

Fig.E6.40

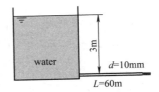

Fig.E6.42

6.43   A liquid with a density of 900kg/m³ flows straight down in a 60mm-diameter cast iron pipe. Estimate the pressure rise over 20m of pipe if the average velocity is 4m/s. Assume $v = 8 \times 10^{-6}$m²/s.

6.44   Find the slope angle of the 10mm-diameter pipe of glass (Fig.E6.44) in which water at 30℃ is flowing with Re = 1800 such that no pressure drop occurs.

6.45   SAE10W oil at 20℃ is pumped through a 20mm-diameter pipe at a flow rate of 12L/min. Calculate the pressure drop in a 20m horizontal section.

6.46   Water at 20℃ flows between the two concentric horizontal pipes of Fig.E6.46 with diameters of 30mm and 36mm. A pressure drop of 180Pa is measured over a 10m section of developed laminar flow. Find the flow rate and the shear stress on the inner pipe.

6.37 用功率 4.5kW 总效率 72% 的电动机驱动水泵从水池内抽水，以 50L/s 的流量输送到水池上方 6m 处的水箱内。不计各项损失，求（1）泵 - 电动机系统总效率；（2）泵进、出口间的压强差。

6.38 用功率 20kW 的电动机驱动油泵，以 $0.1m^3/s$ 的流量输送密度 $\rho = 860kg/m^3$ 的油液。已知输油管直径为 100mm，泵出口油压高于进口油压 250kPa，且电动机效率为 90%。求泵的机械效率。

6.39 流量 $1m^3/s$ 的水通过直径 50cm 的管道进入涡轮机，并经直径 40cm 的管道流出，出口水压下降了 180kPa。已知涡轮 - 发电机组的总效率为 80%，取动能修正系数为 1，求净发电量。

6.40 如图 E6.40 所示，水泵的效率为 85%，进、出水管直径分别为 d_1=60mm 和 d_2=120mm，求水泵的驱动功率。

6.41 水平渐扩管内流量为 $0.025m^3/s$，水管直径从 60mm 扩大到 120mm。若渐扩段水头损失为 0.5m，且进、出口动能修正系数均为 1.05，求渐扩管两端的压强差。

6.42 如图 E6.42 所示，水罐内初始水位高度为 3m，且顶部敞开，罐底有直径 10mm、开口锐利的出水管，水平出水管长为 60m，开口至大气。若系统总损失水头为 1.2m，不计各项损失、动能修正系数为 1，求初始时刻出水口的流速。

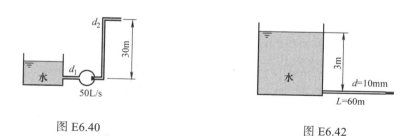

图 E6.40　　　　　　　　　　　图 E6.42

6.43 密度为 $900kg/m^3$、黏度 $v = 8 \times 10^{-6}m^2/s$ 的液体沿直径 60mm 的铸铁管竖直流下。若液体平均流速为 4m/s，计算管内 20m 长度的压强增加值。

6.44 直径 10mm 的玻璃管内水温为 30℃，水流雷诺数 Re=1800，若管内水压保持不变，求玻璃管的倾斜角度（见图 E6.44）。

6.45 用直径20mm的水平管道输送20℃的SAE10W机油，已知流量为12L/min。计算 20m 长管道内的压降。

6.46 水平同心圆管如图 E6.46 所示，内径为 30mm、外径为 36mm，管内水温为 20℃，为充分发展层流，且在 10m 长度内压强下降了 180Pa。求（1）管内水流量；（2）内侧壁面上的剪切应力。

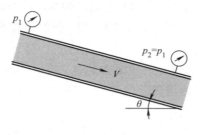

Fig.E6.44

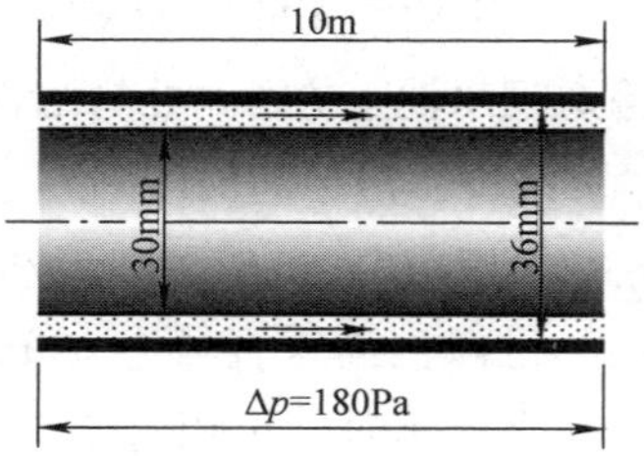

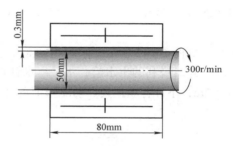

Fig.E6.46

6.47 A board 1m×1m that weighs 60N moves down the incline of 30° shown in Fig. E6.47 with a velocity of $V = 0.3$m/s. Estimate the viscosity of the fluid.

6.48 Calculate the torque T necessary to rotate the rod shown in Fig.E6.48 at 300r/min if the fluid filling the gap is SAE30W oil at 20℃. Assume a linear velocity profile.

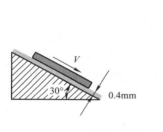

Fig.E6.47

Fig.E6.48

6.49 SAE30 oil at 20℃ is transported in a 100mm diameter pipe at an average velocity of 6m/s. What is the largest size allowed for the roughness element if the pipe is hydraulically smooth?

6.50 A 120mm-diameter horizontal pipe transports SAE10W oil at 20℃. Calculate the wall shear, the average velocity, and the flow rate if the pressure drop over a 10m section of pipe is measured to be 20kPa.

6.51 Water at 15 ℃ flows up a 30° incline in a 60mm diameter plastic pipe with a flow rate of 8.5L/s. Find the pressure change over a 90m length of pipe.

6.52 For the water flow at 20℃ shown in Fig.E6.52, find p_2 if $q = 0.02$m³/s and $p_1 = 200$kPa.

6.53 For the water flow in the pipe as shown in Fig.E6.52, has an abrupt expansion from $d_1 = 80$mm to $d_2 = 160$mm. The velocity in the smaller section is 10m/s and the flow is turbulent. The pressure in the smaller section is $p_1 = 410$kPa. Taking the kinetic energy correction factor to be 1.06 at both the inlet and the outlet, determine the downstream pressure p_2.

6.54 If the flow rate is measured to be 5L/s from the pipe shown in Fig.E6.54. Find the loss coefficient of the valve. Neglect wall friction.

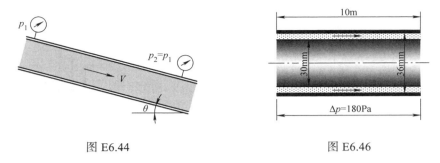

图 E6.44　　　　　　　　　　　　　图 E6.46

6.47　边长 1m × 1m、重 60N 的平板如图 E6.47 所示沿 30° 斜面下滑,速度 V=0.3m/s。求平板与斜面间液体的黏度。

6.48　如图 E6.48 所示,轴颈与轴承间隙中充满 20℃ 的 SAE30W 号机油,假设间隙内油液流速线性分布。求轴以 300r/min 的速度转动时所需的转矩 T。

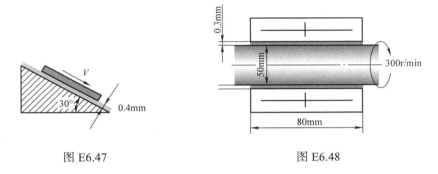

图 E6.47　　　　　　　　　　　　　图 E6.48

6.49　用直径 100mm 的油管以 6m/s 的速度输送 20℃ 的 SAE30 机油。若希望保持油液做水力光滑流动,求油管壁面允许的最大粗糙度平均高度。

6.50　用直径 120mm 的水平管道输送 20℃ 的 SAE10W 机油,若在 10m 长一段管道上压降为 20kPa,求:(1)管壁剪切应力;(2)管内平均流速;(3)流量。

6.51　15℃ 的水沿直径 60mm、向上 30° 倾斜的塑料管内流动,流量为 8.5L/s。求管道中相距 90m 两截面的压强差。

6.52　如图 E6.52 所示,水温为 20℃、流量 q = 0.02m³/s,且测得 p_1 = 200kPa,求 p_2。

6.53　水流经如图 E6.52 所示的突扩管道,若两段直径分别为 d_1 = 80mm 和 d_2 = 160mm,小直径段为紊流且流速为 10m/s,测得该处压强 p_1 = 410kPa。取两端动能修正系数均为 1.06,求下游截面压强为 p_2。

6.54　如图 E6.54 所示,管内水流量为 5L/s。不计管壁摩擦损失,求阀门处局部损失系数。

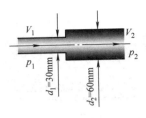

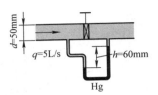

Fig.E6.52   Fig.E6.54

6.55　Water at 20 ℃ is transported in a 250mm diameter concrete pipe with a flow rate of $0.1m^3/s$. Find the head loss over a 300m section of the pipe.

6.56　Water is flowing through a Venturi meter whose diameter is 70mm at the entrance part and 40mm at the throat. The pressure is measured to be 380kPa at the entrance and 150kPa at the throat. Neglecting frictional effects, determine the flow rate of water.

6.57　Gasoline with a density of $680kg/m^3$ and kinematic viscosity of $0.43\times10^{-6}m^2/s$ is being discharged by a 20mm-diameter, 10m-long horizontal pipe from a storage tank open to the atmosphere. The least flow rate is to be 1L/s, disregarding the minor losses, determine minimum height of the liquid level above the center of the pipe.

6.58　Water is to be withdrawn from a 5m-high water tank by drilling a 10mm-diameter hole at the bottom surface. Disregarding the effect of the kinematic energy correction factor, determine the flow rate of water through the hole if (a) the entrance of the hole is well-rounded and (b) the entrance is sharp-edged.

6.59　Water reservoirs A and B are connected to each other through a 20m long, 50mm-diameter cast iron pipe with a sharp-edged entrance as shown in Fig.E6.59. The pipe also involves a fully open gate valve. The water level in both reservoirs is the same, but reservoir A is pressurized by compressed air while reservoir B is open to the atmosphere. If the initial flow rate through the pipe is 5L/s, determine the air pressure on top of reservoir A. Take the water temperature to be 10℃ .

6.60　An orifice with a 50mm-diameter opening is used to measure the mass flow rate of water at 15℃ ($\rho = 999\ kg/m^3$ and $v = 1.139\times10^{-6}m^2/s$) through a horizontal 100mm-diameter pipe (Fig.E6.60). A mercury manometer is used to measure the pressure difference across the orifice. If the differential height of the manometer is 200mm, determine the volume flow rate of water through the pipe, and the head loss caused by the orifice meter.

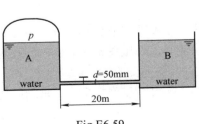

Fig.E6.59

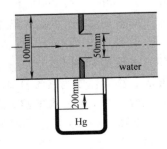

Fig.E6.60

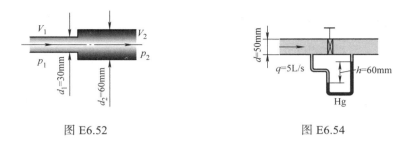

图 E6.52   图 E6.54

6.55 直径 250mm 的水泥管道内水温为 20℃、流量为 0.1m³/s，求长度 300m 段管道内的水头损失。

6.56 用文丘里流量计测量某水管内的流量。已知流量计入口管径 70mm，喉部直径 40mm。两处分别测得压强为 380kPa 和 150kPa。不计摩擦损失，求管内水流量。

6.57 从敞口储油罐内用直径 20mm、长 10m 的水平油管输送汽油，油液密度为 680kg/m³、运动黏度为 0.43×10^{-6}m²/s，若要保证流量不小于 1L/s，不计局部损失，求罐内汽油液面至出油管中心的最小高度。

6.58 在液面高度为 5m 的水罐底部钻一个直径 10mm 的排水孔，不计动能修正系数，计算以下两种情况下的排水流量:（1）排水孔边缘圆滑；（2）排水孔边缘锐利。

6.59 储水罐 A 和 B 通过一直径 50mm、长 20m 的铸铁管道相连，如图 E6.59 所示，管道入口边缘锐利，且管道中安装有全开的闸阀。初始时两水罐内液面高度相同，水温均为 10℃，但 A 罐内部加压而 B 罐敞开。若测得最初时刻水管内的流量为 5L/s，求 A 罐上部空气的压强。

6.60 在 100mm 直径的水平管路中安装直径 50mm 的孔板测流量（见图 E6.60）。已知水温为 15℃（$\rho = 999$ kg/m³ 且 $\nu = 1.139 \times 10^{-6}$m²/s），水银测压管在孔板两端测得液柱高度差为 200mm。求（1）管内水流量；（2）孔板流量计产生的局部损失。

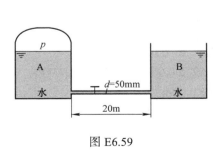

图 E6.59

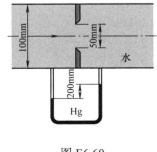

图 E6.60

Appendix A/ 附录 A　ANSWERS/ 习题答案

1.35　$V=0.233$m/s

1.37　0, 0.08, 0.2

1.38　4×10^{-3}

1.39　0.116N · m

1.40　0.64N

1.41　0.0032Pa

1.42　0.05517N · m

1.43　4×10^{-3}Pa · s

1.44　222.4N, 46.67%

1.45　0.565N

1.46　16N

1.48　1.81N

1.50　26.62N · m, 3.19N · m

1.51　8.65×10^{-3}Pa · s

1.52　12.9mm

1.54　9.9mm

1.56　0.0136N/m

1.57　14.38m

1.58　3.37mm

2.1　66.7kPa

2.2　100570pa

2.3　161626Pa

2.17　40.17kPa

2.18　12.2cm

2.19　13.95kPa

2.20　14.9kPa

2.21　20.7kPa

2.22　7.6kPa

2.23　35.12kPa

2.24　2.96m

2.25　62.86kPa

2.26　441.45N, 0.2m; 3384.45N, 1.157m

2.27　941.76kN, 0. 65m

2.28　24.62kPa

2.29　168.7N, 0.27m

2.30　(a) 1060kN, (b) 265kN, (c) 306kN

2.31　205.4kN(1.48, 2.91)

2.32　25.3kN

2.33　43.36kN, 30.66kN

2.34　51.56kN · m

2.35　0.993m

2.36　577.5kN

2.37　2.11m

2.38　365kN

2.39　21.6×10^{3}kg

2.40　3.46m

2.41　5.44m

2.42　37.67kN

2.43　158.9kN, 101kN

2.44　45.17kN, 29.7°

2.45　45.8kN, 15.5°

2.46　19.62kN, 11.64kN

2.47　739.3kN, 164.3kN

2.48　470kN

2.49　58.86kN

2.50　0.628N

2.51　67.62kPa

2.52　6.02kPa, 20.73kPa, 8.73kPa

2.53　50m^{3}

2.54　12.5kPa, 0

3.8　0.94,0.78

3.9　17.89m/s^2

3.10　3.61m/s, 9.22m/s^2

3.11　3.61m/s, 11.70m/s^2

3.12　3 m/s^2

3.17　1.5,−1.5

3.19　14 m/s^2

3.21　0.009m^3/s, 14.47kg/s

3.22　40.7s, 57.5s, 81.3s

4.25　17.98kW

4.26　1cm

4.32　0.0285m^3/s

4.35　3.85m/s, 5.45m/s

4.36　2.23m/s

4.37　1.1m/s

4.38　49m/s, 114m

4.39　31.3m/s

4.44　3.8 × 10^{-4}m^3/s, −1.4m

4.45　6.28 × 10^{-4}m^3/s

4.46　0.28 m^3/s

4.48　32.8%

4.49　47.54kPa

4.50　14.13kN

4.51　5.73kN

4.52　1.41kN

4.53　229N

4.56　7.85N

4.57　14.15kN

4.59　2.26kN

4.63　7.85kPa, 298N

4.64　2.76kN

5.22　200m^3/s, 4.5MPa

5.23　0.4m^3/s, 10MPa

5.24　10m/s, 122.88kN

5.25　16mm, 1.11

5.26　0.0056m^3/s

6.21　0.02m^3/s, 652.4m

6.22　0.6375N/m^2

6.23　0.01m/m

6.26　0.1199m/m, 50W/m

6.27　0.255m, 75W

6.28　1.57m

6.29　9.9MPa

6.30　78.5%

6.31　4.18 m

6.32　8.77m/s, 0.077

6.33　17.96m, 5.29kW

6.34　99kW

6.35　0.01m^3/s

6.36　12.29m, 3.01kW

6.41　33.6kPa

6.42　5.94 m/s

6.43　107kPa

6.44　0.215°

6.45　91kPa

6.47　40 × 10^{-3}Pa・s

6.53　432kPa

6.55　5m

6.56　0.0285m^3/s

6.57　0.63m

6.59　55.88kPa

Appendix B/ 附录 B
The Moody chart/ 莫迪图

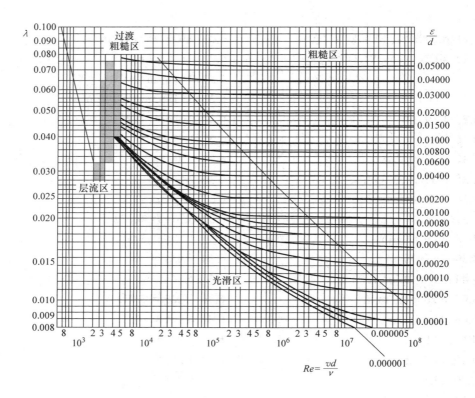

BIBLIOGRAPHY

[1] MO N R. Engineering fluid mechanics [M]. Wuhan: HUST PRESS, 2019.

[2] WANG J W, ZHANG H J, etc. Hydraulic Transmission [M]. Beijing: CHINA MA-CHINE PRESS, 2017.

[3] REN H L, LIN T L, etc. Hydraulic Transmission [M]. Beijing: CHINA MACHINE PRESS, 2019.

[4] LI Y Z, HE W Z, etc. Engineering fluid mechanics: Vol. 1 [M]. Beijing: Tsinghua University Press, 2017.

[5] HUANG Y J, LIU X M. The applications of technique of Electro-Rheological and Magneto-Rheological fluid in mechanical engineering [M]. Xiangtan: Xiangtan University Press, 2015.

[6] POTTER M C, WIGGERT D C. RAMADAN B H . Mechanics of Fluids [M]. 4th ed. Stamford: Cengage Learning, 2012.

[7] CENGEL Y A, CIMBALA J M. Fluid Mechanics: Fundamentals and Applications [M]. New York: McGraw-Hill Education, Inc, 2010.

[8] FINNEMORE E J, JOSEPH F B. Fluid Mechanics with Engineering Applications [M]. New York:McGraw-Hill Education, 2012.

参考文献

[1] 莫乃榕. 工程流体力学 [M]. 武汉：华中科技大学出版社，2019.

[2] 王积伟，章宏甲，等. 液压传动 [M]. 北京：机械工业出版社，2017.

[3] 任好玲，林添良，等. 液压传动 [M]. 北京：机械工业出版社，2019.

[4] 李玉柱，贺五洲，等. 工程流体力学：上册 [M]. 北京：清华大学出版社，2017.

[5] 黄宜坚，刘晓梅. 电磁流变液及其在机械工程中的应用 [M]. 湘潭：湘潭大学出版社，2015.

[6] 波特，威格特，等. 流体力学 [M]. 4 版. 斯坦福：圣智学习出版公司，2012.

[7] 森哲尔，辛巴拉. 流体力学：基础及应用 [M]. 纽约：麦格劳 - 希尔教育出版公司，2010.

[8] 芬纳莫尔，弗朗兹尼. 流体力学及其工程应用 [M]. 纽约：麦格劳 - 希尔教育出版公司，2012.